The Orthopoxviruses

The Orthopoxviruses

Frank Fenner

The John Curtin School of Medical Research
The Australian National University
Canberra, ACT, Australia

Riccardo Wittek

Institut de Biologie Animale
Batiment Biologie
Université de Lausanne
Lausanne, Switzerland

Keith R. Dumbell

Department of Medical Microbiology
Medical School
University of Cape Town
South Africa

Academic Press, Inc.
Harcourt Brace Jovanovich, Publishers
San Diego   New York   Berkeley   Boston
London   Sydney   Tokyo   Toronto

ACADEMIC PRESS, INC.
San Diego, California 92101

United Kingdom Edition published by
ACADEMIC PRESS LIMITED
24-28 Oval Road, London NW1 7DX

Library of Congress Cataloging-in-Publication Data

Fenner, Frank, Date
    The orthopoxviruses / Frank Fenner, Riccardo Wittek, Keith R.
Dumbell.
        p.   cm.
    Includes index.
    ISBN   0-12-253045-4   (alk. paper)
    1. Orthopoxviruses.   I. Wittek, Riccardo.   II. Dumbell, Keith R.
III. Title.
QR405.F46   1988
616.9′12—dc19   88-10362
                                                            CIP

PRINTED IN THE UNITED STATES OF AMERICA
88 89 90 91   9 8 7 6 5 4 3 2 1

Contents

Preface

On May 8, 1980 the World Health Assembly announced that smallpox had been eradicated from all countries in the world and resolved that this unique and historical event should be fully documented. In large part, this resolution has been fulfilled by the publication of "Smallpox and Its Eradication," produced by the World Health Organization under the authorship of Drs. Fenner, Henderson, Arita, Jezek, and Ladnyi. During the writing of this book dealing with the virology, immunology, and pathogenesis of smallpox, it became apparent to one of the authors (Dr. Fenner) that over the decade of the Intensified Smallpox Eradication Program (1967–1977) and since, a great deal of information had been accumulated about variola virus and other orthopoxviruses and the diseases they cause. Most of this information was of too detailed a nature and too peripheral to the eradication of smallpox to be included in the WHO book. After consultations with several other virologists, we decided to produce a book on the genus *Orthopoxvirus*, which would incorporate this new knowledge and provide a comprehensive and integrated picture of this family of viruses.

During its rather prolonged gestation, a new and potentially important use of vaccinia virus was discovered, namely, its use as a vector for genes from other viruses, bacteria, or protozoa that specify antigens important in stimulating the production of a protective immune response. This discovery has greatly increased interest in the molecular biology of vaccinia virus and the pathogenesis of vaccinia, and a chapter on this new development is included.

Because of the time taken for its writing and production, a book cannot hope to be completely up-to-date in such a rapidly advancing field as molecular biology. However, we have endeavored to provide an overview of the molecular biology of the orthopoxviruses as current as mid-1987. We believe that a comprehensive account of all aspects of the biology of all known species of orthopoxviruses, placed in historical context, will be useful to molecular biologists, virologists, immunologists, pathologists, and researchers in the veterinary sciences.

We are grateful to the staff of Academic Press for their assistance in the production of this book.

Frank Fenner
Riccardo Wittek
Keith R. Dumbell

CHAPTER 1

Historical Introduction and Overview

Smallpox, the first human infectious disease to have been rendered extinct by a deliberate campaign of eradication (World Health Organization, 1980a; Fenner *et al.*, 1988), was once the most serious scourge of mankind. Unlike malaria, it was not limited by climate, and unlike plague, it was always present. Before any preventive measures were developed it was associated with case-fatality rates of about 25%, and when large communities were first exposed to it, as were the Amerindians of Mexico and Peru in the sixteenth century, it exerted a major influence on geopolitics. Hopkins (1983) has described the complex and fascinating role of smallpox in history. Its interest to us is that it was

1

caused by an orthopoxvirus, and its importance gave a special emphasis and urgency to the study of the orthopoxviruses from the earliest days of microbiology.

During the nineteenth century the importance of smallpox and the fact that at that time a related orthopoxvirus was the only known vaccine for any infectious disease combined to make the vaccine virus a favored agent for the laboratory investigation of animal viruses. In consequence, vaccinia virus dominated the newly developing field of virology during the latter part of the nineteenth century and the first three decades of the twentieth century, rivaled only by the viruses of rabies, yellow fever, and poliomyelitis. The fact that the vaccinia virus particle was among the largest of virions and lent itself to physical and chemical study using the relatively primitive biophysical instrumentation available before 1950 allowed the early workers to make several observations of fundamental importance to virology (Fenner, 1979). Likewise vaccinia virus was widely used in the early attempts to grow viruses in cultured cells.

Subsequently another orthopoxvirus, mousepox, was used as a model for the study of the pathogenesis of generalized viral infections and the role of cellular immunity in recovery from viral infections. Other orthopoxviruses have provided interesting insights into the natural history of viral infections. The historical development of these studies will be briefly reviewed here as a background for the detailed description of the present state of knowledge covered in later chapters of the book.

VARIOLATION AND VACCINATION

As well as causing many deaths, smallpox left its mark on most of those who survived; the pock-marked face was indisputable evidence of past infection and it was widely recognized that second attacks of smallpox were virtually unknown. Long before the elaboration of the germ theory of infectious disease, people in India and China, and possibly also in Africa, discovered that they could transfer the disease with material extracted from the vesicles and scabs of smallpox patients. What is more, although a few deaths occurred, they observed that such inoculated persons usually suffered a mild disease that protected them from natural smallpox. The practice of "inoculation," or "variolation," as it was later called, spread to western Asia. During the early eighteenth century, when smallpox had become a terrible scourge in Europe, knowledge of the efficacy of variolation was spread from

Turkey to northern Europe. Colonists in North America learned about it from their African slaves as well as from publications in the *Philosophical Transactions of the Royal Society of London,* which was the most important international journal of science in the early eighteenth century.

Variolation flourished in most countries of Europe throughout the eighteenth century, and in the hands of some of its practitioners it was a very effective and relatively safe operation (Razzell, 1977b). But it caused very severe local reactions, there were usually some pustules elsewhere on the body and often a generalized rash, and some of those inoculated died of smallpox. More important, the disease could spread from inoculated subjects to cause severe "natural" smallpox in uninoculated persons. For these reasons, and because smallpox itself was then such a universal and severe disease, Jenner's discovery proved very attractive. He showed (Jenner, 1798) that inoculation of material from lesions of a pox disease of cows produced only a local lesion in the inoculated individual, but provided protection from a smallpox inoculation made some weeks later, and also from disease following exposure to cases of smallpox. Vaccine inoculation, as the practice was named, swept Europe within a few years of Jenner's publication (Baxby, 1981) and the cowpox material was carefully transported, by serial inoculation of children, to the Spanish colonies in the Americas and in China and the Philippines early in the nineteenth century (Smith, 1974).

As far as the science of virology was concerned, the importance of vaccination was that it made an excellent model virus available for study in the bacteriological laboratories that developed in the latter part of the nineteenth century. Not only was the vaccine virus of great practical importance, but it had two very important advantages over variola virus: it was safe to handle and did not cause a severe or transmissible disease and it produced lesions in a wide range of laboratory animals. In consequence the vaccine virus became a popular object of study in laboratories where scientists had become interested in noncultivable and apparently invisible infectious agents—what were then called the "filterable" viruses.

CLASSIFICATION OF THE ORTHOPOXVIRUSES

The demonstration that the viruses that caused smallpox and cowpox were closely related was extended by clinical observations of diseases with similar clinical signs in other domestic animals: horse, sheep, goat, and pig, all of which were grouped together as "pox" diseases. It subsequently transpired that poxviruses caused severe infectious diseases in all these animals, but that the viruses that caused sheep pox, goatpox, and swinepox belonged to different genera of the family *Poxviridae*. The clinical classification was spread even wider to include other diseases of man that produced lesions in the skin: the "great pox" (syphilis) and chickenpox, which are not produced by poxviruses at all. The early workers had suggested a classification of diseases, not of microbes; the latter step was impossible before more was known about the nature of the infectious agents that caused these "pox" diseases.

The large size of the poxvirus particle and the characteristic cytoplasmic inclusion bodies that pathologists observed in sections of poxvirus-infected cells provided criteria for a classification which was based on the nature of the causative agent rather than the disease. As early as 1886, Buist had observed poxvirus particles in stained smears, and early in the twentieth century several methods were developed for staining these minute particles (called "elementary bodies" to distinguish them from inclusion bodies) so that they could be readily seen with a high-power optical microscope (see Plate 1-1). For this reason a classification of viruses based on properties of the particle rather than the nature of the lesions was first developed with the poxviruses, and the morphology of the virion remains the first criterion for the inclusion of an unknown virus in the family *Poxviridae*.

In 1927 Aragão, the Brazilian scientist who first recognized the potential of myxomatosis as a method of biological control of the rabbit (Fenner and Ratcliffe, 1965), grouped together the viruses of "myxoma, variola, bird-epithelioma (fowlpox), molluscum contagiosum, etc.," on the basis of the morphology of their inclusion bodies and elementary particles. A few years later Goodpasture (1933), the distinguished pathologist who introduced chick embryo inoculation into virology (Woodruff and Goodpasture, 1931), formally proposed that vaccinia–variola, fowlpox, horsepox, sheep pox, swinepox, and molluscum contagiosum viruses should be grouped together as a genus for which he proposed the name *"Borreliota."*

These were isolated proposals to deal with the classification of poxviruses only. It was not until 1948 that the first comprehensive classification of viruses was proposed by Holmes in an appendix to the sixth edition of Bergey's *Manual of Determinative Bacteriology* (Holmes, 1948). Although a comprehensive classification was a step forward conceptually, the means used was a step backward, for like the prescientific philosophers Holmes, a plant virologist by training, placed major reliance on the clinical signs of disease. In consequence, he reversed the proposals of Aragão and Goodpasture and placed molluscum contagiosum and myxoma viruses into a genus termed *Molitor*, and other poxviruses into Goodpasture's genus *Borreliota*. Holmes' "nomenclatorial bomb-shell" (Andrewes, 1953) stimulated others, especially the animal virologists, to take classification more seriously. Since to be useful, classification must be subject to international agreement, Andrewes (1951, 1952) promoted interest in viral classification through the International Association of Microbiological Societies, which organized regular triennial international congresses for microbiology. Following the establishment of a Poxvirus Subcommittee by the Sixth International Congress for Microbiology in 1953, Fenner and Burnet (1957) produced a short description of the poxvirus group that has remained the basis of subsequent classifications in respect of the criteria used and the subdivisions adopted.

An International Committee for the Nomenclature of Viruses (ICNV) was established at the Ninth International Congress for Microbiology in 1966, and its name was changed to International Committee for the Taxonomy of Viruses (ICTV) in 1974 (Matthews, 1983). Successive reports from this committee set out the accepted views on the classification and nomenclature of viruses at the times indicated (Wildy, 1971; Fenner, 1976, Matthews, 1979, 1982). Since no changes have been made in the official classification and nomenclature of the family *Poxviridae* since 1979, we have made a few additions to the official ICTV names in order to reflect present knowledge of the poxviruses of vertebrates more accurately (Table 1-1).

TABLE 1-1

Classification of Poxviruses of Vertebrates

Family: *Poxviridae;* Subfamily: *Chordopoxvirinae*

Genus	Prototype virus
Orthopoxvirus[a]	Vaccinia virus
Parapoxvirus[a]	Pseudocowpox virus
Capripoxvirus	Sheep pox virus
Suipoxvirus	Swinepox virus
Leporipoxvirus	Myxoma virus
Avipoxvirus	Fowlpox virus
Yatapoxvirus[a,b]	Tanapox virus
Molluscipoxvirus[a,b]	Molluscum contagiosum virus

[a] Genus includes viruses that infect humans.
[b] Name not yet approved by ICTV.

SPECIES WITHIN THE GENUS *ORTHOPOXVIRUS*

Formerly called the "vaccinia subgroup," or the "vaccinia–variola subgroup," the genus *Orthopoxvirus* includes a number of viruses that are morphologically identical, show extensive serological cross-reactivity and cross-protection in animals, and are found as naturally occurring infections in a variety of different animals. Initial allocation of a poxvirus to the genus *Orthopoxvirus* is made primarily on the basis of cross-protection in experimental animals and cross-neutralization of infectivity. Traditionally, allocation of an orthopoxvirus to a particular species has been based on a range of biological characteristics, of which the most important were the host range (in intact animals) and the morphology and ceiling temperature of the pock produced on the chorioallantoic membrane of the developing chick embryo (Table 1-2). On these criteria, the Fourth Report of the International Committee for the Taxonomy of Viruses (Matthews, 1982) listed the following viruses as members of the genus *Orthopoxvirus:* vaccinia, buffalopox, camelpox, cowpox, ectromelia, monkeypox, rabbitpox, and variola. Carnivorepox and elephantpox ("related to cowpox") and raccoonpox ("probable orthopoxvirus") were listed as other members of the family *Poxviridae* not yet allocated to genera. Tatera poxvirus and Uasin Gishu poxvirus were not mentioned, although both had been described as probable orthopoxviruses a decade earlier, and the California vole orthopoxvirus was not described until 1987. Both rabbitpox and buffalopox viruses are now recognized as strains of vaccinia virus.

Biological characteristics do not always provide an unequivocal way of classifying isolates of orthopoxviruses, especially those with a broad host range like cowpox and vaccinia viruses. The discovery of bacterial enzymes that cleaved DNA at specific sites—site-specific restriction endonucleases (Smith and Wilcox, 1970; see Smith, 1979)—provided another approach to the classification of viruses with DNA genomes. Gangemi and Sharp (1976) showed that two closely related strains of vaccinia virus could readily be distinguished by a small difference in the restriction patterns of their DNAs, and Wittek *et al.* (1977) used overlapping fragment analysis to map the DNAs of rabbitpox and vaccinia viruses, and showed that the internal 60% of the genomes had identical restriction patterns. Mackett and Archard (1979) then showed that the DNAs of all strains of each of five different species of *Orthopoxvirus* showed distinctive patterns after digestion with restriction endonucleases, but shared a large conserved central segment of their genomes. This approach has provided another and a better criterion for the subdivision of the genus into species. Restriction endonuclease

TABLE 1-2

Biological Characteristics of Different Species of Orthopoxvirus[a]

Character	Camelpox virus	Cowpox virus	Ectromelia virus	Monkeypox virus	Raccoon poxvirus	Tatera poxvirus	Uasin Gishu poxvirus	Vaccinia virus	Variola virus	Vole poxvirus
Pocks on CAM[b]	Small, opaque, white	Large, hem.[c]	Very small, opaque, white	Small, opaque, hem.	Very small, opaque, white	Small, opaque, white	Medium size, opaque	Strains vary; large, opaque, white, or hem.	Small, opaque, white	Very small, opaque, white
Ceiling temperature (CAM)	38.5°C	40°C	39°C	39°C	?	38°C	?	41°C	37.5–38.5°C	40°C
Rabbit skin lesion	Transient, non-transmissible	Indurated, hem.	Transient, non-transmissible	Indurated, hem.	Small nodule	Transient, non-transmissible	No lesion	Strains vary; indurated nodule, sometimes hem.	Transient, non-transmissible	Small nodule
Disease in monkeys	Large lesion, localized	Large lesion, localized	?	Generalized rash		?	Susceptible; no rash	Large lesion, localized	Generalized rash	?
Lethality for Mice	Low	Variable	Very high	High	High	Low	Pocks in baby mice	Strains vary; high to very high	Low	High in baby mice
Chick embryos	Low	High	High	Medium	?	Low	?	Very high	Low	?
Type A inclusion bodies	−	+	+	−	+	?	−	−	−	?
Thymidine kinase sensitivity to TTP[d]	−	−	−	−	?	−	?	−	+	?

[a] Data on raccoon poxvirus, tatera poxvirus, Uasin Gishu poxvirus, and vole poxvirus based on a few papers on single viral isolates.
[b] CAM, Chorioallantoic membrane; examined at 48 hours for vaccinia virus; 72 hours for all others.
[c] hem., Hemorrhagic.
[d] TTP, Thymidine triphosphate.

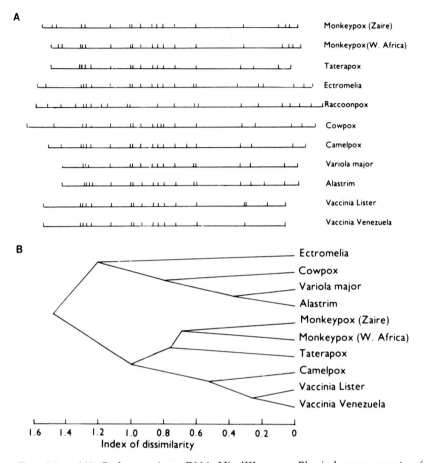

FIG. 1-1. *(A)* Orthopoxvirus DNA HindIII *maps. Physical arrangements of fragments of the genomic DNA of one or two isolates of each species except Uasin Gishu poxvirus and the vole orthopoxvirus, the DNAs of which have not been examined. The consecutive order of digest fragments was determined by cross-hybridizations with cloned DNA fragments of orthopoxviruses with known HindIII maps. From top: Monkeypox (Zaire), monkeypox virus, Congo-8 strain; monkeypox (W. Africa), monkeypox virus, Copenhagen strain; taterapox, tatera poxvirus, Dahomey strain; ectromelia, ectromelia virus, Hampstead strain; raccoonpox, raccoon poxvirus, Maryland strain; cowpox, cowpox virus, Brighton strain; camelpox, camelpox virus, Somalia strain; variola major, variola virus, Harvey strain; alastrim, variola virus, Butler strain; vaccinia Lister, vaccinia virus, Lister strain; vaccinia Venezuela, vaccinia virus, Venezuela strain. (Data from Esposito and Knight, 1985.) (B) Dendrogram showing dissimilarities between these orthopoxviruses (except for raccoon poxvirus), using centroid sorting strategy (Gibbs and Fenner, 1984).*

mapping of the viral DNA was applied by Esposito and Knight (1985) to all known members of the genus *Orthopoxvirus* except Uasin Gishu poxvirus, which was not available in the United States, and the vole orthopoxvirus, which was not described until 1987 (Fig. 1-1). Restriction mapping will be used in this book as the final criterion for allocating a poxvirus to the genus *Orthopoxvirus*, and within the genus for allocating an isolate to a particular species of *Orthopoxvirus*. Table 1-3 sets out the current classification of the species of *Orthopoxvirus*, based on biological characters and DNA maps, and indicates the animals found naturally infected and the breadth of host range in laboratory animals.

TABLE 1-3
Species of the Genus Orthopoxvirus

Species	Host range in laboratory animals	Animals found naturally infected	Geographic range of natural infections
California vole poxvirus	(?) Broad	Voles	United States
Camelpox virus[a]	Narrow	Camel	Africa and Asia
Cowpox virus	Broad	Numerous: carnivores, cow, elephant, gerbils, man, okapi, rats	Europe and USSR
Ectromelia virus	Narrow	Mice, (?) voles	Europe
Monkeypox virus	Broad	Numerous: anteater, great apes, man, monkeys, squirrels	Western and central Africa
Raccoon poxvirus[a]	(?) Broad	Raccoon	United States
Tatera poxvirus	Narrow	*Tatera kempi* (a gerbil)	Western Africa
Uasin Gishu poxvirus	Medium	Horse	Kenya, Zambia
Vaccinia virus	Broad	Numerous: buffalo, cow, man, pig, rabbit	(?) India (buffalopox)
Variola virus	Narrow	Man (disease now eradicated)	Formerly worldwide

[a] Viruses are called "camelpox virus," etc., when they are known to cause a disease with rash (pocks) in the animal mentioned; as "raccoon poxvirus," etc., when no such disease is known.

SPECIES OF *ORTHOPOXVIRUS:* AN OVERVIEW

In the pages which follow the history of the recognition of each of the species of *Orthopoxvirus* is briefly reviewed. The viruses are discussed in alphabetical order.

Camelpox Virus

This virus shares many biological properties with variola virus (Baxby, 1972, 1974), but has a distinctive genome structure (Fig. 1-1). It was first isolated in tissue culture by Ramyar and Hessami (1972) and its affinities with the genus *Orthopoxvirus* were first recognized by Baxby (1972). The camel appears to be the only natural host, and the disease it causes in camels is described in Chapter 10.

Cowpox Virus

Cowpox has been recognized for several hundred years as a disease of cows that is transmissible to man, producing ulcers on the cow's teats or the milker's hands. What is now recognized as cowpox virus was probably the virus originally used for human vaccination (Jenner, 1798); it is not clear when and how it was eventually supplanted by vaccinia virus as the virus used for vaccination. The distinction between cowpox and vaccinia viruses was first made by Downie (Davies *et al.*, 1938; Downie, 1939a,b).

Recently a large number of strains of *Orthopoxvirus* recovered from a variety of animal species in zoos and circuses, and from rodents (Table 1-4), have been recognized as being very similar to the strains of cowpox virus that have been recovered from cows and man in both their biological properties (Baxby and Ghaboosi, 1977; Baxby *et al.*, 1979a,b; Marennikova, 1979) and their genome maps (Mackett, 1981). We have allocated all of these strains to the cowpox virus species (see Chapter 6).

Ectromelia Virus

Infectious ectromelia, later called mousepox, was described in England by Marchal (1930) and subsequently recovered from laboratory mice in many parts of the world (Fenner, 1982; see Chapter 9). Serological studies (Kaplan *et al.*, 1980) show that voles in Britain are a reservoir host of an *Orthopoxvirus*, which is probably either ectromelia virus or cowpox virus.

TABLE 1-4

Probable Affinities of Certain Orthopoxviruses Recovered
from Various Animal Sources[a]

Viral species	Animal from which isolated	Reference
Cowpox	Domestic cat	Baxby *et al.* (1979a)
	Elephant ("elephantpox")	Baxby and Ghaboosi (1977)[b]
	Large carnivores	Marennikova *et al.* (1977)[b]
		Baxby *et al.* (1979a)[b]
	Okapi	Zwart *et al.* (1971)
	White rat ("ratpox")	Marennikova *et al.* (1978a)[b]
	Wild rodents (Turkmenia)	Marennikova *et al.* (1978b)
Monkeypox	Anteater	
	Cercopithecus hamlyni	
	Chimpanzee	
	Gibbon	Peters (1966)[b]
	Gorilla	
	Marmoset	
	Orangutan	
	Squirrel monkey	
	Cynomolgus monkey cells[c]	Gispen and Kapsenberg (1966)
	Squirrel (captured in wild)	Khodakevich *et al.* (1985)
Vaccinia	Buffalo ("buffalopox")	Baxby and Hill (1971)
	Camel	Krupenko (1972)
	Cow	El Dahaby *et al.* (1966);
		Dekking (1964)
	Monkey (MK 10)[e]	Shelukhina *et al.* (1975)
	Pig	Maltseva *et al.* (1966)
	Rabbit ("rabbitpox")	Jansen (1946);
		Rosahn and Hu (1935);
		Christensen *et al.* (1967)
	Cynomolgus monkey cells[c]	Gispen and Kapsenberg (1966)
Variola	Cynomolgus monkey cells[d]	Gispen and Kapsenberg (1966)
("whitepox")	Chimpanzee[e]	Marennikova *et al.* (1972b)
	Sala monkey[e]	
	Sun squirrel[e]	Marennikova *et al.* (1976a)
	Multimammate rat[e]	

[a] The recovery of the viruses listed from the particular animal species (usually zoo or domestic animals) does not preclude the possibility that these animals may be natural hosts for other orthopoxviruses.

[b] Outbreaks associated with zoological gardens or circuses.

[c] Possible laboratory contaminants (Dr. J. G. Kapsenberg, personal communication, 1983).

[d] Proved laboratory contamination (Dumbell and Kapsenberg, 1982).

[e] Probable laboratory contaminants (see Chapter 7).

Monkeypox Virus

This virus was first recovered from cynomolgus monkeys that had been captured in Malaysia in 1958 and shipped by air to Copenhagen, where they were housed together for several weeks before the disease was seen (von Magnus *et al.*, 1959). Several other isolations of the virus were subsequently made from captive primates (Arita and Henderson, 1968). It has a wide host range (Table 1-4). In 1970 monkeypox virus was isolated from skin lesion material from four cases of human disease diagnosed clinically as smallpox, and occurring in Zaire, Liberia, and Sierra Leone (Foster *et al.*, 1972; Ladnyi *et al.*, 1972; Lourie *et al.*, 1972; Marennikova *et al.*, 1972a). Human monkeypox has now been recognized as a rare zoonosis occurring in many parts of western and central Africa (see Chapter 8).

Raccoon Poxvirus

The first indigenous orthopoxvirus to be recovered from the Americas, this virus (see Chapter 10) was isolated from raccoons in the eastern United States (Herman, 1964; Alexander *et al.*, 1972) and unequivocally characterized as an orthopoxvirus by Thomas *et al.* (1975). Esposito and Knight (1985) found that only about half its *Hind*III restriction fragments cross-hybridized with those of vaccinia virus DNA, but Parsons and Pickup (1987) showed that all of them shared some nucleotide sequences with cowpox virus DNA.

Tatera Poxvirus

This virus (see Chapter 10) was recovered from pooled liver/spleen material obtained from small naked-soled gerbils (*Tatera kempi*) captured in Dahomey (present-day Benin) in 1964. It was characterized as an orthopoxvirus on biological criteria by Lourie *et al.* (1975) and shown to have a distinctive genome map by Esposito and Knight (1985) (see Fig. 1-1).

Uasin Gishu Poxvirus

Uasin Gishu disease is a skin disease long known to occur in horses in various parts of Kenya (see Chapter 10). Kaminjolo *et al.* (1974a,b) showed that it was caused by a poxvirus which cross-reacted serologically with vaccinia and cowpox viruses. The horse, a relatively recent arrival in Kenya, is clearly merely a sentinel indicator animal of a virus occurring in some unknown African wildlife reservoir.

Vaccinia Virus

This is the virus that has been most widely used for vaccination (see Chapter 5). Baxby (1977b, 1981) has summarized speculations about its origins, and added some of his own. Many strains were supposed to have been derived from variola virus (Wokatsch, 1972). However, its DNA is so different from that of variola virus (and all other orthopoxvirus species) that "transformation" of one species into another would be impossible without a long and highly unlikely sequence of nucleotide changes (see Chapter 2).

Because of its long history of use in vaccine production laboratories and for research purposes, there are many strains of vaccinia virus, but all strains have many biological features in common, and all have a distinctive genome map (see Chapter 5). Since vaccinia virus has a broad host range and has been very widely used for many decades, accidental infections of domestic animals were not uncommon when human vaccination was practiced on a widespread scale (Table 1-4). Sometimes serial transmission occurred naturally in animals (cows, buffaloes, rabbits); viruses derived from such sources were sometimes called "buffalopox virus" or "rabbitpox virus."

Variola Virus

This virus, which caused human smallpox, has a restricted host range in experimental animals. Monkeys were infected with variola virus as early as 1874 (Zuelzer, 1874) and were extensively used by early pathologists (e.g., Brinckerhoff and Tyzzer, 1906). Variola virus was subsequently grown in the rabbit cornea, a test was developed to differentiate variola from chickenpox (Paul, 1915), and later it was grown in chick embryos (Torres and Teixeira, 1935). North *et al.* (1944) and Downie and Dumbell (1947) showed that the pocks produced by variola virus on the chorioallantoic membrane were sufficiently distinctive to allow its differentiation from "vaccine virus" (both vaccinia and cowpox viruses) by this method. Variola virus DNA has a distinctive restriction map (Mackett and Archard, 1979).

Vole Orthopoxvirus

The most recent addition to the genus *Orthopoxvirus* is a virus recovered from California voles (*Microtus californicus*) by Regnery (1987). Serological surveys showed that it is enzootic in several vole populations in the San Francisco area.

PHYSICAL PROPERTIES OF THE *ORTHOPOXVIRUS* VIRION

The circumstances described in the opening section of this chapter led to the extensive use of vaccinia virus for laboratory studies during the late nineteenth and early twentieth centuries. The large size of its virion and the stability of its infectivity combined to allow scientists of that time to make observations on the physical and chemical structure of vaccinia virus that were of fundamental importance to virology as a whole.

The Particulate Nature of Virus Particles

The brilliant discoveries of Pasteur and Koch in the latter half of the nineteenth century had demonstrated the particulate nature of the bacterial causes of many infectious diseases. Following the demonstrations that the agents of foot-and-mouth disease (Loeffler and Frosch, 1898) and tobacco mosaic disease (Beijerinck, 1899) passed through filters that retained bacteria, there were serious discussions up to the 1930s, especially in relation to plant and bacterial viruses, as to whether such agents were particulate. Medical and veterinary pathologists had less doubts about the particulate nature of viruses, largely because of studies made with the orthopoxviruses.

The virions of both variola and vaccinia viruses were first seen and described by Buist (1886; see Gordon, 1937; Mackie and van Rooyen, 1937). Subsequently, and apparently unaware of Buist's work, Calmette and Guérin (1901) drew attention to the presence of many very small refringent granules in vaccine lymph which they thought might be the "virulent elements." This opinion was endorsed by Prowazek (1905), who demonstrated that such "elementary bodies" could be stained by either Giemsa or Victoria Blue stains. Paschen (1906, 1924) applied improved staining methods (Plate 1-1) and became the champion of the view that these particles (which came to be known as "Paschen bodies") were indeed the viral particles (see Wilkinson, 1979).

At this point it is relevant to remark that viruses need to be thought about as a two-phase system. The virus particle, or virion as Lwoff *et al.* (1959) called it, about which work on the physical and chemical structure of "viruses" is concerned, is an inert phase whose significance is that it provides a way in which viruses can pass from one cell to another and from one host to another. In their dynamic phase, the "vegetative" stage, the virion ceases to exist; its nucleic acid interacts with components of the cell that the virion enters to produce a new entity, the virus-infected cell, in which viral replication proceeds. During this stage viruses are not particulate; progeny virions are produced at the end of the replication cycle.

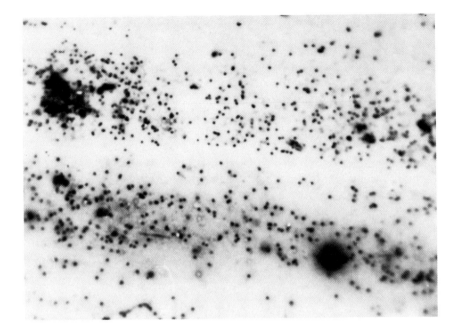

PLATE 1-1. *Virions ("elementary bodies") of variola virus in vesicle fluid, as seen with an optical microscope. Gutstein's stain, ×2500. (Courtesy Professor A. W. Downie.)*

As we have already remarked, the particulate nature of the virions of poxviruses was important in providing a basis for the rational classification of viruses. It also provided a means for monitoring the purification of viral suspensions, which made possible studies on the chemical composition of the virions of vaccinia virus (see below) and experiments by Parker (1938) and Smadel *et al.* (1939), which led to the conclusion that under the proper conditions a single active elementary body of vaccinia virus could initiate infection. This does not imply that all particles in a suspension have equal potency.

The Morphology of the Virion

The elementary bodies, as visualized by the light microscope, were too small for details of their structure to be seen. Radiation of shorter wavelength than visible light was needed if their substructure was to be studied. Barnard developed an "ultraviolet microscope" with which he demonstrated that the virions of ectromelia virus were oval bodies with a clearly defined refractile border and that the inclusion bodies (A type) contained myriads of these tiny particles (Barnard and Elford, 1931; see Plate 9-4).

The next step forward was the development of electron microscopy, which has had impacts on virology and cellular biology comparable with the effects of Leeuwenhoek's development of the optical microscope on bacteriology and protozoology. A tool was now available for the structural analysis of virions, and a comparison was soon made by Ruska and Kausche (1943) of the size and shape of the virions of several poxviruses.

As the resolution of electron microscopes has increased and new methods (metal shadowing, negative staining, thin sectioning, freeze-etching, scanning electron microscopy) have been developed, they have been applied to the understanding of the structure of the virion and viral replication and maturation in the poxvirus-infected cell (see Chapters 2 and 3). Enzymatic digestion as a method for demonstrating structure within virions was first developed in experiments with vaccinia virus (Dawson and McFarlane, 1948).

CHEMICAL COMPOSITION OF THE *ORTHOPOXVIRUS* VIRION

Chemical studies of virions were impossible until they could be purified, hence the significance of Stanley's (1935) "crystallization" of tobacco mosaic virus. Vaccinia virus was the first animal virus of which the virions were purified sufficiently to make accurate chemical analysis possible. This was done by T. M. Rivers and his colleagues at the Rockefeller Institute in New York between 1935 and 1943 (Parker and Rivers, 1935; Benison, 1967; for review, see Smadel and Hoagland, 1942). The quality of these chemical analyses is attested by the fact that Zwartouw's (1964) work over 20 years later was largely confirmatory of the earlier findings.

The method of purification was derived from a technique developed by Craigie (1932); the shaved skin on a rabbit's back was scarified with a selected strain of virus and the product harvested 3 days later by gentle scraping. The material thus obtained was subjected to several cycles of differential centrifugation, a technique first used by MacCallum and Oppenheimer (1922) and by Ledingham (1931) in experiments with vaccinia virus. New methods of purification of poxvirus virions were not introduced until Joklik (1962a) described the use of density gradient centrifugation for the purification of orthopoxviruses grown in the chorioallantoic membrane; this method is still widely employed.

The Rockefeller Institute workers established a number of facts about the chemical composition of vaccinia virions: they contained 5.6% DNA (Hoagland *et al.*, 1940b) but no RNA (Smadel *et al.*, 1940), about 5% lipid, part of which was phospholipid (Hoagland *et al.*, 1940a), and all but traces of the carbohydrate present could be accounted for as the sugar component of the DNA. They also found that suspensions of virions displayed a number of enzymatic activities, but it was shown that alkaline phosphatase, catalase, and lipase were probably adsorbed impurities (Hoagland *et al.*, 1942). Much later, a range of enzymes was shown to be associated with the virions of vaccinia virus (see Table 2-1).

Serological methods were used for analyzing the virus-specific antigens in poxvirus virions and poxvirus-infected cells. These studies revealed that the virions were antigenically very complex and that poxvirus-infected cells contained, in addition to the components of the virion, several nonstructural virus-specific antigens.

The major advances in the chemical characterization of viruses that have occurred since the 1950s have depended upon a number of techniques of physical biochemistry, such as purification by gradient centrifugation (rate-zonal and isopycnic), separation of proteins and nucleic acids by gel electrophoresis, characterization of homology between RNAs and DNAs by hybridization, the analysis of DNAs by use of restriction endonucleases, and combinations of these and other methods.

One-dimensional gel electrophoresis of disrupted vaccinia virions, first used by Holowczak and Joklik (1967a,b), was subsequently developed to the stage that 48 different polypeptides could be distinguished (McCrae and Szilagyi, 1975); two-dimensional gel electrophoresis has increased this number to over 110 (Essani and Dales, 1979), the combined molecular mass of which accounts for about two-thirds of the genetic information in the viral genome. There are, in addition, some 50 "early" nonvirion polypeptides present in extracts of virus-infected cells.

While some of the enzymes found in "purified" preparations of poxvirions by the Rockefeller Institute workers were adsorbed impurities, it has now been shown that many different enzymatic activities do occur within the cores of vaccinia virions. Historically the most important aspect of the discovery of the virion-associated enzymes of poxviruses was that the demonstration of a virion-associated DNA-dependent RNA polymerase (Kates and McAuslan, 1967b; Munyon *et al.*, 1967) stimulated other research workers to look for transcriptases in the virions of other viruses. These were soon found, including an RNA-dependent RNA polymerase in reoviruses and the viruses whose genome consisted of $(-)$ sense single-stranded RNA, and the reverse transcriptase of the retroviruses (for review, see Baltimore, 1971).

Since Hirst's (1941) discovery of hemagglutination by influenza virions, this reaction has been demonstrated with a wide range of viruses, including poxviruses of the genus *Orthopoxvirus* (Nagler, 1942). It was the discovery that ectromelia virus caused hemagglutination that led Burnet and Boake (1946) to investigate its antigenic relationship to vaccinia virus and to classify it as an orthopoxvirus. Viruses of many families produce hemagglutination; in all families except the orthopoxviruses the virion itself induces hemagglutination. However, as first shown by Chu (1948), the virion of vaccinia virus is separable from the hemagglutinin, some of which is derived from the plasma membrane of the infected cell (Blackman and Bubel, 1972). A minority of the progeny virions produced by cells infected with vaccinia virus—those released from infected cells before they lyse—have an envelope which contains the hemagglutinin and several other virus-specified polypeptides (Payne, 1979), but this is not necessary for infectivity. However, the envelope polypeptides play an important role in the protection provided by vaccination (Boulter, 1969; Boulter and Appleyard, 1973) and the envelope itself may be important in the spread of vaccinia virus around the body (Payne, 1980).

Since the genetic information which directs the synthesis of new virions resides in the viral DNA, the chemical study of poxvirus DNA is being pursued with increasingly sophisticated tools. The vaccinia virus genome is a double-stranded DNA molecule with terminal cross-links forming covalently "closed" ends, so that when denatured it is converted into a single-stranded circular configuration encompassing the complementary halves (Geshelin and Berns, 1974). Its molecular mass is about 120×10^6, comprising 180 kbp; it contains no unusual bases and has a very low G plus C content (35%).

By cloning fragments of the vaccinia virus genome (Wittek *et al.*, 1980a) it has been possible to study the replication cycle of vaccinia virus at the molecular level (see Chapter 3). A further development of considerable practical importance has been the insertion into vaccinia virus DNA of cloned genes from other sources—viruses, bacteria, protozoa, and vertebrates (Mackett *et al.*, 1984). This procedure and its applications are described in Chapter 12.

THE USE OF LABORATORY ANIMALS FOR POXVIRUS RESEARCH

Variolation was the first example of the deliberate propagation of a virus, in that variolators obtained crusts from smallpox patients, ground these up, and used the powder for the inoculation of their subjects, a practice that dates back many hundreds of years in China and India. Likewise, arm-to-arm transfer of vaccine virus, originally suggested by Jenner, was employed extensively for maintaining virus for vaccination during the first half of the nineteenth century.

Deliberate large-scale production of vaccine virus in animals was practised in Italy from about 1840. In the 1860s scarification of the skin of calves as a method of production of vaccine spread to France, Germany, Holland, Belgium, Austria, and Japan. Production of vaccine in calves did not begin in England, Jenner's own country, until 1881, and arm-to-arm vaccination was continued there until 1898 (Dudgeon, 1963). Since the beginning of the twentieth century vaccinia virus has been progagated for the production of smallpox vaccine by the scarification of the skin of cows, sheep, and buffaloes. Although chick embryos and cultured cells have been used in a few vaccine production laboratories, the scarification of large animals remained the commonest method of vaccine production throughout the global smallpox eradication campaign.

For experimental purposes, smaller laboratory animals were required, and much work was done by workers at the Pasteur Institute at the end of the nineteenth century and during the first three decades of this century on various aspects of the pathology of vaccinia virus infection of rabbits. Following early experiments on rabbit inoculation by Calmette and Guérin (1901), Noguchi (1915) used intratesticular inoculation to obtain bacteria-free suspensions of virus. Subsequently the Pasteur Institute group (for references, see van Rooyen and Rhodes, 1948) carried out extensive experiments on the serial passage of vaccinia virus in rabbits, and showed that variants that differed substantially in their virulence and pathogenic behavior could be selected by passing the virus either by scarification of rabbit skin ("dermovaccine") or by intratesticular or intracerebral inoculation ("neurovaccine"). As related earlier, scarification of rabbit skin with a dermal vaccinia strain was subsequently developed as a method of obtaining high yields of purified virions, which made their chemical study possible. Paul's test for differentiating between smallpox and chickenpox (Paul, 1916) consisted of the scarification of the rabbit cornea and subsequent examination for opacities and the presence of Guarnieri bodies. Inoculation into the rabbit skin later proved to be a useful method for the biological differentiation of different species of *Orthopoxvirus*.

Monkeys were also used from an early period, and were of particular value for the study of the pathogenesis of variola virus infections, since they are the only experimental animals which develop a generalized infection with a rash after inoculation with variola virus (Hahon, 1961). Monkeys were also useful for the study of the pathogenesis of monkey-pox (Cho and Wenner, 1973), although many other laboratory animals are susceptible to infection with monkeypox virus. Mice proved useful for comparative studies of different species of orthopoxviruses and have been extensively used for a variety of laboratory investigations of infectious ectromelia (mousepox; see Chapter 9).

The developing chick embryo was introduced into virological research for the study of poxviruses (Woodruff and Goodpasture, 1931; Good-pasture et al., 1932). Burnet (1936) exploited the technique of chorioal-lantoic inoculation, using the appearance of the individual pocks pro-duced on the chorioallantoic membrane as a diagnostic tool and for quantitative studies. The appearance of the pocks produced on the chorioallantoic membrane by different poxviruses remains the single most useful biological test for differentiating between various species of orthopoxvirus.

GROWTH OF ORTHOPOXVIRUSES IN CULTURED CELLS

The first attempts to grow a virus in cells removed from the animal body were made with vaccinia virus inoculated onto rabbit cornea (Aldershoff and Broers, 1906). A few years later Steinhardt et al. (1913) showed that vaccinia virus survived for several weeks in fragments of guinea pig and rabbit cornea, and may have multiplied. The first unequivocal evidence that viruses could multiply in cultured cells was the demonstration by Parker and Nye (1925) that vaccinia virus could be carried through 11 passages in rabbit testicular tissue and that the last culture contained over 10^4 times as much virus as the initial preparation. Maitland and Maitland (1928) then showed that vaccinia virus could be grown in fragments of kidney tissue suspended in a mixture of serum and inorganic salts. Some years later, vaccinia virus was the first virus to be grown in cells cultivated in roller tubes (Feller et al., 1940).

Vaccinia virus was the first agent grown in cell culture to be used for the immunization of humans (Rivers, 1931), but this method was never widely used for smallpox vaccine production. With the introduction of a wide range of primary and continuous cell lines into virological re-search, cell culture has become an essential part of research with orthopoxviruses. Differential cellular susceptibility is a useful adjunct to the reactivity of experimental animals for the differentiation of different species of orthopoxvirus, e.g., camelpox and variola viruses (Baxby, 1974).

Orthopoxviruses such as vaccinia and monkeypox viruses, which produce disease in a wide range of laboratory animals, also replicate in a wide range of cell types; others, like variola and camelpox viruses, have a restricted range of susceptible animal hosts and cell types.

ASSAY OF VIRAL INFECTIVITY

Closely connected with methods of propagation was the development of methods of assay of viral infectivity—a basic requirement for quantitative virology. Here again, vaccinia virus, as the most readily available and best known "laboratory" animal virus, was important in many of the early investigations. The earliest attempts at the assay of viral infectivity were made to determine the potency of vaccinial "lymph" used for smallpox vaccination. In the 1920s this material was tested in rabbit skin (Gordon, 1925), either by scarification or intradermal inoculation at a 1 : 1000 dilution, a positive reaction being required before it was considered satisfactory for human vaccination. Assay in rabbit skin was elaborated by Parker (1938), who compared the lesions produced at "limit dilution" by several strains of vaccinia virus. Vaccinia virions are large enough to be counted in a Petroff–Hausser chamber (Parker and Rivers, 1936), and several workers have compared particle counts with infectivity, a problem further investigated by Sharp (1963), using the electron microscope.

Next, with the development of the chorioallantoic membrane for the cultivation of poxviruses, Burnet introduced an assay analogous to the colony counts of bacteria, viz. pock counting (for review, see Burnet, 1936). Keogh (1936) demonstrated that the pocks produced by dermal vaccinia and the neurovaccine strains were different and characteristic of the strain. He also showed the value of pock assay for the titration of neutralizing antibodies. However, soon after the introduction of the plaque assay technique for animal viruses by Dulbecco (1952), this technique displaced others for most infectivity assays of orthopoxviruses, although pock counting on the chorioallantoic membrane remains the official assay procedure for smallpox vaccines (World Health Organization, 1965, 1980b).

THE REPLICATION CYCLE OF VACCINIA VIRUS

Much modern laboratory research in animal virology is concerned with the structure and chemical nature of the virion and the process of viral replication. For the orthopoxviruses most such research has been carried out with vaccinia virus (including rabbitpox virus) as a model for the family *Poxviridae*. Looking back over the scientific literature, it is somewhat surprising to realize how recent has been the development of the modern concepts of viral replication. For example, in a symposium on "The Nature of Virus Multiplication" held in England in April 1952 (Society of General Microbiology, 1953) a paper was presented supporting the notion of replication of poxviruses, as of prokaryotes, by fission, and even in 1959 (Society of General Microbiology, 1959) doubt was expressed about the existence of an "eclipse phase" in the replication cycle of vaccinia virus. It is significant of the profound change in our understanding of viral replication during the last 20 years that the term "eclipse phase" has now disappeared from the vocabulary of virologists.

Development of the modern picture of the replication of poxviruses began with experiments with radioactively labeled virus or its products (Cairns, 1960; Joklik, 1962b; Salzman *et al.*, 1962). The morphogenesis of vaccinia virus was followed by thin-section electron microscopy (Morgan *et al.*, 1954; Dales and Siminovitch, 1961; Dales, 1963). Gel electrophoresis was applied to the analysis of viral nucleic acids and viral polypeptides, both those of the virion and as they appear sequentially in the infected cell (Holowczak and Joklik, 1967a,b). The detailed understanding of poxvirus replication was further developed by the dissection of the viral genome with restriction endonucleases and the molecular cloning of DNA fragments, thus providing probes for the precise analysis of the time sequence of transcription in the infected cell (Wittek *et al.*, 1980b).

GENETIC STUDIES OF ORTHOPOXVIRUSES

Genetic studies of viruses began, and for a long time remained, much more advanced with the bacterial viruses (see Stent, 1963). Although special problems and opportunities were presented by the RNA viruses with segmented genomes (e.g., influenza virus and reovirus) and later with the retroviruses, genetic studies with animal viruses have in the main been derivative, following the lines developed earlier by bacterial virologists.

Apart from Pasteur's work with rabies, the earliest demonstration of the development of stable "variants" of an animal virus by serial passage in particular animals, by particular routes of inoculation, was the work of the Pasteur Institute group with "dermovaccine" and "neurovaccine" in the early 1920s (for references, see van Rooyen and Rhodes, 1948). The next major advance was based on pock assays on the chorioallantoic membrane, when Downie and Haddock (1952) demonstrated that suspensions of cowpox virus contained two kinds of particles that "bred true"; a majority that produced red hemorrhagic pocks and a minority (about 1%) that produced white opaque pocks. This discovery was not exploited until about 10 years later, when similar findings with rabbitpox virus, a strain of vaccinia virus, were used for "genetic mapping" (Gemmell and Fenner, 1960), recombination between different strains of vaccinia virus having been demonstrated 2 years earlier (Fenner and Comben, 1958; Fenner, 1959).

Later genetic studies with the poxviruses made use of conditional lethal mutants, introduced into bacteriophage genetics by Edgar and Epstein (Edgar, 1966). The first experiments exploited the discovery by M. E. McClain in 1961 (McClain, 1965) that some white pock mutants of rabbitpox virus failed to replicate in PK-2a cells. This work was later developed with a large range of PK-negative mutants (Fenner and Sambrook, 1966). At about the same time temperature-sensitive mutants of rabbitpox virus were isolated (Sambrook et al., 1966), and used for recombination experiments (Padgett and Tomkins, 1968). Since then extensive suites of temperature-sensitive mutants have been assembled and used for studies of viral biosynthesis (Dales et al., 1978). Analysis of the orthopoxvirus genome by mapping restriction endonuclease cleavage sites has made it possible to characterize many of the white-pock mutants of cowpox virus (Archard et al., 1984), rabbitpox virus (Moyer and Rothe, 1980), and monkeypox virus (Dumbell and Archard, 1980; Esposito et al., 1981) as having large terminal deletions and/or transpositions.

Although not "genetic" in the sense of involving changes in the genome, a rather interesting phenomenon was discovered with the poxviruses and appears to be peculiar to them. In 1935, influenced by Griffith's (1928) discovery of pneumococcal transformation, Berry and Dedrick (1936) performed what they regarded as an analogous experiment with two leporipoxviruses, active fibroma virus and heat-killed myxoma virus. They obtained an analogous result, viz., the "heat-killed" myxoma virus was "transformed" and acquired infectivity. Using orthopoxviruses, it was shown independently by Fenner et al. (1959) and Hanafusa et al. (1959a) that his phenomenon was not "transformation" in the sense of the transfer of DNA from one virus to another, but was essentially a reactivation of the "heat-killed" virus. It is a general phenomenon among poxviruses (Hanafusa et al., 1959b; Fenner and Woodroofe, 1960); virus "killed" by methods involving damage to its proteins rather than its nucleic acid can be rescued by growing it with another poxvirus whose proteins are intact but whose nucleic acid is lethally damaged (Joklik et al., 1060a,b,c).

PATHOLOGY, PATHOGENESIS, AND IMMUNOLOGY

Pathology

One of the earliest histopathological features of viral infections to be recognized by pathologists was the association of particular types of intracellular "inclusion bodies" with particular viral infections. Indeed, descriptions of inclusion bodies in poxvirus-infected cells proceded the differentiation of viruses from bacteria, the inclusions sometimes being described as protozoa (e.g., Guarnieri, 1892). With little else to distinguish viral from bacterial infections, other than the absence of polymorphonuclear leukocytes and the presence of "small round cells," early pathologists wrote a great deal about these inclusions, which they categorized as intracytoplasmic or intranuclear, and acidophilic or basophilic. Poxviruses were associated with intracytoplasmic inclusion bodies, herpesvirus with intranuclear inclusions (for review, see van Rooyen and Rhodes, 1948).

Subsequent histological and election microscopical work has shown that all orthopoxviruses produce B-type inclusion bodies (Kato et al., 1959), which are viral "factories" (Cairns, 1960). Cowpox, ectromelia, and raccoonpox viruses produce, in addition, strongly eosinophilic A-type inclusion bodies, which may or may not contain large numbers of mature virions (Ichihashi et al., 1971; see Chapter 4).

Pathogenesis

Although the pathological histology of smallpox was extensively studied during the first half of the twentieth century (Lillie, 1930; Bras, 1952a), little was known of its pathogenesis or the role of the immune system in promoting recovery. In particular, what happened during the incubation period of such generalized infections remained obscure. The sequence of events that occurred during the incubation period of generalized poxvirus infections was analyzed with mousepox by Fenner (1948b) and his findings were subsequently confirmed in rabbits infected with rabbitpox virus (Westwood, 1963) and monkeys infected with monkeypox virus (Cho and Wenner, 1973). New pecision was given to studies on pathogenesis by the use of fluorescent antibody staining; Mims (1964, 1966) and Roberts (1962a,b,c) made good use of this method to unravel the cellular events in mousepox infection, studies which emphasized the important role of macrophages.

Observations by Briody et al. (1956) and experiments by Schell (1960a,b) were greatly extended by Wallace et al. (1985) and Bhatt and

Jacoby (1987a), and demonstrated the role of the genotype of the mouse in determining the result of mousepox infections. Studies by Blanden and his colleagues (1970, 1971a,b; for review, see Cole and Blanden, 1982) showed that cell-mediated immunity played an important role in determining recovery from infection.

Immunology

Looking broadly at the subject of immunology, it is perhaps not altogether an exaggeration to suggest that Jennerian vaccination, introduced at the close of the eighteenth century, founded that science. Certainly it was to honor Jenner's discovery that Pasteur (1881) introduced the terms "vaccine" and "vaccination" to indicate the agent and the operation of protective inoculation against diseases other than smallpox.

Concepts of humoral immunity to infectious diseases developed toward the end of the nineteenth century, and attempts were then made to associate immunity in viral diseases with humoral factors. Vaccinia virus was the favorite object for research in this field and serious study of humoral immunity began with the work of Béclère et al. (1898), who demonstrated the passive transfer of immunity to vaccinia virus to heifers that had been inoculated with serum from immunized animals. Subsequently they reported that virus and immune serum reacted *in vitro* (Béclère et al., 1899). Keogh (1936) quantified the analysis of virus neutralization reactions using neurovaccinia virus and the pock-counting technique on the chorioallantoic membrane. In the next year, in the first comprehensive review of the immunology of viruses, Burnet et al. (1937) noted that almost all experimental studies of the immunological reactions of animal viruses to that time had been done with vaccinia virus.

Jenner (1798) made an important observation that relates directly to hypersensitivity and thus to cellular immunology. He observed that in the case of Mary Barge, who had had cowpox, an extensive inflammation was produced by variolation: "It is remarkable that variolous matter when the system is disposed to reject it should excite inflammation . . . more speedily than when it produces the small pox . . . . It seems as if a change, which endures through life, had been produced in the action, or disposition to action, in the vessels of the skin . . ." More recently, Blanden (for review, see Cole and Blanden, 1982) has used mousepox as a model to study the cellular immunology of generalized viral infections.

EXPERIMENTAL EPIDEMIOLOGY

Epidemiology is in general an observational science. Understanding of the epidemiology of smallpox underpinned its control and subsequent eradication, and other orthopoxviruses infectious for man (e.g., cowpox and monkeypox viruses) provide interesting examples of zoonotic diseases.

In the early 1920s the bacteriologist W. W. C. Topley developed the idea of carrying out experimental epidemiology, working with a mouse model of typhoid fever (Topley, 1923). When ectromelia virus was discovered in 1930, Topley and his colleagues immediately began experimental studies of its epidemiology (Greenwood et al., 1936). This work was revived and extended in the late 1940s (for review, see Fenner, 1949d; see Chapter 9). Both of these experimental studies provided data for the more recent theoretical analysis of the population biology of epidemics (Anderson and May, 1979).

THE GLOBAL ERADICATION OF SMALLPOX

Study of the orthopoxviruses was greatly stimulated by the campaign to eradicate smallpox from the world, successfully concluded a decade ago. Although there have been attempts to eradicate other human diseases (yellow fever, malaria, yaws), smallpox was in fact the first disease for which eradication was explicitly proposed, by none other than Edward Jenner, who wrote in 1801 "it now becomes too manifest to admit of controversy that the annihilation of the smallpox, the most dreadful scourge of the human species, must be the result of this practice [of vaccination]." October 1977 saw the successful completion of an 11-year global campaign of eradication conducted by the World Health Organization (Arita, 1979; World Health Organization, 1980a; Fenner et al., 1988).

Much of the research on orthopoxviruses carried out during the decade 1970–1980, apart from that on the molecular biology of viral replication, was stimulated by the global smallpox eradication campaign. Diagnostic procedures were improved and streamlined (World Health Organization, 1969); a new zoonosis, human monkeypox, was recognized (Marennikova et al., 1972a; Jezek and Fenner, 1988); problems relating to a possible animal reservoir of smallpox were raised (Gispen and Brand-Saathof, 1972; Marennikova et al., 1972b, 1976a, 1979) and subsequently disposed of (Dumbell and Archard, 1980; Dumbell and Kapsenberg, 1982; Esposito et al., 1985); a great stimulus was provided for restriction endonuclease analysis of the genomic DNA of orthopoxviruses and their mutants; and a much better understanding was gained of the epidemiology and control of both smallpox and human monkeypox (Fenner *et al.*, 1988).

TABLE 1-5

Historical Events in the Discovery of Various Species of Orthopoxviruses

Date	Event	Reference
Fourth century AD	Recognition of smallpox as a distinct disease in China, by Ko Hung	Needham and Lu (1988)
Seventh century AD	Probable recognition of smallpox as a distinct disease in India	Jolly (1901)
Ninth century AD	Clinical descriptions of smallpox and measles by Rhazes	Al-Razi (1939)
1796	Demonstration by challenge inoculation that poxvirus from cows gave protection against smallpox	Jenner (1798)
1904	Description of variola minor as a distinct disease	de Korté (1904)
1930	Description of infectious ectromelia of mice	Marchal (1930)
1939	Recognition of differences between cowpox and vaccinia viruses	Downie (1939a,b)
1945	Recognition of ectromelia virus as an orthopoxvirus	Burnet (1945)
1959	Discovery of monkeypox virus among captive monkeys	von Magnus *et al.* (1959)
1964	Recovery of raccoon poxvirus	Herman (1964
1970	Recognition of first case of human monkeypox	Ladnyi *et al.* (1972); Marennikova *et al.* (1972a)
1970	Cultivation of camelpox virus in chick embryos	Sadykov (1970)
1972	Recognition of camelpox virus as an orthopoxvirus	Baxby (1972)
1974	Isolation of poxvirus from *Tatera kempi*	Kemp *et al.* (1974)
1974	Recognition of cause of Uasin Gishu disease as an orthopoxvirus	Kaminjolo *et al.* (1974a)
1975	Recognition of Tatera poxvirus as an orthopoxvirus	Lourie *et al.* (1975)
1975	Demonstration that raccoon poxvirus was an orthopoxvirus	Thomas *et al.* (1975
1987	Discovery of orthopoxvirus of California vole	Regnery (1987)

VACCINIA VIRUS AS A VECTOR FOR
VACCINE ANTIGENS

Within 2 years of the recommendation of the World Health Organization that vaccination with vaccinia virus should be discontinued worldwide, it was suggested that vaccinia virus might be used for vaccination of man or domestic animals against any of a wide range of infectious diseases (Mackett *et al.*, 1982; Panicali and Paoletti, 1982). This suggestion was based on the discovery that any foreign gene, or even several different foreign genes, could be introduced into the vaccinia virus genome by homologous recombination, and the relevant gene products would be expressed during replication of the recombinant vaccinia virus. This strategy would enable the great practical advantages of vaccinia virus as a vaccine vehicle (cheapness, heat stability, ease of administration) to be applied to a range of different diseases. This topic is discussed in Chapter 12.

SUMMARY OF HISTORICAL EVENTS IN
ORTHOPOXVIRUS RESEARCH

Table 1-5 lists the dates of discovery of orthopoxvirus diseases and the papers in which the causative agents were shown to be orthopoxviruses.

CHAPTER 2

Structure and Chemical Composition of the Virion

By the year 1900, strains of vaccinia virus had been selected that replicated rapidly and to high titer in a wide range of laboratory animals, whereas very little work had been done with other species of *Orthopoxvirus* or poxviruses of any other genus. Vaccinia virus was therefore early established as the model virus for the genus *Orthopoxvirus* and for the family of viruses now called the *Poxviridae*. Vaccinia virus retains to this day its preeminence as the prototype *Orthopoxvirus* and the prototype poxvirus. We shall therefore use it as our model in this chapter and the next, which are concerned with the molecular biology of the orthopoxviruses, including the replication cycle.

STRUCTURE OF THE VIRION

Historical Development

Virions of vaccinia virus were among the earliest objects to be examined with the electron microscope (Borries *et al.*, 1938). Further understanding of their structure depended on the increasing resolution of the electron microscope and its use with virus particles treated in a variety of ways. The first use of enzymatic digestion as a method of studying the ultrastructure of viruses was carried out with vaccinia virus, to demonstrate the existence of substructure within the brick-shaped virions seen with the electron microscope (Dawson and McFarlane, 1948). In a series of classical studies, Peters (1956) and Stoeckenius and Peters (1955), using enzymatic digestion with deoxyribonuclease and metal shadowing, demonstrated the viral components now known as the outer membrane, the lateral bodies, and the core (see Fig. 2-1). Two other important technical advances in electron microscopy, thin sectioning and negative staining, were first applied to study of the poxviruses shortly after they were introduced. Thin sections have been particularly valuable in elucidating the morphogenesis of the virions of vaccinia virus (Dales and Siminovitch, 1961; see Chapter 3). Negative staining, first used for poxviruses by Nagington and Horne (1962), was subsequently used on virions of vaccinia virus degraded in various ways as a method of analyzing their substructure (Easterbrook, 1966; Pogo and Dales, 1969; Medzon and Bauer, 1970), and for demonstrating the distinctive surface tubular structures of the outer membrane. The negative staining method, applied to material from skin lesions, was the cornerstone of the laboratory diagnosis of smallpox, as it developed during the Intensified Smallpox Eradication Programme of the World Health Organization (Nakano, 1979; Fenner et al., 1988).

Components of the Virion

The virions of all species of *Orthopoxvirus* have an identical structure (Fig. 2-1). Four major structural elements can be distinquished: core, lateral bodies, outer membrane, and envelope. The core consists of an "inner membrane" which encloses the viral DNA and associated proteins; on each side of the core there is an oval mass called the lateral body. The core and lateral bodies are enclosed within an outer membrane, which has a characteristic ridged surface structure (Plate 2-1A).

Two forms of virions can be seen in negatively stained preparations, which have been designated by Westwood et al. (1964) as "M" (mulberry) and "C" (capsule) forms. The M forms are undamaged nonenveloped virions (Plate 2-2A), whereas the C forms (Plate 2-2B) are damaged virions, in which the stain penetrates the particle, revealing a collapsed outer membrane and some of the internal structure. In clinical material, M forms are found in vesicular fluid, whereas most particles in dried scabs are C forms (Fenner and Nakano, 1988).

Virions released spontaneously from cells, rather than by cellular

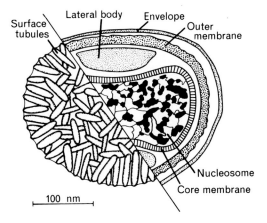

FIG. 2-1. *The structure of the vaccinia virion. Right-hand side, section of enveloped virion; left-hand side, surface structure of nonenveloped particle. The viral DNA and several proteins within the core are organized as a "nucleosome." The core has a 9-nm-thick membrane, with a regular subunit structure. Within the virion, the core assumes a dumbbell shape because of the large lateral bodies. The core and lateral bodies are enclosed in a protein shell about 12 nm thick—the outer membrane, the surface of which consists of irregularly arranged tubules, which in turn consist of small globular subunits. Virions released naturally from the cell are enclosed within an envelope which contains host cell lipids and several virus-specified polypeptides, including the hemagglutinin. Most virions remain cell associated and are released by cellular disruption. These particles lack an envelope, so that the outer membrane constitutes their surface. Both forms are infectious. (From Fenner et al., 1988.)*

disruption, are enclosed within a lipoprotein envelope (Plate 2-1B) which contains several virus-specified polypeptides (Payne and Norrby, 1976; Payne, 1978). Virions released by cellular disruption lack this envelope (Appleyard *et al.*, 1971), their outer surface being composed of the outer membrane.

Core. Easterbrook (1966) obtained isolated cores form a purified suspension of virions by treatment with 2-mercaptoethanol and a nonionic detergent. Within the virion, the core is an oval biconcave disk, dumbbell shaped in a cross-section (Plate 2-1C), but after release the concavities disappear and the core adopts a rounded rectangular shape about the same size as a mature virion (Plate 2-1D). Its coat is composed of an inner smooth membrane about 5 nm thick and an outer layer ("palisade") of regularly arranged short cylindrical subunits about 10 nm long and 5 nm in diameter. When virions were negatively stained at alkaline pH (Peters and Muller, 1963) or when sections were stained

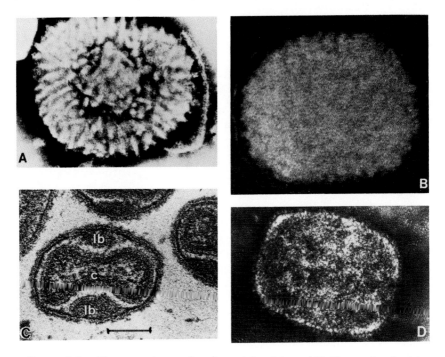

PLATE 2-1. *Electron micrographs of vaccinia virions. (A) Nonenveloped virion, showing the surface tubular elements that make up the outer membrane. (B) Enveloped virion, as released from the infected cell and found in the extracellular medium. (C) Thin section of nonenveloped virion showing the biconcave core (c) and the two lateral bodies (lb). (D) Viral core, released after treatment of virions with Nonidet 40 and 2-mercaptoethanol. Bar = 100 nm. [A, from Dales, 1963; B, from Payne and Kristensson, 1979; C, from Pogo and Dales, 1969; D, from Easterbrook, 1966.]*

with uranyl chloride or indium chloride, the core could be seen to contain two or three broad cylindrical elements which appeared to be connected in a tight S shape.

Treatment of purified virions with sodium dodecyl sulfate results in the release of the viral DNA, which is associated with four major polypeptides, M_r 90K, 68K, 58K, and 10K (Soloski and Holowczak, 1981). The viral DNA is combined with polypeptides in a superhelical conformation. Electron microscopy of this complex fixed immediately after isolation revealed that the supercoiled DNA occurred in globular structures 20–60 nm in diameter, connected to each other by DNA–protein fibers, the overall structure resembling that of the nucleosomes of eukaryotic cells (Fig. 2-1). The core contains many other polypeptides, some of which are enzymes or subunits of enzymes.

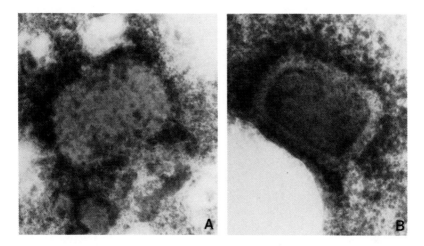

PLATE 2-2. *Virions of variola virus, as seen in negatively stained preparations of material obtained from a patient with smallpox. (A) "M" form, showing tubules of the outer membrane. (B) "C" form, penetrated by phosphotungstate. (Courtesy Dr J. H. Nakano.)*

Lateral Bodies. Within the intact virion, the concavities of the core accommodate the lateral bodies, which are seen by freeze-etching to be oval structures (Medzon and Bauer, 1970). After treatment of suspensions of virions with different detergents the lateral bodies may remain attached either to the core (Easterbrook, 1966) or the outer membrane (Medzon and Bauer, 1970).

Outer Membrane. The structure of the outer membrane has been studied by thin sectioning, negative staining, and freeze-etching. The randomly arranged "surface tubules" (Plate 2-1A) are seen by freeze-etching to consist of double ridges composed of spherical subunits about 5 nm in diameter (Medzon and Bauer, 1970). The double ridge which constitutes the tubule is 15 nm wide, with a center-to-center spacing of the ridges of about 10 nm. The spherical subunits are spaced at 6–7.5 nm centers along the ridges, which vary in size but are commonly about 70–100 nm long. They appear to consist of a single polypeptide, M_r 58K (Stern and Dales, 1976a).

Envelope. There is a general agreement in the scientific literature about the structure of virions of vaccinia virus as described so far, i.e., the core, lateral bodies, and outer membrane. Until comparatively recently, most workers have regarded this as the complete virion, since in suitable preparations such particles have a high particle to infectivity ratio. However, there have been persistent reports that sometimes these particles are enclosed within an envelope (Nagington and Horne, 1962; Noyes, 1962; Dales, 1963; Ichihashi et al., 1971; Morgan, 1976a), but these workers and other commentators (Fenner et al., 1974) tended to regard this additional membrane as an adventitious element of cellular origin that had no particular biological significance. However, an investigation of the efficacy of inactivated vaccinia virus for human vaccination led Boulter (1969) to suspect that the envelope was an important viral component in relation to protection (see Chapter 4). Subsequent experiments (Appleyard et al., 1971; Turner and Squires, 1971) confirmed this view and led these workers (Boulter and Appleyard, 1973) to emphasize the important differences between nonenveloped and enveloped virions. The biological significance of the envelope was conclusively demonstrated in a series of papers by Payne (1978, 1980), who used the terms "intracellular naked vaccinia virus" and "extracellular enveloped vaccinia virus" to describe the two forms of virions of vaccinia virus, both of which are infectious. As seen in thin sections, the envelope is a bilaminar membrance 5–10 nm thick, often closely applied to the outer membrane (Plate 2 1B)

CHEMICAL COMPOSITION OF VIRIONS OF VACCINIA VIRUS

All studies of the chemical composition of the virions of orthopoxviruses, other than those of Payne (1978, 1979), Hiller et al. (1981a), and Hiller and Weber (1985), have been carried out with nonenveloped virions. We shall summarize the results of these studies before describing the chemical composition of the viral envelope.

Composition of Nonenveloped Virions

Vaccinia virus was the first animal virus the virions of which were purified sufficiently to make reasonably accurate chemical analysis possible, and pioneering chemical studies were carried out by T. M. Rivers and his collaborators at the Rockefeller Institute for Medical Research in New York between 1935 and 1943 (Parker and Rivers, 1935; for review, see Smadel and Hoagland, 1942). Comparable studies on other animal viruses were not possible until the mid-1950s, and the quality of the chemical analyses made by the Rockefeller Institute group is attested by the fact that the only subsequent chemical studies of the virions of vaccinia virus (Zwartouw, 1964; Joklik, 1966) were largely confirmatory of the earlier findings. The method of purification used by the Rockefeller Institute workers was derived from a technique developed by Craigie (1932); the shaved skin of a rabbit's back was inoculated by scarification with a selected strain of virus and the virions were harvested 3 days later by scraping the skin gently with a blunt scalpel. The resultant material was then subjected to several cycles of differential centrifugation. More recently this method has been replaced by growth of the virus on the chorioallantoic membrane or in cultured cells, and purification is carried out by density gradient centrifugation of virus released from disrupted cells (Joklik, 1962a).

The most important finding of the Rockefeller Institute group was that the virion of vaccinia virus contained about 5.6% DNA and no RNA. At that time the only other viral genome nucleic acids that had been characterized were the DNA of a bacteriophage (Schlesinger, 1936), and the RNA of tobacco mosaic virus (Bawden et al., 1936).

Early studies of the viral proteins relied on serological techniques, of which the most discriminative were gel diffusion and gel electrophoresis. However, these methods revealed only a small fraction of the polypeptides that are components of the virion or are produced during infection. Gel electrophoresis of detergent-disrupted virions has provided a much more detailed picture of its protein components, and a recent study of vaccinia virus-infected cells by two-dimensional gel electrophoresis has revealed 279 virus-specified proteins (Carrasco and Bravo, 1986); for technical reasons even this figure is regarded as an underestimate.

Purified nonenveloped virions of vaccinia virus contain about 6% lipid (wet weight), which is a component of the outer membrane. The envelope contains additional lipid.

Composition of the Envelope

Compared with the extensive studies that have been made of the chemical composition of nonenveloped virions, there have been few studies of the envelope. Payne (1978) reported that purified enveloped virions contained eight polypeptides (210K, 110K, 89K, 42K, 37K, 21.5K, 21K, 20K) that were not present in nonenveloped virions, and that the same polypeptides could be detected in purified envelope material. All except the 37K polypeptide were glycosylated. In enveloped virions there were also alterations in six polypeptides located in the outer membrane of enveloped particles, compared with nonenveloped particles obtained by disrupting infected cells. The 89K polypeptide was reported to be the vaccinia virus hemagglutinin (Payne, 1979). Hiller and

Weber (1985) considered that the nonglycosylated 37K polypeptide was the major component of the envelope. It has a very low methionine content, which may explain why Payne failed to recognize how abundant it was. Using an antiserum specific for the 37K polypeptide as a probe, Hiller and Weber showed that it accumulated in the Golgi region and these modified membranes were then wrapped around vaccinia virus particles to form the double-walled envelopes that characterize intracellular enveloped virions. The outer membrane is lost when the virion is released from the cell by exocytosis (see Chapter 3). Enveloped virions contain about twice as much phospholipid as nonenveloped virions (Hiller *et al.*, 1981a).

STRUCTURE OF THE GENOME

Size and Base Composition

The vaccinia virus genome consists of a linear double-stranded DNA molecule which has several characteristic features. Geshelin and Berns (1974) estimated a relative molecular mass (M_r) of 122 ± 2.2 million from contour length measurements of DNA molecules obtained by electron microscopy. As restriction enzymes began to be used extensively to analyze orthopoxvirus genomes, several investigators reported M_r values obtained by summation of the sizes of DNA restriction fragments, which are in close agreement with the figure obtained from electron micrographs (Gangemi and Sharp, 1976; Wittek *et al.*, 1977; McCarron *et al.*, 1978; Müller *et al.*, 1978; Mackett and Archard, 1979; De Filippes, 1982; Esposito and Knight, 1985). An M_r value of about 125 million, equivalent to 190 kbp, thus appears to be a good estimate for the size of the vaccinia virus genome.

In addition to its large size, the vaccinia virus genome is characterized by a relatively low G+C content: 36–37 mol% (Joklik, 1962a; Pfau and McCrea, 1963).

Restriction Maps of *Orthopoxvirus* Genomes

It was first thought that restriction enzyme analysis of poxvirus genomes would not be possible, due to the great size and complexity of the DNA molecules. However, Gangemi and Sharp (1976) and Müller *et al.* (1978) showed that if the restriction endonucleases were carefully chosen, the genome could be dissected into a small enough number of fragments to permit their separation by gel electrophoresis. The initial work also showed that closely related orthopoxviruses, such as two strains of vaccinia virus (Gangemi and Sharp, 1976), could readily be distinguished on the basis of their restriction patterns. More distantly related orthopoxviruses yielded fragment patterns the degree of similarity of which closely reflected their biological relatedness, as established by other means (Müller *et al.*, 1978; see Fig. 1-1 and Table 1-2). These findings clearly showed that restriction enzymes were potent new tools with which to classify orthopoxviruses.

Restriction maps of a large number of orthopoxvirus genomes have now been published (Mackett and Archard, 1979; Esposito and Knight, 1985). They show that DNA sequences from the internal part of the genomes display a very similar distribution of restriction sites, i.e., they are highly conserved between different orthopoxviruses. The differences in the restriction patterns observed between species of orthopoxvirus are mainly due to sequence divergence near the ends of the genomes. It is therefore not surprising that the genes encoding structural polypeptides (Weir and Moss, 1984; Wittek *et al.*, 1984a,b; Hirt *et al.*, 1986) and enzymes involved in nucleic acid metabolism (Hruby and Ball, 1982; Weir *et al.*, 1982; Bajszar *et al.*, 1983; Jones and Moss, 1984; Traktman *et al.*, 1984) map within the conserved central portion of the genome. Genes affecting host range and pock morphology, on the other hand, are located toward the ends of the DNA molecules (Archard and Mackett, 1979; Dumbell and Archard, 1980; Lake and Cooper, 1980; Moyer and Rothe, 1980; Drillien *et al.*, 1981; Esposito *et al.*, 1981; Gillard *et al.*, 1986).

Strain Differences in Vaccinia Virus DNA

The most detailed studies of the orthopoxvirus genome have been made with the WR strain of vaccinia virus. However, restriction endonuclease maps have been made of DNA from several other strains of vaccinia virus; the available data for *Hin*dIII cleavage sites is illustrated in Fig. 2-2. Visual inspection of Fig. 2-2A confirms the close similarity between the large central regions of all strains; differences are found mainly at the two ends. Figure 2-2B is a dendrogram produced from restriction data for these strains of vaccinia virus obtained with several enzymes, after analysis as described by Gibbs and Fenner (1984). It is clear that all strains of vaccinia virus have very similar DNA molecules, which are distinctly different from those of monkeypox and variola viruses, and in a broader comparison (Fig. 1-1) from those of all other species of *Orthopoxvirus*.

DNAs of Different Species of *Orthopoxvirus*

The M_r values of the genomes of orthopoxviruses range from 120 million for variola virus DNA to 145 million for cowpox virus DNA, with

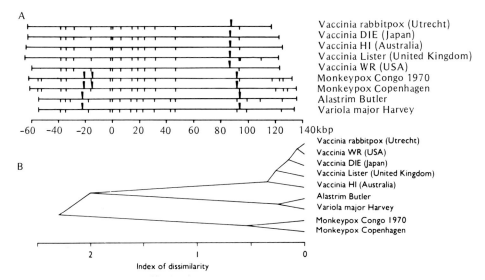

FIG. 2-2. *(A) Restriction maps of five strains of vaccinia virus from different countries, two strains of monkeypox virus, and two strains of variola virus, after digestion with* HindIII *(I) and* Sma1 *(▼). (B) Dendrogram illustrating the similarities and differences between these viral DNAs, as revealed by digestion with* HindIII, Sma1 *and* XhoI, *derived as described in Gibbs and Fenner, 1984). (A, from Mackett and Archard, 1979.)*

some strain variation within each species. As described below, deletion mutants of some strains occur quite commonly; the size of such mutant genomes may therefore be considerably less than that of the wild-type virus. The restriction maps of all species of *Orthopoxvirus* for which data are available are shown in Fig. 1-1; maps for different strains of each species are illustrated in the appropriate chapters.

Structure of the Ends of the Genome

The structure of the genome of vaccinia virus is illustrated diagrammatically in Fig. 2-3. Most of the differences between strains and mutants of vaccinia virus and between species of *Orthopoxvirus* occur in the terminal 20–30 kbp of the molecule. For this reason, and because of their putative role in DNA replication, the ends of the molecule have been studied in particular detail. They contain a number of features of interest.

Long Inverted Terminal Repeats.   Strong cross-hybridization between the terminal restriction fragments suggested that vaccinia virus DNA

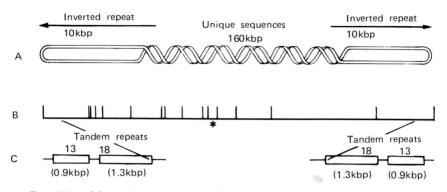

FIG. 2-3. *Schematic representation of the DNA of vaccinia virus (WR strain). (A) Linear double-stranded molecule with terminal hairpins and inverted repeats (not to scale). When denatured it forms a very large single-stranded circular molecule. (B) HindIII cleavage sites. The asterisk indicates the fragment containing the thymidine kinase gene, which is used in experiments with vaccinia virus as a vector for hybrid vaccines. (C) Each 10-kbp terminal portion includes two groups of tandem repeats of short sequences rich in AT.*

contains similar sequences at both termini (Wittek *et al.*, 1977; Mackett and Archard, 1979). Detailed restriction site mapping of the ends of the DNA of rabbitpox virus and Lister vaccinia virus showed that these genomes contained inverted terminal repetitions of about 5.3 and 11.6 kbp, respectively (Wittek *et al.*, 1978b). The presence of an inverted terminal repetition has been demonstrated by electron microscopy with the WR strain of vaccinia virus (Garon *et al.*, 1978). Self-annealing of denatured DNA molecules at low DNA concentration resulted in the formation of large single-stranded circles with a double-stranded projection ("panhandle"), the length of which indicated an inverted terminal repetition of 10.5 kbp.

Cross-hybridization between the termini has been observed with all *Orthopoxvirus* DNAs, including that of raccoon poxvirus (Parsons and Pickup, 1987), but with the notable exception of variola virus DNA, which may lack an inverted terminal repetition or have a very short one (Mackett and Archard, 1979; Esposito and Knight, 1985).

Cross-Links. Denatured vaccinia DNA rapidly reforms the duplex structure when it is allowed to reanneal. This "snap-back" property has been explained by the presence of covalent links between the complementary DNA strands. Based on a sedimentation analysis, Berns and Silverman (1970) concluded that vaccinia virus DNA contained a small number of such cross-links which prevent the two strands from separat-

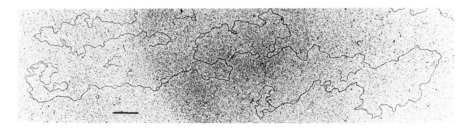

PLATE 2-3. *Electron micrograph of a partially denatured molecule of vaccinia virus DNA, showing the single-stranded loops at each end. Bar = 1 μm. (From Geshelin and Berns, 1974.)*

ing upon denaturation. Electron microscopy was used to localize the cross-links (Geshelin and Berns, 1974). Upon complete denaturation, large single-stranded circles were observed. Partial denaturation resulted in a single-stranded loop at each end of the DNA molecule (Plate 2-3), indication that vaccinia virus DNA is cross-linked at each end.

Nucleotide sequence analysis of the ends of the vaccinia virus genome (Baroudy *et al.*, 1982, 1983) revealed further interesting features. A fragment of 250 nucleotides derived from the very end of the genome was shown to exist as 2 electrophoretic variants: as a slow migrating (S form) and a fast migrating (F form) species, which were present in approximately equimolar amounts. Nucleotide sequencing around the presumed hairpin loop revealed a stretch of 73 nucleotides which were identical in both forms, followed by a sequence of 104 bases which differed in the S and F forms. After this divergence, there was a sequence complementary to the first 73 nucleotides (Fig. 2-4). When the 104-nucleotide divergent sequences are aligned in opposite polarity, it is obvious that one is the inverted and complementary copy of the other. When the two sequences are written with maximal base pairing, several mismatched bases are observed, suggesting that the ends of the vaccinia virus genome exist as single-stranded loops. The vaccinia virus genome can thus be considered as a single continuous polynucleotide chain which is base paired except for 104 nucleotides at each end, which exist as single-stranded loops in two isomeric forms, one of which is the inverted complementary counterpart of the other.

The terminal cross-link is a very characteristic feature of poxvirus genomes, since all DNAs so far examined, including those of avipoxviruses (Szybalski *et al.*, 1963) and the parapoxviruses (Menna *et al.*, 1979) are cross-linked at both termini.

Tandem Repeats. Evidence for repetitive DNA in the vaccina virus genome was first obtained by analysis of DNA reassociation kinetics

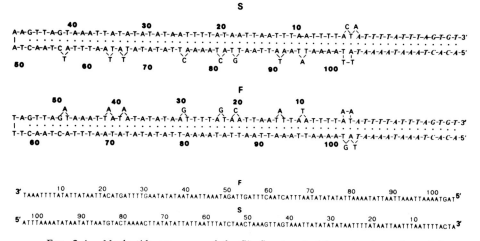

FIG. 2-4. *Nucleotide sequences of the flip-flop terminal-loop structure of vaccinia virus DNA (WR strain). Slow (S) and fast (F) fragments were labeled at the 3' end with ^{32}P cordycepin monophosphate and sequenced according to the Maxam–Gilbert technique. Top: Sequences arranged in the maximum base-paired structures. Bottom: 104-nucleotide sequences arranged in a linear fashion, with F in 3' → 5' direction and S in 5' → 3' direction. (From Baroudy et al., 1983.)*

(Grady and Paoletti, 1977; Pedrali-Noy and Weissbach, 1977). Subsequent studies revealed a very complex arrangement of reiterated DNA sequences comprising approximately the first 3.5 kbp at each end of the genome (Wittek and Moss, 1980; Baroudy and Moss, 1982; Baroudy *et al.*, 1982, 1983). A first repetitive element of 70 bp starts at 87 bp from the terminal loop and is then repeated 13 times in tandem (Fig. 2-5). After an unique intervening sequence of 325 bp, the 70-bp element is again repeated 18 times. The last repeating unit is immediately followed by two consecutive 125-bp repeats and then eight more 54-bp repeats. After

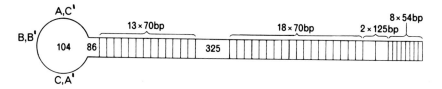

FIG. 2-5. *Schematic representation of the terminal 3.5 kb of the vaccinia virus (WR) genome. The 104-nucleotide flip-flop sequence at the very end of the DNA is shown in its single-stranded form and in both possible configurations (A, B, C; C', B', A'). The arrangement of the repeat sequences is shown by the vertical lines. Unique sequences at the beginning (86 bp) and between the two sets of 70-bp repeats (325 bp) are also shown.*

incomplete copies of the 54- and 70-bp elements, the repetitive structure ends with four copies of a 6- to 7-bp repeat. The 70-, 125-, and 54-bp elements have considerable sequence homologies and redundancies.

Comparison of the Termini of Vaccinia and Cowpox Virus DNAs.   The nucleotide sequence of the ends of cowpox virus DNA was obtained from a fragment which was cloned after removal of the terminal cross-link (Pickup *et al.*, 1982), so that little information is available for the loop structure of cowpox virus DNA. The organization of the sequences of repeat DNA is even more complex than in vaccina DNA. However, as far as the organization and base composition of the repeats at the ends of cowpox and vaccinia virus DNA are concerned, there is a striking similarity in the two viruses, stongly suggesting that the repeat units have evolved from a common ancestral subunit by unequal crossing over, which is probably a rather frequent event in the replication of poxvirus DNA.

Limited information on the fine structure of other orthopoxvirus DNAs has been obtained by restriction enzyme mapping and hybridization analysis (Esposito and Knight, 1985).

Structure of Mutant Genomes

Mutations near the Ends.   A number of orthopoxvirus mutants characterized by alterations of DNA sequences near the ends of the genome have been described. Plausible explanations for this observation are that the DNA at the very end of the genome (about 3.5 kbp) does not contain any genes, and that some genes at the ends are not essential for infectivity.

Terminal restriction fragments from the DNA of a commercially available smallpox vaccine (Lister strain of vaccinia virus) migrated as diffuse bands upon gel electrophoresis, indicating length heterogeneity in the DNA molecule population (Wittek *et al.*, 1978a). DNA from pock-purified virus of the same stock, however, yielded end fragments of discrete length. Similar length heterogeneity has also been observed in rabbitpox virus DNA. Cloning of the virus again resulted in distinct-length end fragments which in one clone were about 200 bp longer than in another clone. From each plaque isolate, however, end fragments of identical length were obtained from both termini when the DNA was cleaved within the inverted terminal repetition, indicating that the length variation affected both termini equally.

Further evidence for variations in the length of the DNA molecule affecting both termini simultaneously has been obtained by McFadden and Dales (1979). Upon analysis of the DNA of a large series of temperature-sensitive (ts) mutants of vaccinia virus, about 20% were shown to have altered genome structures not related to the ts phenotype. These consisted of near-terminal deletions of up to 250 bp which were always observed at both termini of the DNA molecule ("mirror-image deletions").

Such microheterogeneity at the ends of the orthopoxvirus genome is best explained by loss or acquisition of 70-bp repeating elements. The mechanism by which this occurs and the reason why any alteration at one end appears to be copied to the other, thus preserving the overall symmetry of the DNA molecule, remains obscure.

Other variants in genome structure in which the ends of the molecules are involved have been described by Moss *et al.* (1981a). About 20% of all plaque isolates from a serially propagated virus stock yielded unexpected restriction patterns. After cleavage within the inverted terminal repetition, the end fragments did not migrate as a double-molar band but instead produced a "ladder pattern" of regularly spaced bands that differed in size by about 1.6 kbp. A detailed analysis of these mutants showed that the increments in size were due to reiterations of one set of tandem repeats and the unique sequence separating it from the second set. During replication of their DNA, these mutants thus generate a family of molecules characterized by additional blocks of 70-bp tandem repeats. Molecules that contained less than the regular two blocks were not observed, and both ends of the molecules contained the reiterations. The resulting genome structure, however, was very unstable. Upon each cycle of plaque purification, about 20% of such variants reverted to the apparently more stable prototype structure, which contains only two sets of tandem repeats. Unequal crossing over between repeating 70-bp units belonging to different blocks has been proposed to explain acquisition of additional blocks of tandem repeats and reversion to the wild-type genome structure (Moss *et al.*, 1981a).

The variant genomes that have been described seem to result from instability of near-terminal DNA sequences. Other mutants in which the ends of the genomes are affected by very large deletions have been described. Two of these, in which there were deletions of about 10 kbp of DNA from the left-hand end of the genome, were isolated in two laboratories from a serially propagated stock of the WR strain of vaccinia virus (Moss *et al.*, 1981b; Panicali *et al.*, 1981). In both mutants, the deletion did not extend into the inverted terminal repetition but mapped just beyond the junction of terminally repeated and unique DNA sequences. Detailed transcriptional maps for a large portion of the DNA that is deleted showed that at least seven early and possibly two minor late polypeptides are encoded within the deleted part of the genome. It is therefore surprising that in the particular cell lines tested each of these mutants gave yields of progeny virus similar to that of wild-type virus. A mutant of the Copenhagen strain of vaccinia virus with a deletion of almost 20 kbp of the left-hand part of the DNA was isolated and characterized by Drillien *et al.* (1981). Analysis of this mutant showed that it had retained a large portion, it not all, of the inverted terminal repeat sequences, including the terminal cross-link structure, but was unable to replicate in human cells.

White Pock Mutants. In addition to the vaccinia virus mutants just described, deletion mutants of the rabbitpox strain of vaccinia virus, cowpox virus, and monkeypox virus have been isolated on the basis of their white pock phenotype. Wild-type strains of these three orthopoxviruses produce red or ulcerated pocks on the chorioallantoic membrane of embryonated chicken eggs. Between 0.1 and 1% of the pocks, however, have a white appearance. Viruses isolated from such lesions produce exclusively white pocks and consequently these mutants have been designated "white pock" mutants. The genome structure of many such mutants has been analyzed.

Rabbitpox Virus. The white pock mutants of rabbitpox virus were divided into two classes depending on whether they were able to replicate in a pig kidney cell line or not (Fenner and Sambrook, 1966). Mixed infections between host range and non-host range mutants yield progeny virus with wild-type pocks and no host range restriction. However, no wild-type virus is obtained from mixed infections between either host range or non-host range mutants alone (Lake and Cooper, 1980). All mutants with an impaired host range on pig kidney cells have deleted between 4.5 and 30 kbp from the left-hand end of the genome (Lake and Cooper, 1980; Moyer and Rothe, 1980). The deletion comprises predominantly unique sequences, but in three mutants distal portions of the inverted terminal repeat are also deleted. In each case, however, portions of the repeat sequences, including the terminal cross-link, are preserved.

Most non-host range white pock mutants are characterized by large deletions of about 15 kbp at the right-hand terminus of the genome (Lake and Cooper, 1980; Moyer *et al.*, 1980). In two mutants the deletion was seemingly compensated by the copy round of 19.5 and 22.5 kbp of DNA from the extreme left-hand end of the genome, which resulted in a net increase in genome size of 4.5 and 7.5 kbp, respectively. A third non-host range mutant had a deletion of about 9 kbp of DNA from the

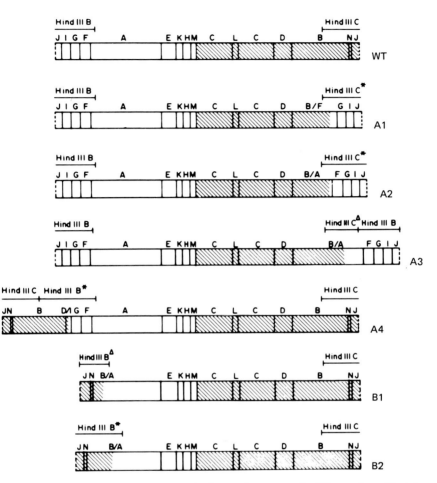

FIG. 2-6. *Restriction maps (XhoI) of white pock mutants of rabbitpox virus. The left and right halves of the parental rabbitpox virus genome (WT) are arbitrarily distinguished by crosshatching to demonstrate more readily the original source of the terminal sequences transposed in the variants. Maps both of white pock non-host-range mutants (A) and of white pock host range mutants (B) are shown. The terminal fragments resulting from HindIII digestion are also shown. Mutant A4 has the unimolar cross-linked HindIII C fragment of rabbitpox virus in two-molar amounts. Similarly, A3 has the unimolar cross-linked HindIII B fragment of rabbitpox virus present in two-molar amounts. The maps are drawn to scale to emphasize the net insertions of DNA in mutants A1, A2, A3, and A4 and the net deletions of DNA in mutants B1 and B2. Parental HindIII fragments in mutants that contain deletions (△) or insertions (*) are also indicated. (From Moyer and Graves, 1981.)*

left-hand terminus, apparently not including the host range function, which was compensated by the copy round of 42 kbp of DNA from the right-hand terminus. The result was a net increase in genome size of about 33 kbp and in the acquisition of huge inverted terminal repeats of 42 kbp. This mutant was rather unstable, and a segregant was isolated in which there was a deletion of the transposed sequences in addition to unique sequences from the left-hand end, which resulted in the host range mutant phenotype.

White Pock Mutants of Cowpox Virus. The white pock mutants of cowpox virus were the first white pock mutants of an orthopoxvirus to be isolated (Downie and Haddock, 1952). They are characterized by large deletions of between 31.5 and 40.5 kbp of the right-hand end of the genome (Archard *et al.*, 1984; Pickup *et al.*, 1984). Except for one mutant described by Pickup *et al.* (1984), which appears to be a simple deletion mutant, all others are characterized by a copy round of left-hand sequences to the deleted right-hand terminus. The transposed sequences range in size between 4.5 and 40.5 kbp in the case of the mutants described by Archard *et al.* (1984), and between 21 and 50 kbp in the mutants described by Pickup *et al.* (1984).

White Pock Mutants of Monkeypox Virus. Special attention has been given to white pock mutants of monkeypox virus since reports from one laboratory had suggested that monkeypox virus may segregate mutants that are indistinguishable biologically (and in their restriction maps) from variola virus (see Chapter 7). Restriction maps of several white pock mutants isolated in another laboratory, however, revealed that the distribution of restriction sites was typical of monkeypox virus in a large internal region of the DNA (Dumbell and Archard, 1980; see Chapter 8). The ends of the genome showed complex sequence rearrangements which in two mutants can best be explained by deletion of right-hand sequences followed by transposition of left-hand sequences to the deleted end, thus restoring genome symmetry. Other mutants showed more complex genome structures involving both ends of the DNA molecule.

Two monkeypox white pock mutants isolated by Esposito *et al.* (1981) also revealed sequence rearrangements at the right-hand terminus of the genome. Both mutants were segregated from a pock-purified virus of the red pock phenotype showing transposition of left-hand sequences to a deleted right-hand end. One of the white pock mutants isolated from it had a genome structure which was indistinguishable from the parent; in the other almost all of the transposed DNA had been deleted, leaving intact an inverted terminal repetition.

POLYPEPTIDES OF ORTHOPOXVIRUSES

Vaccinia Virus Polypeptides

Craigie and Wishart (1936a,b) and subsequently Shedlovsky and Smadel (1942) studied an antigen found in filtrates of dermal pulp which they designated "LS," and a nucleoprotein ("NP") antigen extracted from suspensions of purified virions (Smadel $et\ al.$, 1942). Later workers, who used immunodiffusion to differentiate many more antigens in poxvirions and in the extracts of poxvirus-infected cells ("soluble antigens"), succeeded in distinguishing about 10 different antigens in the virion (e.g., Marquardt $et\ al.$, 1965) and up to 20 in extracts of infected cells (Westwood $et\ al.$, 1965). Immunological methods were then used to follow the synthesis of "early" and "late" proteins during the replication cycle and relate these to the structure of the virion (Cohen and Wilcox, 1968).

The size of the orthopoxvirus genome suggested that it coded for many more polypeptides than could be recognized as antigens. In the late 1960s polyacrylamide gel electrophoresis and electrofocusing by ampholytes provided new and powerful methods for separating polypeptides from a complex mixture according to their M_r or net electric charge.

The most recent study of the number of virus-specific polypeptides synthesized in vaccinia virus-infected cells (Carrasco and Bravo, 1986) utilized two-dimensional electrophoresis, and revealed a total of 279 polypeptides. Since the total estimated M_r of these proteins exceeds the coding capacity of the vaccinia virus genome, some are undoubtedly products of the processing of others. On the other hand, the figure is an underestimate in that some polypeptides may not have entered the gels, and small polypeptides (<10K) would not have been detected.

Structural Polypeptides.   The first studies of the structural polypeptides of nonenveloped virions, using one-dimensional polyacrylamide gels, were initiated by Holowczak and Joklik (1967a,b) and expanded by Sarov and Joklik (1972), Obijeski $et\ al.$ (1973), and McCrae and Szilagyi (1975). Estimates of the number of different virion-specific polypeptides were raised progressively from 17 to 48. It was demonstrated that some large polypeptides found in extracts of infected cells were precursors that were cleaved to become virion proteins during maturation, and the location of some of the major polypeptides in the virion was determined. Using two-dimensional gel electrophoresis, Essani and Dales (1979) distinguished at least 110 polypeptides in the vaccinia virion, many of which are present in very small amounts. Several are enzymes (see Table 2-1), most of which are involved in nucleic acid metabolism.

Virus-specified polypeptides can be defined either according to their time of snythesis in relation to viral DNA replication, i.e., as the products of early genes, which are synthesized before DNA replication, or of late genes, which are transcribed during or after DNA replication (see Chapter 3). Specific functions have been assigned for some polypeptides.

Products of Early Genes

The bulk of early RNA sediments as 10–12S species, indicating an average size of about 1000 nucleotides (Oda and Joklik, 1967). Since hybridizaton studies (Oda and Joklik, 1967; Kaverin et al., 1975; Paoletti and Grady, 1977; Boone and Moss, 1978) showed that about one-half of one strand-equivalent of the viral DNA is transcribed early in infection, one can estimate that there are between 50 and 100 early genes. This estimate is in good agreement with the number of early proteins detected by two-dimensional gel electrophoresis of infected cell extracts. Two studies of *in vitro* translation of RNA selected by hybridization to DNA restruction fragments (Cooper and Moss, 1979; Belle Isle et al., 1981) together provide a survey of the entire genome. Approximately 100 early genes were mapped to the viral DNA, on which they are more or less evenly distributed.

Several early mRNAs have been mapped precisely on the viral genome and in most cases the corresponding *in vitro* translation product has been identified, but a biological function has been determined for only a few of them. Some early genes have been sequenced; those of particular interest are discussed below.

Early Genes Encoded within the Inverted Terminal Repetition.   The inverted terminal repetition contains four early genes that yield *in vitro* translation products of 7.5K, 42K, 19K, and 21K (Wittek et al., 1980b, 1981; Cooper et al., 1981a). Figure 2-7 shows the map positions of these genes and the direction of transcription of their mRNAs. Partial sequences of the gene encoding the 7.5K polypeptide (Venkatesan et al., 1981) and the complete sequences for the genes encoding the 19K and 42K proteins have been reported (Venkatesan et al., 1982).

The 7.5K Gene. This gene is of particular interest for two reasons. First, its 5′ flanking region, which has been analyzed in some detail, contains both early and late regulatory signals for gene expression (see Chapter 3). Second, these sequences have been extensively used to express foreign genes inserted into the viral genome (see Chapter 12). Presum-

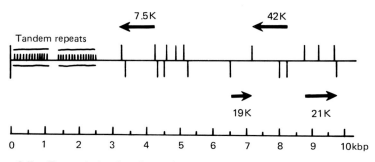

FIG. 2-7. *Transcriptional and translational map of the inverted terminal repeat of vaccinia virus DNA (WR strain). The map positions and directions of the early RNAs are shown, as are the* in vitro *translation products that they encode. Except for the tandem repeats at the far left, the vertical lines above and below the horizontal are HpaII and HincII sites, respectively. (Based on Cooper et al., 1981a).*

ably because of their dual promoter functions, these sequences yield relatively high levels of gene expression. The 7.5K polypeptide is expressed in large amounts in infected cells, but its biological role remains to be determined.

The Vaccinia Virus Growth Factor (VGF). When the sequence of the 19K protein was reported, no information on its possible function was available. Subsequent searches in protein sequence data bases, however, revealed a striking homology to epidermal growth factor (EGF) and α-transforming growth factor (Blomquist *et al.*, 1984; Brown *et al.*, 1985; Reisner, 1985). A growth factor secreted into the culture medium of cells infected with vaccinia virus has been purified to homogeneity and partially sequenced (Stroobant *et al.*, 1985; Twardzik *et al.*, 1985). This showed that the secreted polypeptide is a processed form of the 19K protein which has undergone extensive posttranslational modifications such as removal of a signal and transmembrane sequence, as well as glycosylation.

VGF has several properties similar to those of authentic growth factors: it binds to the epidermal growth factor receptor and stimulates its autophosphorylation, it is mitogenic, and it stimulates anchorage-independent cell growth. Buller *et al.* (1987a) produced VGF⁻ mutants in order to explore its functions. Since the yields of VGF⁻ and wild-type virus were similar in a variety of cell types containing low and high densities of VGF receptor, it is unlikely to play a role in the initiation of infection. However, both in cultured cells and on the chorioallantoic membrane, VGF induced rapid proliferation of cells near the focus of infection (Buller *et al.*, 1987b). Cell growth preceded infection and was a consequence of VGF secretion. VGF⁻ mutants were less virulent than wild-type virus after intracerebral inoculation in mice and intradermal inoculation in rabbits.

The 19K growth factor gene lies within the inverted terminal repeat of the WR strain of vaccinia virus. In other strains of vaccinia virus, and in monkeypox, variola, cowpox, and ectromelia viruses, homologous sequences are present as a single copy only, located just central to the inverted repeat at the left end of the DNA map (Archard *et al.*, 1985).

Products of Late Genes

Marker rescue has proved to be particularly useful for the mapping of early genes, many of which encode enzymes. Alternative procedures had to be devised to map late genes, many of which encode structural proteins. Employing a combination of hybrid selection of RNA, *in vitro* translation and immunoprecipitation, the genes for several major structural proteins have been mapped and some of them have been sequenced. The list includes the genes for the major proteins of viral core (Wittek *et al.*, 1984b; Rosel and Moss, 1985), the gene for a very abundant polypeptide of 11K (Wittek *et al.*, 1984a; Bertholet *et al.*, 1985; Tsao *et al.*, 1986) as well as that encoding a 28K protein (Weir and Moss, 1984). The latter has some sequence homology to the transforming protein of avian erythroblastosis virus and to some extent also to the human epidermal growth factor receptor (Chen and Barker, 1985). The biological significance of this finding remains unclear.

Two further late genes that have been mapped and sequenced are those encoding the 37K protein (Hirt *et al.*, 1986) and the viral hemagglutinin (Shida, 1986; see below). The 37K protein, which is unglycosylated but contains palmitic acid, is the major antigen of the viral envelope and thus, like the hemagglutinin, it is only present on those particles that are naturally released from infected cells.

The early and late genes just described encode polypeptides with more or less well-defined functions. Many additional early and late genes have been identified as open reading frames by sequence analysis of large portions of the viral DNA (Plucienniczak *et al.*, 1985; Niles *et al.*, 1986). Most of these open reading frames encode polypeptides of unknown function and location, and the time of their expression has not been defined. In two cases, however, tight clusters of genes identified by DNA sequencing were found to be expressed either exclusively (Weinrich and Hruby, 1986), or predominantly (Rosel *et al.*, 1986), late in infection. Perhaps the most important conclusion that can be drawn from these studies is that vaccinia virus genes are extremely closely packed along the DNA molecule and in extreme cases may even overlap.

Products of Two Genes of Cowpox Virus

An early and late gene of cowpox virus have also been sequenced. The product of the early gene appears to be responsible for the red pock phenotype of the Brighton strain of cowpox virus (Pickup et al., 1986). The gene was mapped by translocating DNA sequences of wild-type virus that were deleted in white pock mutants into the thymidine kinase gene of the mutant in order to restore the red pock phenotype. The gene is located between 31 and 32 kbp from the right-hand end of the genome and encodes a 38K protein that has considerable homology to inhibitors of serine proteases, which are involved in blood coagulation. It thus appears that interference with the normal pathway of blood coagulation plays a part in the production of hemorrhagic pocks by wild-type cowpox virus.

The late gene encodes the major protein component of the A-type inclusion bodies produced by cowpox virus (Patel and Pickup, 1987). It is probably one of the most strongly expressed genes of any orthopoxvirus, since its product, a 160K protein, may represent up to 4% of the total cell protein. The 5' flanking sequence of this gene may be useful for constructing vaccinia virus recombinants yielding high levels of expression of foreign genes (see Chapter 12).

Important Antigens

Two groups of polypeptides are of particular practical importance: (1) those involved in immunity and protection and (2) antigens that are *Orthopoxvirus* species-specific and therefore useful in viral classificaton and diagnosis.

Protective Antigens. These fall into two classes: epitopes on the surface tubular elements of the outer membrane and antigens in the viral envelope.

A 58K protein polymerizes to form the surface tubules, which have been isolated from virions in a pure form. Such preparations elicit neutralizing antibody against nonenveloped but not against enveloped virions, and block the neutralizing capacity of antibody raised against nonenveloped virions (Stern and Dales, 1976a).

Other protective antigens are located in the viral envelope, which is found only in virions that are naturally released from cells. Antibody to the envelope neutralizes the infectivity of enveloped forms of vaccinia

virus, as demonstrated by the "comet" inhibition test of Appleyard *et al.*
(1971), whereas antibody to inactivated intracellular nonenveloped
virions does not (Payne, 1980; see Chapter 4).

Vaccinia Virus Hemagglutinin. Because of the extensive use of the
hemagglutination-inhibition test in serological studies of vaccinia and
smallpox (see Chapter 4), a special description is required of the
orthopoxvirus hemagglutinin (HA). The production of HA that agglutinated red cells of some but not all chickens was discovered by Nagler
(1942), using extracts from vaccinia virus-infected cells. Subsequently,
all other orthopoxviruses (but no other poxviruses) were shown to
produce HA, although several mutants of vaccinia virus have been
found which have lost this property.

Early studies suggested that vaccinia HA was a lipoprotein, composed
of a viral antigen and a lipid which was responsible for attachment to
the red blood cell (Burnet, 1946; Stone, 1946; Chu, 1948). However,
subsequent work showed that it was a glycoprotein, in which the
carbohydrate moieties were both N- and O-linked, the O-linked oligo-
saccharides being responsible for hemagglutination. A vaccinia virus
mutant that is HA$^-$ (IHD-W) produces a glycoprotein that is serologically
related to the wild-type HA, but its M_r is 41K, rather than 85K (Shida
and Dales, 1982).

From the time of its discovery, the HA was regarded as separable from
the virus particle (Burnet and Stone, 1946) and this was the orthodox
view for many years. The numerous investigations that supported the
dissociation of HA and virion were all carried out with preparations
obtained by the disruption of infected cells. However, Payne and
Norrby (1976) and Payne (1979) reported that vaccinia HA was an 89K
glycoprotein which occurred in the envelopes of enveloped vaccinia
virions and also in the membranes of vaccinia virus-infected cells
(causing hemadsorption).

The gene which specifies the viral hemagglutinin has been mapped
and sequenced by Shida (1986). As deduced from the nucleotide
sequence, the hemagglutinin has an M_r of 35K, whereas the mature
form has an apparent size of 85–89K (Weintraub and Dales, 1974; Payne,
1979; Shida and Dales, 1981). This difference is due at least in part to
extensive posttranslational modifications resulting in the presence of
both N- and O-linked oligosaccharide chains (Shida and Dales, 1981).

Vaccinia Virus LS Antigen. Early workers devoted much attention to a
protein found in large amounts in filtrates of dermal pulp, which they
called the LS antigen (Shedlovsky and Smadel, 1942). Rondle and
Dumbell (1962) reported that cowpox virus produced an LS antigenic

complex that could not diffuse through agar unless it was first treated with trypsin, and Amano and Tagaya (1981) demonstrated that the LS antigen of vaccinia virus cross-reacted with components of the cowpox A-type inclusion body. Patel *et al.* (1986) have brought these findings into sharper focus. They demonstrated that the most abundant viral protein in cells infected with cowpox virus was the 160K protein of the A-type inclusion bodies. This protein cross-reacted with proteins in cells infected by other orthopoxviruses, which occurred as components of A-type inclusion bodies in cells infected with ectromelia virus (130K) and raccoon poxvirus (155K). It also cross-reacts with the LS antigen, which is abundant in cells infected with vaccinia virus (94K) and monkeypox virus (92K), which do not produce A-type inclusion bodies (Dr. D. Pickup, personal communication, 1987).

Nucleoprotein Antigen. Smadel *et al.* (1942) described an antigen obtained by extraction of purified suspensions of virions of vaccinia virus with NaOH, which they called nucleoprotein (NP) because it contained all the viral DNA. The NP antigen is a mixture of molecular species. It appears to include the antigens that show serological cross-reactivity between viruses belonging to all genera of vertebrate pox-viruses that have been tested (Woodroofe and Fenner, 1962). These may include polypeptides of M_r 64–65K and 40K (Ikuta *et al.*, 1979; Kitamoto *et al.*, 1987).

Virus-Coded Enzymes

Because vaccinia virus replicates in the cytoplasm, it cannot utilize the variety of nuclear enzymes that are involved in DNA replication and transcription of most other DNA viruses. In consequence, the vaccinia virus genome codes for a large number of enzymes that subserve such functions (Table 2-1). Many of these virus-specified enzymes are components of the virion and are located within the core, and mRNA is synthesized when cores are incubated under appropriate conditions (see Chapter 3). Others are not components of the virion, but are synthesized during the replication cycle.

Virion-Associated Enzymes. *RNA Polymerase.* The presence of an RNA polymerase associated with purified virions of vaccinia virus was first demonstrated by Kates and McAuslan (1967b) and Munyon *et al.* (1967). The enzyme, which has been purified extensively (Baroudy and Moss, 1980; Spencer *et al.*, 1980), has an M_r of 500K in its native form. It is composed of at least seven subunits ranging in mass from 17K to 140K. An enzyme with a very similar subunit composition has also

TABLE 2-1

Enzymes Isolated from Vaccinia Virus[a]

RNA polymerase[b,c]
  M_r: 500K (140K, 137K, 37K, 35K, 31K, 22K, 17K)
  nNTP → RNA + nPP$_i$
  Requires single-stranded DNA template, Mn^{2+}, role in transcription
Poly(A) polymerase[d]
  M_r: 80K (55K, 30K)
  nATP + N($-$N)m → N($-$N)m($-$A)n + nPP$_i$
  Primer dependent, specific for ATP, role in polyadenylation of mRNA
RNA triphosphatase[e,f]
  M_r: 127K, (95K, 31K)
  pppN($-$N)n → ppN($-$N)n + P$_i$
  Part of complex with RNA guanylyltransferase and RNA (guanine-7-)
    methyltransferase, role in capping mRNA
RNA guanylyltransferase[g,h,i,j,k]
  M_r: 127K (95K, 31K)
  GTP + ppN($-$N)n → GpppN($-$N)n + PP$_i$
  Part of complex with RNA triphosphatase and RNA (guanine-7-)methyltransferase,
    95K polypeptide forms covalent GMP complex, capping enzyme
RNA (guanine-7-)methyltransferase[g,h]
  M_r: 127K (95K, 31K)
  AdoMet + GpppN($-$N)n → m^{7}GpppN($-$N)n + AdoHcy
  Part of complex with RNA triphosphatase and RNA guanylyltransferase
RNA (nucleoside-2'-)methyltransferase[l,m]
  M_r: 38K
  AdoMet + m^{7}GpppN($-$N)m → m^{7}GpppN^m($-$N)n + AdoHcy
  Methylation of cap structures
5'-Phosphate polynucleotide kinase[n]
  2ATP + pN($-$N) → pppN($-$N) + 2ADP
  Partially characterized, role in RNA processing suggested
Adenosine triphosphatase[o,p,q]
  M_r: 61K
  ATP + H$_2$O → ADP + P$_i$
  Specific for ATP or dATP, DNA dependent
Nucleoside triphosphatase[p,q]
  M_r: 68K
  NTP + H$_2$O → NDP + P$_i$
  Ribo- or deoxyribonucleoside triphosphates, stimulated by DNA or RNA,
    antigenically distinct from ATPase
Endoribonuclease[r]
  RNA + H$_2$O → smaller RNA
  Not well characterized, suggested role in RNA processing
Deoxyribonuclease (pH 4.4 optimum)[s,t,u,v]
  M_r: 105K (50K)
  DNA + H$_2$O → mono- and oligodeoxyribonucleotides
  Specific for single-stranded DNA, exo- and endonuclease
Deoxyribonuclease (pH 7.8 optimum)[s,t]
  M_r: 50K

TABLE 2-1 (*Continued*)

DNA + H₂O → oligodeoxyribonucleotides
Specific for single-stranded DNA, endonuclease
Deoxyribonuclease (pH 6.5 optimum)[w]
  Supercoiled closed duplex DNA → duplex linear DNA. Also cross-links DNA.
  Unclear whether a distinct enzyme or another activity of neutral or acid
    deoxyribonuclease
DNA topoisomerase[x,y]
  M_r: 32K
  Type 1, relaxes both left- and right-handed superhelical DNA
Protein kinase[z,aa]
  M_r: 62K
  ATP + protein → P-protein + ADP
  Phosphorylates serine and threonine residues of viral acceptor protein, activated by
    protamine
Alkaline protease[bb]
  Protein + H₂O → peptides
  Detected but not isolated or characterized

---

[a] Based on Moss (1985); [b] Baroudy and Moss (1980); [c] Spencer *et al.* (1980); [d] Moss *et al.* (1975); [e] Tutas and Paoletti (1977); [f] Venkatesan *et al.* (1980); [g] Martin and Moss (1975); [h] Martin and Moss (1976); [i] Martin *et al.* (1975); [j] Shuman and Hurwitz (1981); [k] Shuman *et al.* (1980); [l] Barbosa and Moss (1978a); [m] Barbosa and Moss (1978b); [n] Spencer *et al.* (1978); [o] Gold and Dales (1968); [p] Paoletti and Moss (1974); [q] Paoletti *et al.* (1974); [r] Paoletti and Lipinskas (1978a); [s] Pogo and Dales (1969); [t] Pogo and O'Shea (1977); [u] Rosemond-Hornbeak and Moss (1974); [v] Rosemond-Hornbeak *et al.* (1974); [w] Lakritz *et al.* (1985); [x] Bauer *et al.* (1977); [y] Shaffer and Traktman (1987); [z] Kleiman and Moss (1975a); [aa] Kleiman and Moss (1975b); [bb] Arzoglou *et al.* (1978)

---

been purified from the cytoplasm of vaccinia virus-infected HeLa cells (Nevins and Joklik, 1977). Characteristic features of the purified enzyme include its high resistance to α-amanitin, preference for manganese over magnesium, and poor activity on double-stranded DNA as template.

The genes encoding what are presumably all subunits have been mapped on the viral DNA (Morrison *et al.*, 1985; Jones *et al.*, 1987) and two genes for RNA polymerase subunits of 147K and 22K have been sequenced (Broyles and Moss, 1986), providing rigorous proof that they are virus coded. The deduced amino acid sequence of the 147K subunit has several domains of considerable homology to the *Escherichia coli* RNA polyermase B′ chain, the large subunit of yeast RNA polymerases II and III, and the large subunit of *Drosophila* RNA polymerase II.

Capping Enzyme. The RNA molecules transcribed by the RNA polymerase are modified at the 5′ end by the addition of a cap structure. This complex process is accomplished by enzymes that are contained in the virion and have been purified and characterized in great detail

(Martin and Moss, 1975, 1976; Martin *et al.*, 1975; Tutas and Paoletti, 1977; Barbosa and Moss, 1978a,b; Shuman *et al.*, 1980; Venkatesan *et al.*, 1980; Shuman and Hurwitz, 1981). The first step in the capping reaction consists of removal of the γ-phosphate from the 5' end of the RNA molecule. Guanosine-monophosphate is then added and methylated at the 7-position. These reactions are catalyzed by three enzymatic activities: RNA triphosphatase, RNA guanylyltransferase, and RNA (guanine-7-)methyltransferase, all of which are contained in an enzyme complex consisting of two subunits of M_r 95K and 31K. Finally, the penultimate nucleoside is methylated by RNA (nucleoside-2')methyltransferase to yield the mature cap.

Poly(A) Polymerase. The 3' ends of vaccinia virus mRNAs are polyadenylated (Kates and Beeson, 1970b; see Chapter 3). A poly(A) polymerase has also been isolated from viral cores and characterized (Moss and Rosenblum, 1974; Moss *et al.*, 1975).

Other Enzymes. Besides the enzymes with a well-defined role in mRNA synthesis, vaccinia virus cores contain several other enzymes that appear to be involved in nucleic acid metabolism (see Table 2-1).

Viral Enzymes Not Associated with the Virion. *Thymidine Kinase.* Thymidine kinase (TK) catalyzes phosphorylation of thymidine to thymidine monophosphate. McAuslan (1963a,b) showed that vaccinia and cowpox viruses induced TK synthesis in infected HeLa cells for a limited time, followed by repression. This repression, but not the induction of enzyme synthesis, was inhibited by UV irradiation of the virus prior to infection, and the TK activity then increased to over 30 times that of uninfected cells. Isolation of TK-negative mutants (Dubbs and Kit, 1964) provided further evidence that the enzyme was virus coded, and the gene has been now mapped in the viral genome (Hruby and Ball, 1982; Weir *et al.*, 1982; Bajszar *et al.*, 1983) and sequenced (Hruby *et al.*, 1983; Weir and Moss, 1983). It is located near the center of viral DNA, at the left-hand end of the *Hind*III J fragment. As inferred from the DNA sequence, the viral enzyme has an M_r of 20K, whereas a value of 80K has been found for the native enzyme isolated from infected cells (Kit *et al.*, 1977), indicating that it is composed of four subunits. Vaccinia virus TK has no significant homology to the herpesvirus TK. The sequences of the TK genes of monkeypox and variola viruses, and of fowlpox virus, have been compared (Esposito and Knight, 1984; Boyle *et al.*, 1987). The deduced amino acid sequences of the three orthopoxvirus TKs are very similar, and that of fowlpox TK shows extensive homologies with them (Fig. 2-8), suggesting that the poxvirus TK genes have evolved from a common ancestor gene.

Since the TK gene is not essential for viral replication, this locus has

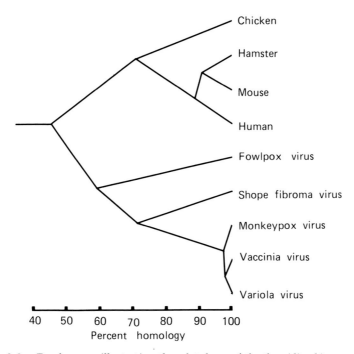

FIG. 2-8. *Dendrogram illustrating the relatedness of the thymidine kinase genes of three orthopoxviruses, an avipoxvirus, a leporipoxvirus, and four vertebrates. (Personal communication from Dr. D. Boyle and Dr. A. J. Gibbs, based on Boyle et al., 1987, Bradshaw and Deininger, 1984, Lewis, 1986, Lin et al., 1985, and Upton and McFadden, 1986.)*

been used extensively to insert foreign genes into the vaccinia virus genome. This site also offers the advantage that recombinant virus carrying insertions in the TK gene can easily be selected for on the basis of the TK⁻ phenotype (see Chapter 12). TK⁻ strains of vaccinia virus are less virulent than the parental wild type (Buller *et al.*, 1985).

DNA Polymerase Gene. The DNA polymerase gene is another early gene that has been mapped (Jones and Moss, 1984, 1985; Traktman *et al.*, 1984) and sequenced (Earl *et al.*, 1986). The M_r of DNA polymerase derived from the gene sequence is 109K, which is in good agreement with that of the purified enzyme (Challberg and Englund, 1979), indicating that the active enzyme is a monomer. Comparison of the deduced amino acid sequence with those of other proteins revealed a long stretch of striking homology to the Epstein–Barr virus DNA polymerase and a shorter region of homology with the adenovirus DNA polymerase. This indicates either coevolution of proteins with similar functions, or common ancestry.

Nonstructural Polypeptides of Vaccinia Virus-Infected Cells

Early attempts to categorize the cell-associated polypeptides by running extracts of vaccinia-infected cells on one-dimensional polyacrylamide gels demonstrated that several polypeptides were produced that were not ultimately incorporated into the virion. The most abundant of these appears to be the LS antigen (see above). The use of two-dimensional polyacrylamide gels considerably improved the definition of these polypeptides and J. A. Hackett (personal communication, 1980) identified about 45 nonstructural virus-coded polypeptides in extracts of infected cells. Some are enzymes such as thymidine kinase or DNA polymerase (see above), and a presumably virus-coded ribonucleotide reductase (Slabaugh *et al.*, 1984; Slabaugh and Mathews, 1986). The functions of others have yet to be determined, but some of them are precursors of virion polypeptides. The best studied example of the latter are the two major core polypeptides 4a and 4b (Sarov and Joklik, 1972), both of which are processed from high M_r precursors (Katz and Moss, 1970a,b; Moss and Rosenblum, 1973; Pennington, 1973). A few polypeptides, such as two DNA-binding polypeptides, occur in much larger amounts in the infected cell than in the virion.

Polypeptides of Different Species of *Orthopoxvirus*

Esposito *et al.* (1977b) and Arita and Tagaya (1980) noted that variola, monkeypox, vaccinia, and cowpox viruses each had distinctive structural polypeptides in the 30K–40K range. Turner and Baxby (1979) found that the core proteins of the several species they examined were indistinguishable on a one-dimensional gel; however, the membrane and subsurface polypeptides showed distinctive differences in polypeptides of the 30K–40K region, and they also found differences in the 20K–25K and 50K–60K regions. Camelpox, cowpox, monkeypox, and vaccinia viruses had distinctive patterns. Some viruses of uncertain affiliations (Lenny, MK-10, and buffalopox) were indistinguishable from vaccinia virus by this test; by restriction mapping all are strains of vaccinia virus. On the other hand, the polypeptides of "elephant virus" and "Moscow rat-carnivore virus" (Marennikova *et al.*, 1977), which by genome mapping are strains of cowpox virus, showed the same polypeptide pattern as ectromelia virus in that all three lacked a 37K polypeptide that was present in cowpox virus.

Harper *et al.* (1979) compared the late intracellular polypeptides produced by 24 strains of orthopoxvirus, including the "whitepox" viruses recovered in Moscow from wild animals shot in Zaire. The patterns of variola, vaccinia, monkeypox, and cowpox viruses in one-dimensional gels were distinctive; two whitepox virus isolates were indistinguishable from those of variola virus. The intracellular polypeptides of Lenny, MK-10, and buffalopox viruses were very similar to those of vaccinia virus, whereas those of the Moscow rat-carnivore virus indicated their affinities with cowpox virus.

Analysis of viral polypeptides by polyacrylamide gel electrophoresis thus confirms the classification of species of *Orthopoxvirus* set out in Tables 1-2 and 1-3, which was based upon biological tests and restriction mapping of their genomes. It is a useful method for classifying viruses of uncertain provenance, but DNA analysis is a simpler and more reliable procedure.

Serological Comparisons of Different Species of *Orthopoxvirus*

Serology should provide a ready means of recognizing species-specific proteins and of deciding what species of *Orthopoxvirus* was responsible for the production of orthopoxvirus antibodies that might be found in naturally infected wild animals or humans. Early work by Downie and his associates (Downie and McCarthy, 1950) showed that there was a high degree of antigenic cross-reactivity between variola, cowpox, vaccinia, and ectromelia viruses when tested by either complement fixation or neutralization tests, although differences could be detected in titers against heterologous and homologous viruses and by absorption of antisera.

Gel diffusion tests, which had previously been shown to allow the recognition of up to 20 different antigens in rabbitpox virus-infected cells (Westwood *et al.*, 1965), provided an obvious method for attempting the serological differentiation of different species of *Orthopoxvirus*. Using absorbed sera, Gispen and Brand-Saathof (1974) and subsequently Esposito *et al.* (1977b) demonstrated species-specific antigens for variola, vaccinia, and monkeypox viruses. By this test whitepox viruses were indistinguishable from variola virus and rabbitpox and buffalopox viruses were indistinguishable from vaccinia virus. Maltseva and Marennikova (1976) used absorption in the test well of gel diffusion plates to distinguish between variola, monkeypox, vaccinia, and cowpox viruses, and showed that isolates from various animals in outbreaks in zoos (okapi, elephant, carnivores) reacted like cowpox virus (see Chapter 6).

The fact that different species of *Orthopoxvirus* produce specific antigens as well as antigens shared with other members of the genus suggests that it should be possible to determine the species of virus responsible for the orthopoxvirus-specific antibodies sometimes found in sera of wild animals. This would be particularly useful in surveys of animal sera carried out in efforts to elucidate the ecology of monkeypox virus (see Chapter 8). Four methods have been successfully used to differentiate variola-, monkeypox-, and vaccinia-specific antisera: assays by gel precipitation (Gispen and Brand-Saathof, 1974), immunofluorescence (Gispen et al., 1976), radioimmunoassay (Hutchinson et al., 1977), and the ELISA test (Marennikova et al., 1981), after absorption of the tested sera with homologous and heterologous antigens. All procedures were effective in allowing specific diagnoses of monkeypox infection to be made with human and certain monkey sera, but the absorption procedure makes the standard radioimmunoassay test, for example, unsuitable for routine use with sera from a range of different animals, such as are usually collected during ecological surveys. This problem has been to some extent overcome by the use of *Staphylococcus* A protein instead of animal-species-specific γ-globulin (Fenner and Nakano, 1988).

LIPIDS OF VIRIONS OF VACCINIA VIRUS

Lipids comprise about 6% of the wet weight of nonenveloped vaccinia virions (Smadel and Hoagland, 1942), and occur in the outer membrane. The phospholipids are cellular in origin but differ somewhat in composition from those found in cytoplasmic membranes, suggesting that virion polypeptides may regulate, at least in part, the amounts of individual phospholipids inserted into the outer membrane (Dales and Mosbach, 1968; Stern and Dales, 1974, 1976b). On the other hand the glycolipids in the outer membrane of the virion are present in the same molar ratios relative to phospholipids as those found in cellular membranes (Anderson and Dales, 1978).

Hiller et al. (1981a) showed that extracellular enveloped virions contained about twice as much phospholipid as intracellular virions. Both forms were characterized by a low ratio of neutral to negatively charged phospholipids (55:45), whereas cellular membranes show a proportion of 80:20.

CHAPTER 3

Replication and Morphogenesis of Vaccinia Virus

As we pointed out in Chapter 1, viruses exist in two phases. The inert "extracellular" phase of vaccinia virus, the virion, has just been described. This chapter is concerned with the vegetative phase of vaccinia virus, when it ceases to be an entity separate from the host cell. We shall consider in turn the way virions enter cells, the very complex series of events that result in the production of new virions, and the way the virions get out of cells. During this process there are important changes in the host cells; although these are intimately related to the replication process we shall for convenience discuss them in the next chapter.

During the 1950s virologists were still uncertain about the existence of an "eclipse phase" during poxvirus replication, because of the high background of cell-associated but nonreplicating infectious virus that occurred in the systems then available. However, with the development of cell culture techniques and the increase in understanding of the nature of viral replication after about 1960 these doubts were dissipated.

The large size and distinctive appearance of the vaccinia virion made electron-microscopic study of uptake, uncoating, and morphogenesis of vaccinia virus particularly rewarding, although these studies remain largely descriptive. The development of recombinant DNA methodology has greatly facilitated studies on DNA replication, transcription, and translation, and made it possible to relate these events to the replication cycle. Most of this chapter is devoted to these aspects of replication.

THE REPLICATION CYCLE

Poxviruses, African swine fever virus, and to some extent iridoviruses are the only DNA viruses which replicate in the cytoplasm of the infected host cell. Poxvirus DNA replication occurs within discrete cytoplasmic foci called factories or virosomes (Cairns, 1960; Kato et al., 1960; Dales and Siminovitch, 1961; Loh and Riggs, 1961; Harford et al., 1966). No radioactive host polypeptides have been detected in purified factories isolated from cells labeled prior to infection, suggesting that exclusively viral proteins are involved in the replication process (Dahl and Kates, 1970; Polisky and Kates, 1972). Several virus-coded enzymes have been isolated either from purified virions or from infected cell extracts (see Table 2-1).

The replication cycle is represented diagrammatically in Fig. 3-1. Upon penetration of the virus into susceptible host cells the particles shed their outer protein layers, leaving the intact core. Virion-associated enzymes are activated by the uncoating process and a first burst of RNA synthesis occurs. These early mRNAs are then translated to yield early proteins, typical examples of which are the thymidine kinase and the DNA polymerase. Following the expression of early genes, the DNA is completely uncoated and replicated. Late genes, most of which code for structural polypeptides, are then transcribed from replicated, or possibly replicating, DNA. Finally, progeny virions are assembled in a complex process of morphogenesis.

INITIATION OF INFECTION

Because of the complex structure of poxvirus virions, the initiation of infection is a somewhat more complex process than with many other viruses. It consists of two stages: adsorption, penetration and "first-stage uncoating," which involves disappearance of the envelope (if

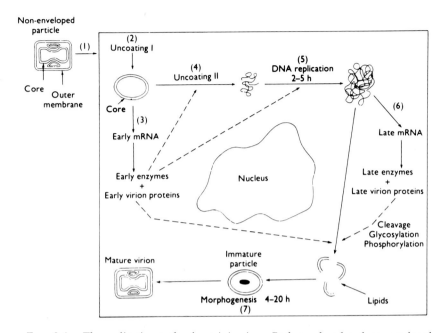

Non-enveloped particle

Core
Outer membrane

(1)

(2) Uncoating I

Core
(3)
Early mRNA

(4) Uncoating II

(5) DNA replication 2–5 h

(6)

Late mRNA

Early enzymes
+
Early virion proteins

Nucleus

Late enzymes
+
Late virion proteins

Cleavage
Glycosylation
Phosphorylation

Mature virion

Immature particle

Morphogenesis   4–20 h
(7)

Lipids

FIG. 3-1. *The replication cycle of vaccinia virus. Both enveloped and nonenveloped particles are infectious but differ in their attachment to cells and mode of entry (not shown). The sequence of events is (1) attachment and entry, (2) uncoating I, whereby the outer membrane is removed by cellular enzymes, leaving the core, (3) immediate early transcription from DNA within the core by the viral transcriptase, leading to the production of early enzymes which include the enzyme that produces (4) uncoating II and release of the viral DNA into the cytoplasm. Simultaneously (5) DNA replication takes place, after which (6) late transcription occurs from the newly synthesized DNA, followed by translation, and cleavage, glycosylation, and phosphorylation of some of the late proteins. Morphogenesis (7), including the appearance of cupules, occurs entirely within a cytoplasmic factory (not shown). (From Moss, 1985.)*

present) and breakdown of the outer membrane, followed by "second-stage" uncoating, during which viral DNA is released from the core.

Adsorption and Penetration

Adsorption depends on two factors: the rate of collision between virions and cells, and the adsorption environment (Joklik, 1964). Virions of vaccinia virus exist in two forms, enveloped and nonenveloped (see Chapter 2), which differ in their mode of adsorption and penetration (Payne and Norrby, 1978). The adsorption of enveloped virions is less dependent than that of nonenveloped virions on the adsorption environment, because of the higher affinity of the envelope for cell receptors. Following adsorption, enveloped virions enter the cell by fusion of the envelope with the cell membranes. The penetration of nonenveloped particles is dependent on cellular metabolic activities and the particles are taken up by viropexis, although it has been reported that in some systems the outer membrane may fuse with the cell membrane, releasing core and lateral bodies into the cytoplasm (Chang and Metz, 1976).

Uncoating

Soon after penetration, the outer membrane is degraded by cellular enzymes, with rapid hydrolysis of some virion protein and all the viral phospholipid. These processes are insensitive to inhibitors of RNA and protein synthesis, and result in the production of subviral cores, within which the viral genome is protected from nucleases.

Release of the viral DNA from the core (second stage uncoating) is prevented by inhibitors of transcription or translation. Although initially ascribed to a host cell-coded activity (Joklik, 1964), it is now clear that second stage uncoating is controlled by the viral genome itself. Experiments by Cairns (1960) with [^{3}H]thymidine-labeled vaccinia virus, showed that in multiply infected cells there was an initial lag period, then all virions were uncoated simultaneously and DNA replication began. Superinfecting particles added at the end of the lag period were uncoated immediately (Joklik, 1964). Dales (1965) provided morphological evidence for the synchronous release of DNA from a multiplicity of cores, the viral DNA being extruded through distinct breaks in the core membrane.

Transcription from the viral DNA occurs initially within the core (see below). One of the products is an "uncoating protein," which has recently been studied in an *in vitro* system (Pedley and Cooper, 1987). The partially purified protein had an M_r of 23K. It was produced within 20 minutes of the addition of virus to cells, reached a maximum level by 2 hours, and maintained this level for the next 24 hours. Heat-inactivated virus failed to produce the uncoating protein, but tests were not done on the capacity of such particles to be uncoated by the 23K protein, in the way in which active poxvirions induce uncoating in the Berry–Dedrick phenomenon (nongenetic reactivation).

EARLY TRANSCRIPTION

Early transcription starts shortly after penetration of the virus particles into the host cell and before second stage uncoating (Kates and

McAuslan, 1967a). Furthermore, early transcription does not require *de novo* protein synthesis, indicating that early mRNAs are made exclusively by the enzymes contained in the virus core. Early mRNAs have methylated caps (Wei and Moss, 1975) and poly(A) tails (Kates and Beeson, 1970b). Thus, although these mRNAs are made in the cytoplasm, they have a structure typical of eukaryotic mRNAs.

In Vitro RNA

Since all the enzymes required for synthesis and modification of mRNA are contained in the virion, it is not surprising that RNA synthesis occurs when purified cores are incubated *in vitro* under appropriate conditions. This RNA is referred to as *in vitro* RNA, and has been characterized in considerable detail. By several criteria, such as sedimentation characteristics (Kates and Beeson, 1970a) and protein-coding capacity (Beaud *et al.*, 1972; Fournier *et al.*, 1973; Jaureguiberry *et al.*, 1975; Nevins and Joklik, 1975; Pelham, 1977; Bossart *et al.*, 1978; Cooper and Moss, 1978; Pelham *et al.*, 1978), this *in vitro* RNA appears to be identical to early *in vivo* RNA. Furthermore, *in vitro* RNA hybridizes to between 14% (Kates and Beeson, 1970b) and 50% (Nevins and Joklik, 1975) of one strand-equivalent of the viral genome and is thus in this respect also similar to early *in vivo* RNA (Oda and Joklik, 1967). The higher values that were reported on the basis of RNA excess hybridizations (Boone and Moss, 1978; Paoletti and Grady, 1977) may have been due to the presence of low amounts of long RNA transcripts. Such long RNA molecules, sedimenting between 20 and 30S (Paoletti, 1977a), are made in rather high amounts when cores are incubated at low ATP concentration (Paoletti and Lipinskas, 1978b) or when they are heated prior to incubation (Harper *et al.*, 1978). The biological significance of the high-M_r RNA remains unclear (see below).

Synthesis of Mature Early mRNAs

The high-M_r RNA made *in vitro* under particular incubation conditions remains core associated but at least part of it subsequently appears to be extruded as 8–12S RNA characteristic of the bulk of early *in vitro* and *in vivo* mRNA (Paoletti, 1977a). Furthermore, the sequences found in high-M_r and messenger-size RNA appear to be related (Paoletti, 1977b). RNAs several thousand nucleotides in length have also been found in early RNA isolated from infected cells (see, for example, Wittek *et al.*, 1980b; Cooper *et al.*, 1981b). These long molecules have not been characterized in great detail, but from the number and size of such RNAs that map to given segments of the viral genome it is clear that they overlap each other and also overlap the mature mRNAs. It is therefore tempting to speculate that the long RNA molecules represent precursors of the mature RNAs, as had initially been proposed for the high-M_r *in vitro* RNAs (Paoletti, 1977b; Paoletti and Grady, 1977). However, ultraviolet target size studies, which showed a linear relationship between the size of the protein and the dose required to inhibit its expression, argue strongly against the high-M_r RNA being an obligatory intermediate in the biogenesis of mature early mRNAs.

Information on the sequence relationship of individual high-M_r RNAs and the major mRNA species transcribed from similar regions of the DNA has come from transcriptional mapping of the thymidine kinase (TK) and DNA polymerase early genes (Bajszar *et al.*, 1983; Traktman *et al.*, 1984; Jones and Moss, 1985). These studies may also provide clues on how such long RNA molecules are made. Functional thymidine kinase is encoded by two mRNA species of 590 and 2380 nucleotides, which have a common 5' end. The long RNA also contains the coding sequences of an additional protein which, however, is not translated from the long RNA. Instead, this product is made by a separate mRNA species that is initiated downstream of the TK coding sequences but which has the same 3' end as the long TK mRNA. This indicates that in vaccinia virus early mRNAs, as in most other eukaryotic mRNAs, the AUG closest to the 5' end is used for translation initiation (Kozak, 1986). Thus, some early mRNAs are structurally polycistronic, but functionally monocistronic. A similar situation also exists for the DNA polymerase, which is encoded by two RNA species of 3400 and 3900 nucleotides that have a common 5' end.

One possibility to explain these multiple 3' ends is to assume that they are generated by cleavage of high-M_r precursor molecules. Such a mechanism would require a site-specific endoribonuclease. An enzyme for which no role has yet been found and which appears to possess the required properties has been isolated from vaccinia virions and partially characterized (Paoletti and Lipinskas, 1978a).

However, a recent study on transcription termination of an early gene (Rohrmann *et al.*, 1986) makes the cleavage hypothesis very unlikely. Soluble extracts from vaccinia virions, previously shown to initiate early gene transcription correctly (Golini and Kates, 1985; Rohrmann and Moss, 1985), also correctly terminate transcription. This also made it possible to identify the DNA sequences involved in the formation of 3' ends. In the particular early gene examined, the translation termination codon is immediately followed by the sequence TTTTTAT, which is essential for correct termination to occur, 50 to 70 bp downstream from this signal. Occasional disregard of this signal, favored under certain experimental conditions, presumably gives rise to the long read-through transcripts with multiple 3' ends. In the *in vitro* system there appeared to be little sequence specificity at the transcription termination site itself (Rohrmann *et al.*, 1986). Analysis of the 3' ends of cDNA clones, however, indicated that *in vivo* transcription terminates on the first or second T of one of several copies of the sequence TATGT found downstream of the early gene coding sequence (Yuen and Moss, 1986). Multiple copies of this sequence, located 50 to 100 nucleotides downstream from the transcription termination signal, may also explain the microheterogeneity of the 3' ends of early RNAs.

The recent isolation from vaccinia virions of an RNA polymerase complex that specifically initiates and terminates transcription (Broyles and Moss, 1987) made it possible to isolate the termination factor. The protein involved was purified to homogeneity and shown to be identical to the viral multisubunit capping enzyme by all criteria examined (Shuman *et al.*, 1987). How the enzyme accomplishes its dual function (the modification of the 5' end and the generation of the 3' termini of mRNA) remains an intriguing question. Interestingly, capping of the RNA is not a prerequisite for the formation of the correct 3' end.

The studies on transcription termination also provided information on the polyadenylation reaction, which appears to differ in important respects from that of higher eukaryotes. In the latter, polyadenylation may be directly coupled to 3' processing and depends on specific polyadenylation signals (Birnstiel *et al.*, 1985). In vaccinia virus, polyadenylation is not coupled to processing events and does not require specific signals. Instead, the viral poly(A) polymerase apparently polyadenylates any free 3'-OH end and exogenous RNA added to the *in vitro* transcription reaction is an efficient competitor for polyadenylation (Rohrmann *et al.*, 1986).

Besides the mechanism by which they are polyadenylated, vaccinia virus early mRNAs differ from most eukaryotic mRNAs transcribed by RNA polymerase II in the absence of splicing. This type of RNA processing was not found with early RNAs encoded within the inverted terminal repetition (Wittek *et al.*, 1980b). Many other studies addressing the question of processing, either directly or indirectly, have since confirmed these observations and so far no evidence for splicing of early vaccinia virus RNAs has been obtained. Processing at the 5' end has also been ruled out by a study on *in vitro* RNAs (Venkatesan and Moss, 1981). Three early mRNAs encoded within the inverted terminal repetition were found to be identical to the corresponding mRNAs made *in vivo* with respect to size, map position, and *in vitro* translation products. When these RNAs were transcribed *in vitro* by cores in the presence of radioactive nucleoside triphosphates, the β phosphate of the initiating ribonucleotide was subsequently found in the cap structure, demonstrating that the cap sites correspond to the sites of initiation of RNA synthesis.

LATE TRANSCRIPTION

RNA isolated at late times in infection, i.e., after DNA replication, hybridizes to at least one entire strand-equivalent of the vaccinia virus DNA (Oda and Joklik, 1967; Kaverin *et al.*, 1975; Paoletti and Grady, 1977; Boone and Moss, 1978). An unusual property of late mRNA is its enormous length heterogeneity. This was demonstrated by *in vitro* translation of size-fractionated total RNA from infected cells (Cooper *et al.*, 1981b; Mahr and Roberts, 1984). Whereas early mRNAs encoding the bulk of early proteins are between 500 and 1500 nucleotides long, late mRNAs are on the average much longer, ranging in size from about 500 to 4000 nucleotides. Furthermore, for early mRNAs there is a direct correlation between the size of the RNA and the size of the *in vitro* translation product. This is not the case for late mRNAs, since a particular *in vitro* translation product appears to be made from a very broad size spectrum of RNA and the longest RNA species may exceed the minimal required coding length manyfold. The recent finding (see below) that late mRNAs undergo unusual processing events at their 5' ends provides a plausible explanation for the length heterogeneity.

Another characteristic feature of late transcription is the production of large amounts of double-stranded RNA (Colby and Duesberg, 1969; Duesberg and Colby, 1969; Colby *et al.*, 1971; Boone *et al.*, 1979; Varich *et al.*, 1979). Relatively little early RNA formed double-stranded structures upon self-annealing. However, late in infection about 15% of the RNA was ribonuclease resistant after self-annealing. The same percentage of double-stranded RNA was also detected by annealing of labeled *in vitro* RNA to an excess of unlabeled late RNA (Paoletti and Grady, 1977). Thus, late in infection sequences from both strands of the DNA are transcribed and these include "anti-sense" early sequences.

REGULATION OF EARLY AND LATE GENE EXPRESSION

Evidence that the large amount of genetic information contained in the vaccinia virus genome is expressed in a temporally regulated fashion was first obtained by labeling infected cell proteins with radioactive amino acids (Holowczak and Joklik, 1967b; Moss and Salzman, 1968; Pennington, 1974). The pattern of newly synthesized polypeptides changes drastically at the time when the viral DNA starts to be replicated, about 3 hours after infection. Transcription of late genes requires protein synthesis and DNA replication, the reason for which is not clear. Inhibitors of protein or DNA synthesis, however, are powerful tools to arrest the viral replication cycle at the early stage. Under such conditions, pure populations of early RNAs may be studied.

Early RNAs made in the presence of inhibitors of either protein or DNA synthesis are indistinguishable. Thus, as compared for instance to herpesviruses, where the early genes can further be subdivided depending whether their expression does not (immediate early genes), or does (early genes) require protein synthesis, there appears to be little justification to adopt a similar classification for vaccinia virus early genes.

5′ Flanking Sequences of Early and Late Genes

The molecular basis for the temporal regulation of vaccinia virus genes is still poorly understood. By analogy to eukaryotic genes, the signals responsible for the regulation of early and late gene expression are believed to reside in the 5′ flanking regions of the genes. Comparison of these regions might therefore reveal sequence elements that are conserved in early and late genes, implying that they represent the signals for the correct temporal regulation of the corresponding genes. Figure 3-2 shows a comparison of the 5′ flanking sequences of several early genes. The most obvious feature of these sequences is their high A + T content. Absent, however, is a TATA box, located at about 30 nucleotides upstream of the transcription initiation site of most cellular and viral genes transcribed by RNA polymerase II (for review, see Breathnach and Chambon, 1981). This is not surprising in view of the fact that vaccinia virus utilizes its own RNA polymerase to transcribe the genes.

Computer analysis of the 5′ flanking sequence of four early genes revealed two stretches of nine and eight nucleotides, centered at about positions −13 and −40, respectively, that are conserved in these genes (Weir and Moss, 1983). Subsequent deletion analysis, however, showed that the distal element is not essential for early gene expression (Cochran et al., 1985b). The significance of the proximal element also remains to be established, since so far this sequence has not been mutagenized and tested functionally. Vassef (1987) recently proposed an alternative consensus sequence that may be involved in early gene expression (Fig. 3-2). The 5′ flanking sequences of several late genes are also shown in Fig. 3-2, where they are aligned on the ATG translation

Early Genes

```
          -50        -40        -30        -20        -10        +1
           |          |          |          |          |          |
 7.5K    AATACAATAATTAATTTCTCGTAAAAGTAGAAAATATATTCTAATTTATTG   a

 19K     TAACAATATATTATTAGTTTATATTACTGAATTAATAATATAAAATTCCCA   b

 42K     TAAAATATAATAAAGCAACGTAAAACACATAAAAATAAGCGTAACTAATAA   b

 TK      TGATGGATATATTAAAGTCGAATAAAGTGAACAATAATTAATTCTTTATTG   c

147K^i   TCGTCCGTCATGATAAAAATTTAAAGTGTAAATATAACTATTATTTTTATA   d

147K^ii  ATAACTATTATTTTTATAGTTGTAATAAAAAGGGAAATTTGATTGTATACT   d

 22K     TTACCAAATCAGACGCTGTAAATTATGAAAAAAAGATGTACTACCTTAATA   d

110K     CTTCCCCAATGTTTGGGATTCATTTAAATGAAAATATATTTCTAAATTCTA   e
```

Late Genes

```
          -50        -40        -30        -20        -10        +1
           |          |          |          |          |          |
 11K     TGTATGTAAAAATATAGTAGAATTTCATTTTGTTTTTTTCTATGCTATAAATG  f

 28K     TTTGTAACATCGGTACGGGTATTCATTTATCACAAAAAAAACTTCTCTAAATG  g

 4a      AAATCGTTGTATATCCGTCACTGGTACGGTCGTCATTTAATACTAAATAAATG  h

 4b      AGATTGGATATTAAAATCACGCTTTCGAGTAAAAACTACGAATATAAATAATG  i

 37K     ATTCTAGAATCGTTGATAGAACAGGATGTATAAGTTTTTATGTTAACTAAATG  j

 HA      ATAGAACAAAATACATAATTTTGTAAAAATAAATCACTTTTTATACTAATATG  k
```

Fig. 3-2. *5' flanking sequences of early and late genes. In early genes the position +1 is the first or the major transcription start site; in the late genes the position +1 is arbitrarily defined as the A of the ATG translation initiation codon. Conserved sequence elements of early genes proposed by Vassef (1987) are underlined, as also are conserved elements in late genes. The 147K^i and 147K^ii sequences are upstream of two distinct mRNA start sites of the same gene. TK and HA represent the thymidine kinase and hemagglutinin gene, respectively; 4a and 4b are the genes encoding the precursors of the*

initiation codon. Again these sequences are very A + T rich. Perhaps the most striking feature of these regions is that in most late genes the sequence ATG is immediately preceded by the sequence TAA. The few exceptions to this rule are the hemagglutinin gene (Shida, 1986; Fig. 3-2) and late genes identified by Weinrich and Hruby (1986). In the case of the gene encoding the precursor of the major core polypeptide 4b (Rosel and Moss, 1985; Fig. 3-2), the ATG is preceded by the sequence TAAATA. Thus the sequence TAAAT, which in most late genes also contains part of the translation initiation codon, is highly conserved, suggesting some important function. Further conserved elements in the 5′ flanking region of late genes include stretches of five to eight T or A residues starting at 16 to 20 bp upstream of the ATG codon.

Functional Analysis of the 5′ Flanking Sequences. Several powerful methods are currently available to study DNA sequences involved in gene regulation in vaccinia virus. Extracts prepared from infected cells were shown to initiate early gene transcription accurately on added DNA templates (Puckett and Moss, 1983; Foglesong, 1985). More recently, soluble extracts have been prepared from virus cores and also shown to terminate early gene transcription correctly (Golini and Kates, 1985; Rohrmann and Moss, 1985). Transient expression systems also provide convenient means to study regulatory sequence elements (Cochran et al., 1985a). This procedure consists of fusing putative vaccinia virus promoter sequences to the coding region of a reporter gene whose product can easily be detected and quantified. The DNA is then amplified in a plasmid vector and transfected into cells that have been infected with vaccinia virus. The reporter gene is thus expressed by enzymes provided by the infecting virus. The disadvantage of the transient expression system is that it is not well suited to study early gene regulation, since the test gene carried on the transfected plasmid becomes accessible to the viral transcription system only after early gene expression, which occurs within viral cores, is already completed.

The ideal assay system to study regulatory DNA sequences is to generate recombinant vaccinia virus in which the DNA sequences to be assayed are stably integrated in the viral genome. Any foreign DNA inserted into such recombinants is expressed in the normal infection

---

major core polypeptides. (a) Venkatesan et al. (1981); (b) Venkatesan et al. (1982); (c) Weir and Moss (1983); (d) Broyles and Moss (1986); (e) Earl et al. (1986); (f) Bertholet et al. (1985); (g) Weir and Moss (1984); (h) E. Van Meir, personal communication; (i) Rosel and Moss (1985); (j) Hirt et al. (1986); (k) Shida (1986).

cycle and thus under conditions that closely reflect the *in vivo* situation. The procedures used to generate such vaccinia virus recombinants is described in detail in Chapter 12, but will be outlined briefly here. A recombinant plasmid containing the viral thymidine kinase (TK) gene and flanking sequences is linearized at a restriction site within the coding sequences. The foreign gene coding sequence, fused to a vaccinia virus promoter fragment, is then ligated into the TK gene and the DNA is amplified. To generate recombinant virus, this DNA is transfected into cells that have been infected with standard virus. When the DNA of the infecting virus is replicated, recombination occurs between the TK gene sequences on the transfected plasmid and those in the viral DNA. The foreign gene thus becomes integrated into the viral genome. Since this insertion disrupts the TK gene, recombinant virus can easily be selected for on the basis of their TK⁻ character.

Vaccinia Virus Promoter Sequences

Early Promoters. Recombinant vaccinia virus was used by Vassef (1987) to identify early promoter sequences. Segments of randomly fragmented vaccinia virus DNA that promoted the expression of the herpesvirus thymidine gene were sequenced and compared with the 5' flanking region of the early genes. This made it possible to identify conserved sequence elements (Fig. 3-2, underlined) that are centered at about 20 bp upstream of the mRNA start sites, and presumably represent early promoter elements.

The 7.5K Promoter. The only early promoter which has been analyzed in any detail is that of the 7.5K gene, which was initially classified as an early gene, since it is transcribed *in vitro* in cores (Venkatesan and Moss, 1981) and *in vivo* in the presence of inhibitors of protein synthesis (Cooper *et al.*, 1981a). However, when cells were infected with a vaccinia virus recombinant carrying the chloramphenicol acetyltransferase (CAT) gene under the control of the 7.5K promoter and treated with DNA synthesis inhibitors, CAT gene expression was markedly reduced as compared to untreated controls (Mackett *et al.*, 1984). This indicated that the 7.5K promoter was active both before and after DNA replication. Nuclease S1 mapping of early and late RNA revealed two distinct RNA start sites for the 7.5K gene (Cochran *et al.*, 1985a). The one used late in infection is located about 55 nucleotides upstream from the one used early. Since there is no ATG between the two start sites, the same polypeptide is made from both RNAs. Vaccinia virus may use such dual promoters to make certain gene products throughout the infection cycle. A few other examples of such dual promoters have been reported

(Wittek *et al.*, 1984b; Rosel *et al.*, 1986). In these cases, the 5' ends of early RNAs were mapped about 35 nucleotides upstream from late RNA start sites.

The closely spaced early and late regulatory signals in the 7.5K gene flanking region made it possible to delimit the minimal sequence requirement for early and late gene expression of the same gene. Deletion analysis demonstrated that first late and then early gene expression was lost when more than about 30 nucleotides upstream of the corresponding RNA start sites were removed (Cochran *et al.*, 1985b). Thus, both early and late gene expression in vaccinia virus appear to require suprisingly short stretches of 5' flanking sequence.

The 11K Gene Late Promoter. Most of the information on the regulation of late gene expression has been gained through the study of the 11K late gene 5' flanking sequence. Insertion of these sequences into the TK early gene of vaccinia virus recombinants showed that all necessary elements required for late gene transcription resided in a maximum of 100 bp that also contained the putative RNA start site (Bertholet *et al.*, 1985). Various mutations were therefore introduced into these sequences and their effects on gene expression were studied after insertion into recombinant virus (Bertholet *et al.*, 1986; Hänggi *et al.*, 1986). The following conclusions with respect to late gene regulation can be drawn from these studies. First, about 20 nucleotides upstream of the ATG translation initiation codon are sufficient to regulate late gene expression. Furthermore, all mutations around the putative mRNA start site, or within the conserved TAAAT motif, abolish late gene transcription. On the other hand, the TAAAT motif by itself is not sufficient to direct late gene expression.

Mechanism of Late mRNA Production

Mapping of the 5' ends of several late mRNAs suggested that they had very short 5' untranslated leader sequences apparently consisting of about 5 nucleotides, or even less. However, in these studies, the nuclease S1 mapping procedure was used, and this method does not detect RNA sequences that are not complementary to the gene coding sequences and flanking regions. Studies of the 5' end of the the 11K late mRNA by primer extension analysis showed that the late gene coding sequences were preceded by a poly(A) stretch that was immediately fused to the late gene AUG translation initiation codon (Bertholet *et al.*, 1987; Schwer *et al.*, 1987). This homopolymer stretch is not encoded in the 5' flanking region of the 11K gene, indicating that these molecules are made by some unusual mechanism. The mRNA of a strongly expressed cowpox virus gene also contains a 5' poly(A) sequence (Patel and Pickup, 1987).

The most attractive model for the mechanism by which the poly(A) leader and the late gene coding sequences become linked is that proposed for transcription from certain late promoters of the bacteriophage T4 (Kassavetis *et al.*, 1986), which has been discussed by Patel and Pickup (1987) to explain the presence of the poly(A) leader in the cowpox virus mRNA. According to this model, the 5' poly(A) sequence is generated by a series of abortive initiation events in the late gene promoter region. These would occur on the first T residue of the sequence ATTTA (coding strand) of the highly conserved TAAAT motif (noncoding strand) characteristic of late genes, and produce the trinucleotide transcript pppApApA. This trimer then slides in a 3' to 5' direction on the DNA template, again exposing one or more T residues on the template which are used for elongation of the trimer. Multiple rounds of sliding and elongation would finally produce a long 5' poly(A) stretch. There is considerable experimental evidence to support this model for bacteriophage T4; it remains to be established whether it also applies to late transcription in poxviruses.

If most or all mRNAs had a poly(A) leader, this could explain the shutoff of early gene expression late in infection, which has been particularly well-studied in the case of the viral TK gene (McAuslan, 1963a,b; Jungwirth and Joklik, 1965; Zaslavsky and Yakobson, 1975; Hruby and Ball, 1981). Synthesis of this enzyme is switched off about 4 hours after infection, when late genes begin to be expressed. Nevertheless, functional TK mRNA, identified by *in vitro* translation, can be isolated from cells that have stopped making the enzyme. This discrepancy could be explained on the basis of structural differences between early and late mRNAs and by assuming that late in infection the translation apparatus of the cells is modified so as to allow the efficient translation only of those mRNAs that have a poly(A) leader, but that *in vitro* systems from noninfected cells would not discriminate between the two types of structures. Thus the unusual structure of the 5' ends of late mRNAs may also provide an explanation for the switch from early to late gene expression at the level of translation.

REPLICATION OF THE GENOME

Viral DNA synthesis in vaccinia virus-infected cells starts at about 2 hours after infection, reaches a maximum by about 3 hours, and then declines rapidly, ceasing several hours before infectious progeny virions are first detectable. Parental DNA molecules containing one nick presumably serve as templates for viral DNA replication (Esteban and Holowczak, 1977b; Pogo, 1977, 1980). This nicking occurs immediately after uncoating of the viral DNA (Pogo, 1977). Replication then proceeds in a semiconservative fashion as revealed by density labeling of replicating DNA molecules (Esteban and Holowczak, 1977a). The high degree of single strandedness of replicating molecules (Estenban and Holowczak, 1977b; Pogo et al., 1981) is compatible with a strand displacement mechanism of DNA synthesis. The elongation of the daughter strands appears to proceed from initiation sites of replication located at or near the ends of the parental DNA molecule (Esteban et al., 1977; Pogo et al., 1981). Furthermore, hybridization of pulse-labeled replicating DNA to the isolated single strands of the vaccinia virus DNA showed that the strands are synthesized asymmetrically, i.e., only one strand is replicated from any given end (Pogo et al., 1981). The concept of asymmetry in the replication of the two daughter strands makes discontinuous DNA synthesis, as previously postulated on the basis of "Okazaki-like" fragments (Esteban and Holowczak, 1977a; Pogo and O'Shea, 1978) a less likely mechanism for vaccinia virus DNA replication.

Evidence for the formation of concatameric forms of vaccinia DNA has been obtained by sedimentation studies (Archard, 1979) and restriction analysis of replicating DNA molecules (Moyer and Graves, 1981; Baroudy et al., 1983).

Cross-linking of the progeny molecules is a postreplicative process that occurs late in infection (Esteban and Holowczak, 1977a; Pogo and O'Shea, 1978).

Models for Replication of Vaccinia Virus DNA

All known DNA polymerases require a free 3'-OH end to initiate DNA synthesis. Such a primer can be provided by first synthesizing short RNA segments complementary to the 3' end of the template strands. However, subsequent excision of the RNA primer leaves a gap at the 5' end of the daughter strands which cannot be filled in by DNA polymerase. As pointed out by Watson (1972), progressively shorter daughter strands would be produced inevitably upon each round of replication.

One possible solution to this problem is to assume that the chromosome ends consist of palindromic sequences which can form hairpin loops (Cavalier-Smith, 1974; Bateman, 1975). Such hairpins could form either transiently during the replication process (Cavalier-Smith, 1974) thus providing a free 3'-OH end or, alternatively, the hairpin could represent the normal condition (Bateman, 1975) and replication would simply proceed around the loop. Terminal palindromic sequences of this kind, which allow self-priming, play a key role in a model proposed for the replication of parvovirus DNA (Tattersall and Ward, 1976).

The presence of hairpin loops at the ends of the vaccinia virus DNA which consist of imperfect palindromes (see Chapter 2) make the essential features of the parvovirus model applicable to the replication of poxvirus genomes. The first self-priming model was proposed by Moyer and Graves (1981) for the replication of rabbitpox virus DNA. Baroudy *et al.* (1982, 1983) subsequently showed that such a model also accounts for the flip–flop inversions of the terminal hairpin sequences of the vaccinia virus DNA. Furthermore, the occurrence of concatemeric intermediates in DNA replication (Moyer and Graves, 1981; Baroudy *et al.*, 1983) in which genomes are fused in alternating head-to-head tail-to-tail orientation (Moyer and Graves, 1981) is readily explained by these models.

Figure 3-3 depicts the essential features of the simplest version of this model. The first step consists of a site-specific nicking of the parental DNA molecule within the inverted terminal repeat but outside the terminal loop. In the next step, the terminal loops are unfolded, exposing free 3'-OH ends which serve as primer to synthesize the complementary strand. This is shown to occur at both ends of the parental molecule. The newly synthesized DNA then folds back on itself by base pairing of the inverted terminal repeat. Again a free 3'-OH end is provided from which replication of the daughter strands proceed. In the resulting daughter molecules, the loops at one end represent inverted complementary copies of those present at the corresponding end of the parental molecule.

This model can also account for the concatemeric intermediates in DNA replication, by assuming that the nick is introduced at only one end of the parental DNA molecule. Self-primed replication would then start at that end, and after two cycles a tetramer would be formed in which the individual unit-length molecules were fused in alternating left/right–right/left orientation. Genomic DNA molecules could be excised from the concatemer by site-specific nicking and the terminal hairpin restored by "folding back" the single-stranded extensions at each end.

A key enzyme in this replication model is a site-specific endonuclease which introduces nicks at defined locations within DNA molecules before and after DNA replication. An excellent candidate for such an enzyme has been isolated from vaccinia virions (Lakritz *et al.*, 1985). The purified enzyme not only introduces site-specific nicks but subsequently

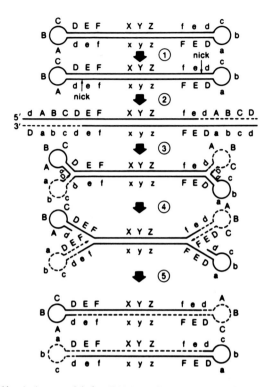

FIG. 3-3. *Self-priming model for DNA replication (Baroudy* et al., *1982). For explanation see text.*

also cross-links the two DNA strands of linear molecules obtained by prolonged incubation with the enzyme, and thus fulfills all requirements for excision and cross-linking of progeny molecules from concatemeric intermediates.

Recent studies strongly suggest the idea that unit-sized genomes are indeed excised from concatemeric replicative intermediates (Delange *et al.*, 1986; Merchlinsky and Moss, 1986). When the junction fragments from such intermediates were cloned into plasmid vectors and transfected into virus-infected cells, the plasmids not only replicated, but were also resolved into linear molecules with vector DNA in the center and viral DNA hairpins at the two ends. However, the fact that apparently any DNA transfected into infected cells is replicated (Delange and McFadden, 1986) argues against specific origins of replication. Thus, presumably the initial stages of DNA replication are less sequence dependent than proposed in the model shown in Fig. 3-3. Resolution of concatemeric intermediates and formation of hairpin ends, on the other hand, require specific DNA sequences which map close to the ends of the genome.

A Possible Mechanism for the Generation of Mutants

The genome structure of the deletion and deletion/transposition mutants of several orthopoxviruses, described in Chapter 2, closely resembles that of incomplete particles of adenovirus (Daniell, 1976). A plausible model has been proposed by Sambrook (cited by Daniell, 1976) to explain the generation of the defective adenovirus genomes. It is tempting to speculate that the corresonding orthopoxvirus mutants may arise by a similar mechanism. The Sambrook model is based on the replication of a single displaced strand which could be generated in the proposed replication scheme of the orthopoxvirus DNA shown in Fig. 3-3 by assuming that nicks occur at both ends of the DNA molecule but that replication proceeds only from one end. The model proposes that the single displaced strand is nicked accidentally within the unique sequences close to either end. The truncated end then folds back and "illegitimate base pairing" (Daniell, 1976) occurs either within sequences of the inverted terminal repeat or within unique sequences. The free 3'-OH end is then used as primer to synthesize the complementary strand. Replication then proceeds by the same self-priming mechanism shown in more detail in Fig. 3-3. If the illegitimate base pairing has occurred within the inverted terminal repetition, a molecule is produced which has deleted some unique sequences and a part of the repetition. If, on the other hand, base pairing has occurred within the unique sequences, a molecule results which has deleted unique sequences of one end but which also has inserted unique sequences from the other end in an inverted orientation. The two types of genome mutants frequently observed in orthopoxviruses are thus readily explained by the same basic mechanism.

The results of detailed analysis of several white pock mutants of cowpox virus further support the proposed model. If such mutants arose by recombination events between recognition sequences, as has been proposed to explain the analogous rabbitpox virus mutants (Moyer et al., 1980), one might expect to find unusual and/or homologous sequences at the junctions of the transposed sequences. Pickup et al. (1984) have sequenced the flanking regions where such transposition has occurred in white pock mutants of cowpox virus. In the four mutants examined, no sequence homology nor any unusual structural features were observed, suggesting that such mutants arise by illegitimate base pairing rather than by site-specific recombination. The fact that the deletions in all cowpox mutants so far examined are of very similar size should not be taken as argument in favor of site-specific recombination, since selection for the white pock phenotype on the one hand and viability on the other could impose rather narrow limits on the size of the deletion one might expect to find in such mutants (Pickup *et al.*, 1984). The proposed model also readily explains the mirror-image deletion in vaccinia virus DNA (McFadden and Dales, 1979), since any aberration close to one end of the genome would inevitably be copied to the opposing terminus.

ROLE OF THE HOST CELL NUCLEUS IN THE REPLICATION OF POXVIRUSES

As already described, poxviruses replicate in the cytoplasm of the infected host cell, and for a long time it was believed that the nucleus played no role in replication. However, several investigators have detected poxvirus DNA in the nuclei of infected cells (LaColla and Weissbach, 1975; Gafford and Randall, 1976; Taylor and Rouhandeh, 1977). In HeLa cells infected with vaccinia virus the nuclear viral DNA represents sequences from the entire genome (Archard, 1983). Minnigan and Moyer (1985), on the other hand, were unable to detect rabbitpox virus DNA in the nucleus of A549 cells by *in situ* hybridization. In enucleated cells viral DNA is produced with normal timing but in reduced amounts (Pennington and Follett, 1974; Hruby *et al.*, 1979), suggesting that the host cell nucleus is not a major site or viral DNA replication. However, some undefined nuclear function(s) are required at a later stage of virus replication in order to obtain virus progeny.

Inhibition of viral replication by the drug α-amanitin (Hruby *et al.*, 1979; Silver *et al.*, 1979), which inhibits host cell RNA polymerase II, but which has no effect on the viral RNA polymerase (Nevins and Joklik, 1977; Baroudy and Moss, 1980; Spencer *et al.*, 1980; Archard *et al.*, 1985), showed that host cell RNA polymerase II was directly involved at some stage of the replication cycle. This view was further confirmed by the observation that α-amanitin-resistant cells treated with the drug produced a normal yield of progeny (Silver *et al.*, 1979). A possible explanation for the requirement for the host cell enzyme was recently provided by Morrison and Moyer (1986), who showed that a large subunit of the host enzyme was translocated from the nucleus to the factory areas and subsequently found associated with the viral enzyme isolated from purified virions. This suggests that the cellular subunit is directly involved in the transcription of the viral genome. However, a new problem for the interpretation of these results emerged when vaccinia virus mutants were isolated that were able to replicate in α-amanitin-treated or enucleated cells (Villareal et al., 1984). Furthermore, a single gene, that was mapped in the viral genome, appeared to be responsible for the α-amanitin-resistant phenotype (Villareal and Hruby, 1986). The subject has been reviewed recently by Moyer (1987); at the present time no clear picture has emerged as to the role of the cellular enzyme in viral replication.

MORPHOGENESIS

Encouraged by the large size of the virus particle and the cytoplasmic site of replication, virologists undertook thin-section electron microscopy of cells infected with vaccinia virus soon after the technique was developed. However, studies on morphogenesis are still at the descriptive stage; the molecular basis of various events has yet to be elucidated. Reviews written by two of the pioneers in this field (Morgan, 1976b; Dales and Pogo, 1981) should be consulted for further details of the process. Little further work has been carried out since then except for studies of envelopment by Payne, Hiller, and their colleagues.

Virion formation occurs entirely within circumscribed, granular, electron-dense areas of the cytoplasm, which have been called viral "factories" (Cairns, 1960). In suitably stained cells, these are seen as B-type inclusion bodies. The first morphological evidence of developing virus particles is the appearance of a 50- to 55-nm wide membrane, initially as "caps" or in three dimensions, as "cupules" (Plate 3-1). No polyribosomes are seen within these factories, all viral proteins being synthesized elsewhere in the cytoplasm and transported to the factory. The appearance of the cupules is soon followed by completion of these membranes to produce roughly spherical "immature particles," which contain a finely granular viroplasm. Although these immature particles resemble isolated viral cores (see Plate 2-1D), the membrane of the immature particle becomes the outer membrane of the virion and the lateral bodies and core differentiate within it. Initially the membrane is covered by a layer of spicules, which consist of an M_r 65K protein (Essani et al., 1982), and confer sufficient rigidity on the membrane to maintain its spherical shape. The spicule-backed membranes are produced as early functions, and if DNA replication is blocked with hydroxyurea immature particles, lacking viral DNA are formed. Removal of the block results in synchronous maturation, which Morgan (1976a,b) has used to follow the insertion of viral DNA into the immature particle.

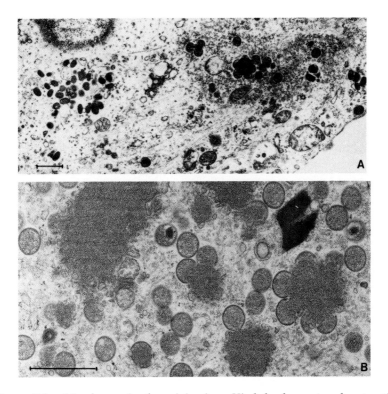

PLATE 3-1. *Morphogenesis of vaccinia virus. Viral development and maturation occur within viral "factories" in the cytoplasm, which contain no polyribosomes, indicating that proteins are synthesized elsewhere and transported to the factory for assembly. (A) Low-power view (bar = 1000 nm) of factory area showing immature particles on right and nonenveloped mature virions on left. (B) High power view (bar = 1000 nm) showing foci of viroplasm, dense DNA paracrystals, and the cupules of developing viral particles, which develop into spherical immature particles. (A, from Payne and Kristensson, 1979, courtesy Dr. L. G. Payne; B, from Stern et al., 1977, courtesy Dr. S. Dales.)*

Treatment of infected cells with rifampicin, on the other hand, interferes with the development of spicules and the developing particles are then irregular in shape and fail to progress to maturation. On reversal of rifampicin inhibition, further transcription and translation occur and within minutes the membrane is covered with spicules and assumes a regular convex shape (Grimley *et al.*, 1970).

During the process of maturation the spicules are replaced by the surface tubules that characterize the surface of the outer membrane of mature nonenveloped virions (Essani *et al.*, 1982; see Plate 2-1A), the protein of the tubules being among the last components to be incorporated into the virion. The outer membrane also contains lipid, which is taken from the cellular pool. While these processes are occurring in the outer membrane, the undifferentiated viroplasmic matrix within it differentiates into the DNA-containing core and the lateral bodies. The expression of late functions is required for the internal differentiation of the virion; Silver and Dales (1982) suggest that a nonvirion protease (M_r 12.5K) is involved in posttranslational cleavage of some of the polypeptides within the immature particle. Internal differentiation of the core, the development of lateral bodies, and the acquisition of infectivity occur as temporally coordinated, close linked events. Mature virions are moved out of the factories and accumulate in adjacent areas of the cytoplasm (Plate 3-1), or with most strains of cowpox and ectromelia viruses are occluded within A-type inclusion bodies.

Envelopment

The process of envelopment was first described by Dales (1963), and confirmed by subsequent studies (see Morgan, 1976b; Dales and Pogo, 1981). The importance of the envelope in generating an adequate immune resonse after vaccination was recognized by Boulter and Appleyard (1973) (see Chapter 4), and its significance in pathogenesis was demonstrated by Payne (1980). It is not an artifact of cell culture, since it occurs in infected mouse brains (Kristensson et al., 1984) and lungs (Payne and Kristensson, 1985). Various stages of the process are illustrated in Plate 3-2. A proportion of the mature virions, which differs with different strains of the virus (Payne, 1980), migrates to the Golgi apparatus, apparently along actin-containing microfilaments of the cellular cytoskeleton (Hiller et al., 1981b). Here they are enclosed with a double-membrane structure comprising Golgi membranes that have been altered by the insertion of several virus-encoded proteins. These double-wrapped virions then migrate to the cell surface, where the outer membrane fuses with the plasma membrane, releasing particles with a single-layered envelope from the cell.

Release of enveloped virions thus involves two cellular membrane systems: the Golgi and plasma membranes. In infected cells both the viral envelope and the plasma membrane contain several virus-specified proteins: 89K (or 85K; Shida, 1986), 42K, 37K, and a group of 20–23K proteins (Payne, 1978, 1979). All these are glycoproteins except the 37K protein, which is a nonglycosylated, acylated protein which is the major protein component of the envelope (Hiller and Weber, 1985). The M_r of the "37K" protein obtained from gels has apparently been somewhat

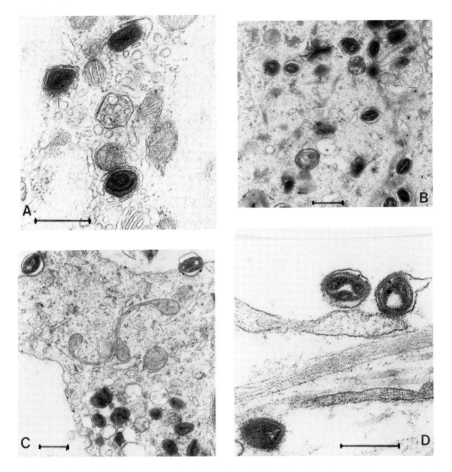

PLATE 3-2. *Envelopment and release of vaccinia virions in the respiratory tract of the mouse. (A) Mature virions acquire double envelopes in vicinity of Golgi complex. (B) Particles with double envelopes in the cytoplasm, en route to the plasma membrane. (C and D) Particles with single envelope released from cell by fusion of outer envelope with plasma membrane. Bars = 500 nm. (From Payne and Kristensson, 1985, courtesy Dr. L. G. Payne.)*

underestimated, since as deduced from the gene sequence its size is close to 42K (Hirt *et al.*, 1986).

Occlusion within Inclusion Bodies

Three species of orthopoxvirus (cowpox, ectromelia, and raccoonpox) produce two types of inclusion bodies in infected cells: irregularly shaped B-type inclusion bodies or viral factories, found in all orthopoxvirus-infected cells, and eosinophilic A-type inclusion bodies (see Plate 4-4). The protein of the A-type inclusion bodies is a single species of M_r 130–160K (Patel *et al.*, 1986). With most strains of the species of orthopoxvirus that produce A-type inclusion bodies, mature nonenveloped virions are incorporated into them (occluded), but some strains of cowpox and ectromelia viruses produce A-type inclusion bodies that are devoid of virions (see Chapter 4). Occlusion does not depend on the production of mature virions, since particles of cowpox virus whose maturation has been arrested by treatment with rifampicin can become occluded (Ichihashi and Dales, 1971).

The Pathogenesis, Pathology, and Immunology of Orthopoxvirus Infections

The extent of investigation of disease processes differs greatly in different orthopoxvirus infections. The pathology and histopathology of smallpox have been intensively studied, mostly many years ago; humoral immunity to vaccinia in humans and rabbits has long been a subject of investigation; and detailed experiments have been carried out on pathogenesis and cell-mediated immunity in mousepox. However, data on pathogenesis and pathology are very limited for infections of the natural host or experimental animals with camelpox virus, raccoon poxvirus, tatera poxvirus, Uasin Gishu disease virus, and the vole orthopoxvirus; what information is available on these diseases is provided in Chapter 10. Each of the other five species of *Orthopoxvirus* and camelpox virus produces a systemic infection with a generalized rash in one or more species of animal; some of them produce localized infections of the skin in certain species (Table 4-1).

TABLE 4-1

Animal Species in Which Different Species of Orthopoxvirus *Produce Either a Localized Skin Lesion or a Generalized Rash*

Virus	Natural host	Localized pustular skin lesion	Generalized pustular rash
Camelpox virus	Camel	Monkey	Camel
Cowpox virus	(?) Wild rodents	Man, cow	Felidae (many species), rat, elephant, rhinoceros, okapi
Ectromelia virus	(?) Wild rodents, laboratory mouse	—	Mouse
Monkeypox virus	African squirrels, African monkeys	—	African squirrels, Asian monkeys, man
Vaccinia virus (dermal)	?	Man, cow, rabbit	Man (rare), water buffalo
Neurovaccinia virus[a]	?	—	Rabbit
Variola virus	Man	—	Man, monkeys

[a] Including rabbitpox virus.

The term pathogenesis is used to describe the mechanisms involved in the production of disease, including the spread of virus through the body and the physiological responses of the host organism to infection. The immune response is such an important component of pathogenesis that it will be described separately. The pathogenesis of localized skin lesions produced by the introduction of cowpox virus or vaccinia virus into the skin of man, rabbit, or cow is simple; virus replicates locally and produces a pustule, some virions move via lymphatics to the draining lymph nodes which may become enlarged and tender, and there may be a transient viremia, but further spread of the virus through the body and replication in other sites is usually aborted by the developing immune response. The pathogenesis of diseases characterized by a generalized rash is more complex; it has been studied most extensively in several model systems: mousepox in mice, neurovaccinia virus (rabbitpox virus) infections in rabbits, and monkeypox in Asian monkeys.

PORTALS OF ENTRY IN ORTHOPOXVIRUS INFECTIONS

Apart from local skin lesions produced after the entry of orthopoxviruses into minute abrasions or by deliberate inoculation, it is difficult to be certain of the portal of entry in infections with orthopoxviruses, in most of which there are several alternative routes (Table 4-2). There is no evidence that any member of the subfamily *Chordopoxvirinae* replicates in invertebrate hosts. Experiments by Regnery (1987) suggest that mechanical transmission by arthropod vectors may occur in infections of California voles with vole orthopoxvirus, but no other examples are known, although it is important in fowlpox (genus *Avipoxvirus*) and myxomatosis (genus *Leporipoxvirus*). Guillon (1970) has suggested that the mite *Ornithonyssus bacoti* may act as a vector of ectromelia virus, infection being transferred when mice eat mites that have ingested blood from viremic mice.

The Skin

Infection via minute abrasions in the skin occurs in cowpox and vaccinia virus infections of calves sucking dams (cows or water buffaloes) with lesions on their teats, and on the hands of humans who milk such animals. The usual result is a localized lesion of the lips or hands, respectively, but systemic infection with a generalized rash may occur in water buffalo calves. Ectromelia virus infection of mice usually occurs via minute lesions of the skin, producing a localized "primary lesion"

TABLE 4-2

Animal Species in Which Systemic Infections with a Generalized Rash Occur after Natural Infection with Various Species of Orthopoxvirus, *and the Suspected Portals of Entry*

Viral species	Animal species	Suspected portals of entry
Camelpox virus	Camel	Respiratory tract
Cowpox virus	Rodents, felines[a]	Respiratory tract, digestive tract
Ectromelia virus	Mouse	Skin, respiratory tract, digestive tract
Monkeypox virus	Man, monkeys, certain squirrels[b]	Respiratory tract, digestive tract
Vaccinia virus	Rabbits[c], water buffaloes[d]	Respiratory tract, skin
Variola virus	Man, monkeys	Respiratory tract, skin,[e] congenital (rare)

[a] Rarely, other zoo animals (elephant, rhinoceros).
[b] In zoo outbreak, several other species (see Table 8-1).
[c] With neurovaccinia (rabbitpox) virus.
[d] Especially in young animals.
[e] Usually as variolation.

which is followed by systemic spread, which produces a generalized rash in mice that do not die from acute hepatitis (see Chapter 9). Roberts (1962b) studied the sequence of events after infection of mice by scarification, using fluorescent antibody to trace events at the histopathological level. He found that scarification needed to be severe enough to introduce virus into the dermis if infection was to be produced with small doses of virus. Primary epidermal lesions consisted of single infected cells in the Malpighian layer; dermal lesions were groups of two or three fibroblasts or histiocytes, which had often migrated some distance from the site of scarification. Subsequently the infection spread contiguously in the epidermis, aided by spread to more distant sites from infected dermal cells. The skin lesion became edematous about the fourth day, probably as a result of cutaneous anaphylaxis. Epidermal hyperplasia occurred in an annulus of normal cells just beyond the advancing margin of infected cells. Spread from the skin lesion to the lymphatics was probably due to movement of infected histiocytes.

The Respiratory Tract

Experimental Studies.   Experimentally, infection by the inhalation of virus aerosols has been demonstrated with ectromelia virus in mice (Edward *et al.*, 1942; Roberts, 1962a), with rabbitpox virus in rabbits (Westwood *et al.*, 1966), and with variola virus in monkeys (Hahon and Wilson, 1960; Noble and Rich, 1969).

Ectromelia Virus. After inhalation of ectromelia virus in an aerosol that contained particles of various sizes, infection of both upper and lower respiratory tracts occurred (Roberts, 1962a). The first evidence of viral replication in the lung was seen in alveolar macrophages and in single mucosal cells of small bronchioles. Two days later the number of infected alveolar macrophages had increased greatly and the mucosal foci had increased in size. By the third day, endothelial cells of the adjacent lymphatics were infected, and occasional infected macrophages were found in the pulmonary lymph nodes. The spleen was also infected at this time. Entry of virus also occurred via cells of the olfactory mucosa and macrophages associated with it. From the third day onward the course of the disease was dominated by viral replication in the lymph nodes, spleen, and liver (see below).

Vaccinia Virus. Westwood *et al.* (1966) showed that rabbits could be infected with very small doses of both a dermal strain of vaccinia virus or with rabbitpox virus (a strain of neurovaccinia virus), when exposed to aerosols produced by the Henderson apparatus, which produced a cloud of dried particles mostly 1 μm or less in diameter. Since their study was limited to infectivity titrations of various tissues and organs, it did not reveal the histological details of the infectious process. However, they showed that susceptible rabbits housed in the same room as rabbits suffering from rabbitpox could be infected over distances of up to 12 feet. Subsequently Thomas (1970), using an adhesive surface sampling technique, readily demonstrated aerosols of rabbitpox virus in these rooms, even though six air changes took place each hour.

Variola Virus. Hahon and Wilson (1960) readily infected Asian monkeys (*Macacus irus*) with aerosolized variola virus, the lungs being the primary and major site of viral replication. Noble and Rich (1969) used special plastic isolation chambers with a controlled airflow to demonstrate that susceptible *Macacus irus* could be infected when exposed upwind of monkeys infected by intranasal inoculation 3 and 4 days earlier. They also showed that cross-infections with variola virus occurred in cynomolgus monkeys placed in contact with other monkeys during the period of rash, probably via respiratory infection. Infection was maintained by this means for six generations of passage (Fig. 4-1).

Natural Infections. Although mice can be infected with ectromelia virus by the respiratory route, epidemiological observations in an outbreak in the National Institutes of Health in 1979 (Wallace *et al.*, 1981)

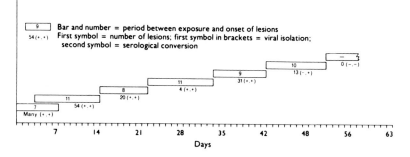

Fig. 4-1. *Contact infection of* Macaca irus philippinensis *through six successive passages before failure in transmission. The first monkey was infected by intranasal inoculation and after 7 days (indicated in bar) developed many lesions, from which variola virus was recovered (first "+" symbol) and subsequently showed serological conversion (second "+" symbol). The second monkey was placed in the same cage 3 days after the inoculation and developed 54 lesions 11 days after exposure; it was then placed in contact with the third monkey, and so on. The penultimate monkey in the series developed 13 lesions, but virus was not recovered from the crusts, although antibodies developed. The last monkey failed to develop lesions or convert serologically. (Based on Noble and Rich, 1969.)*

suggest that this route is not of major importance in mousepox. On the other hand, rabbitpox virus probably spread between laboratory rabbits by the respiratory route in outbreaks in the United States.

Variola Virus. Epidemiological evidence indicates that in human smallpox the usual mode of entry of variola virus was via the respiratory tract and that excretions from the mouth and nose, rather than scab material, were the most important source of infectious virus (Fenner *et al.*, 1988). In theory, infection by inhaled virus could have occurred via the mucous membranes of the mouth, the nasal cavity, or the oropharynx; or via the alveoli of the lungs. However, careful study of these sites in fatal cases of smallpox failed to disclose any evidence of a "primary lesion" there. Nor were patients infectious before the enanthem appeared, at the end of the incubation period. It seems, therefore, that the primary infection in the mouth, pharynx, or respiratory tract did not produce a sizeable lesion nor did the lesion of infection ulcerate and release virions onto the surface.

Usually infection occurred only in persons who were in face-to-face contact with overt cases, which were most highly infectious during the first week of the rash, before virus was released from skin lesions but at a time when there was abundant virus in the oropharyngeal secretions, due to release from ulcers of the enanthem. Long-distance aerosol spread was rare, but was clearly implicated in an outbreak in Meschede Hospital in Germany in 1970 (Wehrle *et al.*, 1970). In this outbreak, an electrician who had just returned from Pakistan was admitted to hospital as a suspected case of typhoid fever. He was confined to an isolation room, and developed a rash 3 days after admission. Smallpox was confirmed by electron microscopy 2 days later and the patient was transferred to the smallpox hospital. In spite of rigorous room isolation of the patient during the 5 days he was in the hospital, 19 further cases of smallpox occurred in patients located on all three floor levels of the building; 17 of them being infected directly from the index case. Even a visitor who had spent only 15 minutes inside the hospital, away from patient care areas, was infected. Tests of air movement with smoke generated in the room occupied by the index case confirmed that air moved to all the rooms in which secondary cases occurred. The infectivity of the index case was enhanced by the fact that he had a dense rash and, an unusual circumstance in smallpox, a severe bronchitic cough.

Monkeypox Virus. Although cases of human monkeypox have a rash and enanthem very like those of cases of smallpox, person-to-person infection is too rare to produce endemic disease in man (see Chapter 8). The mode of transmission of monkeypox virus among wild animals and from animals to man in Africa is not known, but oral infection appears more likely than respiratory infection (see below). Airborne infection probably occurred in an outbreak in a primate house in the zoological gardens in Rotterdam in 1966, in which many primates and a few other animals sustained clinically severe infections (Peters, 1966; see Chapter 8).

The Oral Route

Just as in infection by the respiratory route virus may enter at any level from the pharynx to the alveoli, so oral infection may involve entry via small abrasions of the skin of the lips or the oral mucous membrane, or via the epithelial cells of the small intestine. The oral route may play a role in natural infection in several different orthopoxvirus infections. Fenner (1947b) found that mice could be infected orally with ectromelia virus only if very large doses were used, such as those involved if a mouse ate the carcass of a mouse that had died of acute mousepox. Guillon (1970) suggested that mice may be infected by eating mites (*Ornithonyssus bacoti*) that have fed on viremic mice. Gledhill (1962a) found that mice could be regularly infected by the oral route if large doses were given *per os,* and about 20% were infected if small doses were used. Infections caused by small doses were often inapparent, but ectromelia virus could be recovered for several weeks after infection from the feces, the intestinal tract, and lesions on the base of the tail near the anus. Mahnel (1983) found that a strain of ectromelia virus attenuated by serial passage in chick embryo fibroblasts produced seroconversion and protection when given in drinking water in doses of 10^6 $TCID_{50}$/ml or higher.

Oral infection seems the most likely source of cowpox virus infection of domestic cats, a not uncommon sporadic infection in Britain (Bennett *et al.*, 1986), and in the infection of large species of Felidae, as in the outbreak in the Moscow Zoo, in which consumption of white rats infected with cowpox viruses seemed the most likely source (Marennikova *et al.*, 1977). Monkeypox virus may be spread between squirrels, and possibly from squirrels to monkeys, when partly eaten oil palm seeds are contaminated by one animal and eaten by another (see Chapter 8).

THE SPREAD OF INFECTION THROUGH THE BODY

Early work on the pathogenesis of generalized orthopoxvirus infections, which provided a model that has proved useful in understanding the pathogenesis of other exanthematous diseases, was carried out with mousepox (Fenner, 1948a,b). The overall pattern elaborated in studies of mousepox was confirmed in experiments with rabbitpox in rabbits and monkeypox in Asian monkeys.

Mousepox

Genetically susceptible mice that do not die from acute viral hepatitis develop a generalized pustular rash, and the signs of the naturally occurring disease can be reproduced by the footpad inoculation of mice with a small dose of virus (Fenner, 1948c). The course of events after footpad inoculation was followed by sacrificing mice at frequent intervals and titrating the viral content of certain organs—the inoculated foot, the regional lymph node, the spleen, the skin, and the blood. The results indicated that the sequence of events during the incubation period in mousepox followed a consistent pattern (Fig. 4-2). If the appearance of the primary lesion was taken as the end of the incubation period, it was evident that this symptom-free period was occupied by the stepwise progression of virus through the body: infection, replication, and liberation, usually accompanied by cell necrosis, occurring first at the site of inoculation and then in the regional lymph node, whence virus and/or virus-infected cells entered the bloodstream. Using fluorescent antibody staining to identify virus-infected cells, Mims (1964) showed that the phagocytic cells were the first infected in the liver and spleen. When infection had breached the macrophage barrier the virus replicated to high titer in the parenchymal cells of the liver and in the lymphocytes of the spleen, organs which usually showed semiconfluent necrosis (see Plate 9.5). The bone marrow is also an important early site of viral replication (Bhatt and Jacoby, 1987b). A day or so after infection of these organs, large amounts of virus were liberated into the bloodstream and during this secondary viremia focal infection of the skin, kidneys, lungs, intestines, and other organs occurred. There was again an interval during which the virus replicated to reach a high titer before visible changes were produced, so that a period of about 2 or 3 days usually elapsed between the appearance of the primary lesion and the focal lesions of the secondary rash (Fig. 4-2). Mims (1964) demonstrated by fluorescent antibody staining that circulating lymphocytes and monocytes contained viral antigen; only small amounts of virus could be recovered from plasma or serum. Mousepox appears to be unusual among the generalized orthopoxvirus infections in that the spleen, liver, and bone marrow are early target organs for viral replication.

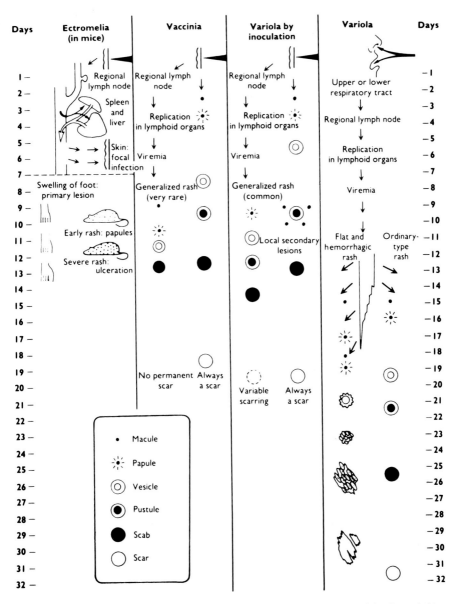

F I G. 4-2. *The spread of virus around the body and the evolution and healing of skin lesions in a model system (mousepox, after Fenner, 1948b), and in vaccinia, inoculation smallpox, and "natural" smallpox in man. (From Fenner et al., 1988.)*

Rabbitpox

Studies on the pathogenesis of rabbitpox confirmed the picture that had emerged from experiments with mousepox. After rabbits were infected by the respiratory route (Bedson and Duckworth, 1963; Westwood et al., 1966) virus spread through the body in a stepwise manner. Virus could always be isolated from the lungs, and very high titers also occurred in the gonads and the adrenal glands. The incubation period, from infection until the onset of fever, was 4–6 days. A rash usually appeared on about the sixth day, although often rabbits died before the rash became evident ("pockless" rabbitpox; Christensen et al., 1967).

In spite of the early replication of rabbitpox virus in the lung, rabbits were not infectious until nasal and conjunctival discharges appeared, and environmental contamination appeared to be maximal 6–7 days after infection. Examination by fluorescent antibody staining revealed that after aerosol infection, primary lesions developed at two distinct sites in the respiratory tract—the bronchioles and the alveoli. As well as being spread to the bloodstream via the lymphatics, rabbitpox virus could spread directly from the bronchiolar or alveolar lesions to adjacent blood vessels. Viremia was mainly leukocyte associated. Animals that died early, before the rash appeared, had viremias which increased exponentially to reach very high titers at the time of death; those that died later had much lower titers (Westwood et al., 1966).

Monkeypox

Cynomolgus monkeys are highly susceptible to monkeypox and usually develop a generalized rash (von Magnus et al., 1959). Wenner and his colleagues (for review, see Cho and Wenner, 1973) studied the pathogenesis of monkeypox in cynomolgus monkeys infected by intramuscular inoculation. These investigations confirmed that in orthopoxvirus infections of primates, as well as in those of mice and rabbits, there was a stepwise progression of infection, a generalized rash developing after viremia had been established owing to the replication of virus in the internal organs. Prior to the development of the rash the virus replicated in the spleen, tonsils, and lymph nodes, without producing extensive or severe lesions. Generalized lymphadenopathy developed in the first week after infection and persisted until the end of the third week. Gross enlargement of the cervical and inguinal lymph nodes is a feature of monkeypox infections in chimpanzees (McConnell et al., 1968) and humans (see Chapter 8).

Smallpox

Experimental studies of variola virus infection of primates (Hahon and Wilson, 1960; Westwood *et al.*, 1966; Noble and Rich, 1969) suggested that the same pattern of spread through the body occurred, but no single organ was clearly identified as the central focus of viral replication. Experimental studies were not possible in human smallpox, but observations on the viremia and the excretion of virus in oropharyngeal secretions, combined with autopsy studies on acutely fatal cases, suggested a similar mode of spread of variola virus (Fenner *et al.*, 1988; see Fig. 4-2).

The Spread of Virus through the Body

Orthopoxviruses are principally disseminated through the body in the lymph and bloodstream as cell-associated virions, although some infectivity is always found free in plasma. In cell cultures enveloped forms of vaccinia virus were more important than nonenveloped forms in the dissemination of virions over a distance (Boulter and Appleyard, 1973), as shown by the "anti-comet" test (see Plate 4-7). Payne (1980) showed that in infections of mice with vaccinia virus, the capacity to spread to distant organs in the mouse was correlated with the production of enveloped virions (Table 4-3). Enveloped virions also occur in infections of mice with cowpox virus (Payne, 1986) and ectromelia virus (Buller, 1986).

THE CONSEQUENCES OF SYSTEMIC INFECTION

The rash is the most characteristic feature of generalized orthopoxvirus infections. The individual lesion is termed a "pock." Its appearance and development differs somewhat in different animals, because of different structures of their skin, so that the lesions of the rash in the thin skin of the mouse differ histologically from those in the thicker skin of man or monkey.

Highly virulent strains of certain orthopoxviruses, in a proportion of animals of certain host species, may cause acute death before a rash develops: e.g., ectromelia virus in mice, rabbitpox virus in rabbits, monkeypox virus in orang utans, or early-type hemorrhagic smallpox in man. Such humans or animals exhibit profound prostration and other signs of severe toxemia.

TABLE 4-3

Relationship of in Vitro Virus Release to in Vitro and in Vivo Parameters of Virus Dissemination[a]

| Vaccinia strain | RK 13 cells | | | Mouse virulence[b] | |
| | Virus production | | | | |
	Enveloped virions[c]	Ratio: $\dfrac{\text{nonenveloped}}{\text{enveloped}}$	Formation of "comets"[d]	Case-fatality rate (%)	Brain titer[e]
South Africa	5.6	250	—	0	—
Cape Town	5.7	100	—	0	—
Venezuela	6.0	160	—	0	—
Lister	6.3	300	—	0	—
Tashkent	6.6	160	—	0	—
WR	6.6	300	—	85	4.7
Lederle 7N	6.7	250	—	0	—
HI-White	6.8	50	—	0	—
Dairen	7.0	16	—	0	2.7
Lafontaine	7.6	40	±	0	3.6
IHD-White	8.0	12	+	25	3.3
Gallardo	8.3	5	+	25	3.3
IHD-J	8.3	6	+	70	4.6

[a] Based on Payne (1980).
[b] After inoculation with 10^6 plaque-forming units intranasally.
[c] Titer in $\log_{10}$ units per milliliter.
[d] In cell monolayers with liquid overlay (see Plate 4-7).
[e] Titer in $\log_{10}$ units per gram of brain.

The Rash

The primary event for the production of the focal lesions of the rash in orthopoxvirus infections is the localization, in the small blood vessels of the dermis, of virus particles that are circulating in the bloodstream, either free in the plasma or more commonly within virus-infected leukocytes. Subsequently, adjacent epidermal cells are infected and the skin lesion develops as described below.

In the infection of laboratory animals with orthopoxviruses, skin areas treated so as to promote a local inflammatory response are associated with an increased density of skin lesions (vaccinia in rabbits: Camus, 1917; ectromelia in mice: Fenner, 1948a). Numerous observations (Ricketts, 1908; MacCallum and Moody, 1921) attest to the effect of mild trauma leading to local vasodilatation on the density of skin lesions in different parts of the body in smallpox (the "garter" effect). However, no satisfactory physiological explanation has yet been provided for the highly characteristic "centrifugal" distribution of the skin lesions in smallpox.

Toxemia

All clinical observers have commented on the "toxic" appearance of patients with variola major, especially those suffering from flat-type or hemorrhagic-type smallpox. Although very severe cases did occur, extremely rarely, in variola minor, there was usually a great contrast in the general appearance and condition of patients with variola major and variola minor who had approximately the same numbers of pustules. The patient with acute variola major was usually very sick, whereas the patient with variola minor of apparently similar severity (in terms of the number of skin lesions) might well be ambulant. No adequate explanation is available for either the toxemia of variola major or the difference in severity between cases of variola major and variola minor with rashes of similar extent. Viral antigen was readily demonstrable in the plasma of patients with severe smallpox (Downie et al., 1953, 1969). Perhaps the formation of immune complexes between such antigen and IgM antibodies, and the associated activation of complement, initiated a series of physiological effects that produced the so-called "toxic" symptoms of variola major.

Seeking an explanation of the cause of death in smallpox, Boulter et al. (1961a) examined various physiological parameters in a model system— rabbitpox. The only consistent physiological changes observed in sick rabbits was extreme hypotension, leading to a shocklike syndrome, decreased urinary output, and a rise in blood urea and plasma potassium levels. Death seemed to be due to lethal concentrations of potassium ion, which occurred possibly as a consequence of the severe hypotension. No information is available on blood pressure or blood potassium changes in severe cases of smallpox.

The other model system, mousepox, provides few clues as to the cause of death in other orthopoxvirus infections, in that it is unique in the extent of viral replication in the liver and spleen, which is enough to account for the occurrence of death within a few hours of the first detectable signs of illness in some mice (see Chapter 9). Some cowpox virus infections in large felines were associated with pulmonary lesions that were sufficient to account for their death.

PATHOLOGICAL CHANGES IN ORTHOPOXVIRUS INFECTIONS

The extent of the pathological changes in orthopoxvirus infections varies according to the species of virus and the species of host. As well, within particular virus-species host-species combinations, the response is influenced by the genetic constitution of both virus and host. In localized infections, e.g., vaccinia or cowpox in man, pathological changes are usually limited to the single lesion at the site of viral entry and the draining lymph node. In most systemic orthopoxvirus infections few lesions are found in the internal organs; the pathology is largely confined to the skin lesions of the generalized rash. Acute mousepox in susceptible strains of mice provides an exception; in these infections there is extensive necrosis of the liver, spleen, and lymph nodes and changes in many other organs. However, the characteristic lesion of orthopoxvirus infections is the skin lesion of the rash, the pathological histology of which has been most fully investigated in human smallpox.

Before describing the pathological changes in animals, it will be useful to review briefly the changes seen in cultured cells and the chorioallantoic membrane after infection with orthopoxviruses.

Changes in Cultured Cells

Within 1 or 2 hours of infection with most species of orthopoxvirus, "toxic" changes may occur in the infected cells, which in monolayer cultures become rounded and retract from each other (Fig. 4-3). Very soon after infection, new virus-coded antigens are found in the plasma membrane (Ueda *et al.*, 1969, 1972), and by the fourth hour there is cytological evidence of viral replication; basophilic areas appear in the cytoplasm, then the B-type inclusions of Kato *et al.* (1959) or viral

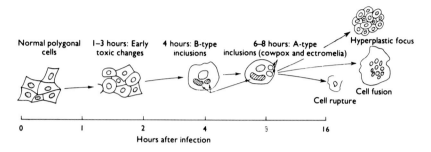

FIG. 4-3. *Cellular changes seen at various times after infection of a monolayer of cultured cells. (From Fenner et al., 1988.)*

"factories" (Cairns, 1960). Eventually gross changes occur in the cells; depending upon the particular virus–host cell combination there may be an aggregation into hyperplastic foci (variola virus), adjacent cells may fuse to form large karyocytes, or cell necrosis and rupture may occur, with release of the cell-associated virions. Late in infection, A-type inclusions develop in cells infected with cowpox, ectromelia, or raccoonpox viruses, although Lees and Stephen (1985) found that cultured hepatocytes infected with ectromelia virus did not become rounded up, and as in the liver of infected mice, they developed B-type but not A-type inclusion bodies.

Some of the earliest changes, and the most significant in relation to the immune response (see below), are the virus-induced alterations in the plasma membranes of infected cells. Some virus-coded antigens are expressed on the surface of the cell within 2 hours of infection (Ueda *et al.*, 1972; Amano *et al.*, 1979); other polypeptides which develop late in infection, including the hemagglutinin and several other envelope glycoproteins, are also incorporated into the plasma membranes of infected cells (Payne, 1979). The development of membrane-associated hemagglutinin can be followed by hemadsorption tests, which have been used in the analysis of differences between variola major and alastrim viruses (Dumbell and Wells, 1982; Dumbell and Huq, 1986; see Chapter 7). Some of these viral antigens promote cell fusion and thus cell-to-cell spread of virions.

Inclusion Bodies. Depending on the species of orthopoxvirus, one or two types of inclusion bodies may be found in infected cells. Irregular, weakly staining inclusion bodies were first described in cells infected with variola and vaccinia viruses by Guarnieri (1892), and used to be called "Guarnieri bodies." Much more prominent, eosinophilic inclusion bodies (Plate 4-1A) are found in cells infected with ectromelia virus

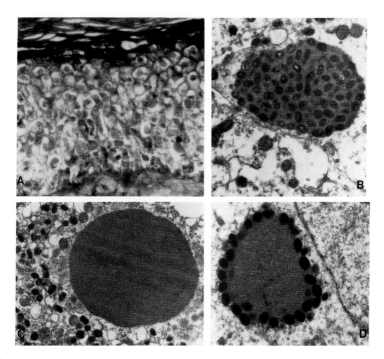

PLATE 4-1. *A-type inclusion bodies in the cytoplasm of cells infected with ectromelia virus or cowpox virus. (A) Section of skin of mouse after infection with ectromelia virus, stained with Mann's stain. (B,C, and D) Electron micrographs of inclusion bodies of three types (see Baxby, 1975): (B) V$^+$ inclusion bodies, with numerous occluded virions; (C) V$^-$ inclusion bodies which completely lack occluded virions; (D) V^i inclusion bodies with virions attached to the periphery. (B and C, courtesy Dr. Y. Ichihashi; D, from Baxby (1975), courtesy Dr. D. Baxby.)*

(Marchal, 1930), cowpox virus (Downie, 1939a), and raccoon poxvirus (Patel *et al.*, 1986). In a systematic study, Kato *et al.* (1959) proposed that the Guarnieri-type inclusion bodies should be called "B-type" and the Marchal–Downie inclusion bodies "A-type." B-type inclusion bodies are found in all poxvirus-infected cells, and are the sites of poxvirus replication ("viral factories," see Chapter 3).

With most strains of cowpox and ectromelia virus, A-type inclusion bodies can be seen by electron microscopy to contain large numbers of mature virions. Such strains of virus are designated V$^+$ (Ichihashi and Matsumoto, 1968a,b; Ichihashi *et al.*, 1971; Plate 4-1B). However, with some strains of these viruses the A-type inclusion bodies are devoid of virions (V$^-$; Plate 4-1C), or rarely, the virions are clustered around the periphery (V^i; Baxby, 1975; Plate 4-1D). Some strains of cowpox virus fail to produce A-type inclusion bodies, but produce instead a protein that gives a reaction of identity in gel diffusion tests with the LS antigen of vaccinia virus (Amano and Tagaya, 1981).

Using a V^- strain of cowpox virus, Patel et al. (1986) showed that the inclusion bodies consisted of a single species of a 160K protein which is synthesized in large amounts late in the replication cycle. Antibodies to the cowpox virus 160K protein react specifically with a 130K protein found in cells infected with ectromelia virus and a 155K protein found in cells infected with raccoon poxvirus (in each of which it is the A-type inclusion body protein), a 94K protein found in cells infected with vaccinia virus (the LS antigen), and a 92K protein found in cells infected with monkeypox virus (Patel et al., 1986; D. J. Pickup, personal communication, 1987). Conservation of these proteins in all species of orthopoxvirus tested has also been demonstrated at the gene level. Thus, vaccinia virus DNA contains a gene whose 5' end is almost identical with that of the cowpox virus gene encoding the 160K protein. Furthermore, DNA–DNA hybridizations showed that the DNAs of other orthopoxviruses, but not of tanapox virus or myxoma virus, contain consequences that hybridize to a probe consisting of the coding region of the cowpox virus gene (D. J. Pickup, personal communications, 1986, 1987; J. J. Esposito, personal communication, 1987).

The factor necessary for inclusion of virions in the A-type inclusion body (occlusion) has been designated "VO," and is present on the surface of the virion (Ichihashi and Matsumoto, 1968b), into which it is incorporated a short time before occlusion takes place (Shida et al., 1977). However, expression of VO can occur in the presence of rifampicin, which inhibits maturation (see Chapter 3), and immature particles of cowpox virus may then be incorporated into A-type inclusion bodies (Ichihashi et al., 1971).

In cells doubly infected with V^+ and V^- strains of cowpox virus, the dominant V^+ strain provides the VO factor for integrating V^- virions into the inclusion body (Ichihashi and Matsumoto, 1968a; Ichihashi and Dales, 1973). Other orthopoxviruses, such as vaccinia virus, which do not produce A-type inclusion bodies, express the VO factor and can supply the deficiency for occlusion of V^- cowpox virus strains in doubly infected cells.

Occlusion within inclusion bodies affects the resistance of infectivity of poxviruses that occur in this form, since the matrix protein provides protection and ensures that the "infectious unit" consists of a large number of viral particles. However, single virions which are released from cells as enveloped particles are also produced in cowpox virus-infected cells (Payne, 1986), and by analogy with vaccinia virus (see below), probably play a role in the spread of virions from one cell to another, and around the body. A-type inclusion bodies are not formed in some tissues, such as the liver in ectromelia virus-infected mice (see Chapter 9).

Pocks on the Chorioallantoic Membrane

The fact that dilute suspensions of vaccinia virus would produce individual lesions (pocks) on the chorioallantoic membrane was first demonstrated by Keogh (1936), who noted the differences between pocks produced by dermal strains of vaccinia virus and neurovaccinia virus, and used the pock-counting method to assay viral suspensions and neutralizing antibodies. This method of cultivation has remained an important technique for studying orthopoxviruses, the lesions produced by different species of orthopoxvirus that infect humans being sufficiently distinctive for pock morphology to have been used as a diagnostic method throughout the Intensified Smallpox Eradication Program (see Chapter 11). Most species of orthopoxvirus produce white, non-ulcerated pocks that vary in diameter from about 0.2 to 3 mm; the pocks produced by other species (cowpox, monkeypox, and neurovaccinia viruses, including rabbitpox virus) are ulcerated or hemorrhagic (Plate 4-2; see Table 1-2).

Histology of Pocks.   The pocks produced by all species of orthopoxviruses begin as areas of proliferation of the ectodermal cells, which is caused by a diffusible protein that is produced very early after infection (vaccinia growth factor, Buller et al., 1987b, or more generally, "poxvirus growth factor"). Infection of cells spreads centrifugally, proliferation preceding infection. In parallel with the proliferation of the ectoderm there is edema and inflammatory changes in the mesoderm. Within 24 hours, eosinophil leukocytes are conspicuous, and when extensive necrosis occurs, as in the pocks produced by vaccinia virus, the mesoderm becomes packed with cells of various kinds including leukocytes, many of them necrotic. B-type inclusion bodies are not readily demonstrable, but A-type inclusion bodies are prominent in cells of the pocks produced by cowpox and ectromelia viruses. The variations seen in lesions produced by different species of orthopoxvirus depend upon several factors: the extent to which proliferation occurs, the stage at which cell necrosis begins and the extent of the necrotic change, the invasion of the mesoderm with leukocytes, and the extent of leakage of erythrocytes from small vessels in the mesoderm (Fig. 4-4). The endoderm is usually slightly hyperplastic.

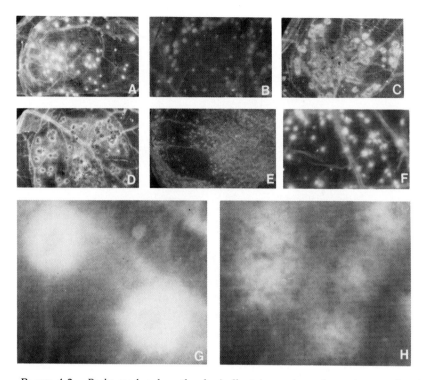

PLATE 4-2. *Pocks produced on the chorioallantoic membrane by various species of orthopoxvirus. (A) Variola virus; (B) monkeypox virus; (C) vaccinia virus (Lister strain); (D) cowpox virus; (E) ectromelia virus; (F) camelpox virus; (G) variola virus (×11). (H) monkeypox virus (×11). (A–D, F–H, courtesy Dr. J. H. Nakano.)*

In lesions produced by cowpox virus there is a conspicuous accumulation of erythrocytes just beneath the thickened ectoderm (Plate 4-3). In pocks produced by monkeypox virus and neurovaccinia virus, on the other hand, the hemorrhagic appearance of the pock is due to the presence of erythrocytes in the surface tissues of the ulcerated pock.

Pocks produced by variola virus consist of greatly proliferated ectoderm and the mesoderm is heavily infiltrated with leukocytes (Downie and Dumbell, 1947). Those produced by ectromelia virus are smaller, but almost every cell in the ectoderm contains an A-type inclusion body (Burnet and Lush, 1936b).

Effects of Temperature on Pock Morphology. Leukocytic infiltration and hemorrhage in pocks on the chorioallantoic membrane is temperature dependent. At an incubator temperature of 38°C there is marked

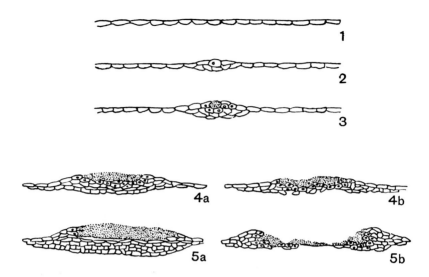

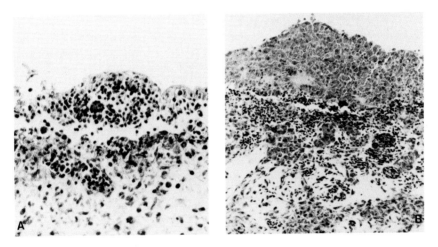

Fig. 4-4 *Diagram illustrating the response of the ectoderm of the chorioallantoic membrane to infection with orthopoxviruses (Modified from Burnet, 1936). (1, 2, and 3) Initial proliferation of the ectoderm, produced by all orthopoxviruses. The lesion may then develop so as to be predominantly proliferative, with varying degrees of necrosis (4a, 5a; characteristic of variola and ectromelia viruses and white pock mutants), or necrosis and ulceration may be more severe (4b, 5b; characteristic of vaccinia and monkeypox viruses).*

Plate 4-3. *Sections of the hemorrhagic pocks produced by monkeypox and cowpox viruses, showing the location of the dark-staining nucleated erythrocytes. (A) Monkeypox virus (×142). The erthrocytes are at the surface and in the tissue of the pock. (B) Cowpox virus (×71). The erythrocytes lie beneath the greatly thickened ectodermal layer of the pock. (Courtesy Dr. J. H. Nakano.)*

infiltration and even wild-type cowpox virus lesions are grayish white. Wild-type monkeypox virus pocks are white at an incubator temperature of 36°C or above, and only exhibit the hemorrhagic center at 35°C. One white pock mutant of monkeypox virus produced a white pock at 35°C but had a hemorrhagic center at 34°C (J. H. Nakano, personal communication, 1981).

Ceiling Temperature of Viral Growth. When carrying out experiments on the growth of ectromelia virus on the chorioallantoic membrane, Burnet and Lush (1936b) noticed that ectromelia virus (not then recognized to be an orthopoxvirus) did not produce pocks at 39.5°C, but did if the temperature was lowered to 37°C. This idea was developed by Bedson and Dumbell (1961), who introduced the concept of the "ceiling temperature" as the highest temperature at which pock formation by orthopoxviruses would occur. The ceiling temperature has proved to be a useful method for distinguishing between species of orthopoxvirus (see Table 1-2), and for distinguishing American strains of variola minor virus (alastrim virus) from variola major virus (see Chapter 7). Cultured cells can also be used for measuring the ceiling temperature (Porterfield and Allison, 1960).

Pocks Produced by White Pock Mutants. In addition to the opaque white pocks produced by most species of orthopoxvirus, the species that produce hemorrhagic pocks (cowpox, monkeypox, and neurovaccinia viruses, including rabbitpox virus) always produce a small proportion of "white" (nonhemorrhagic) pocks; the variants that produce them are designated white pock mutants. These were first observed with cowpox virus. Histologically, the main features distinguishing the pocks produced by the white pock mutants of cowpox virus from those produced by wild-type cowpox virus are the absence of erythrocytes beneath the thickened ectoderm and the intense infiltration of the mesodermal tissue with leukocytes. All white pock mutants of cowpox virus that have been studied have deletions of variable lengths of DNA from the right-hand end of the genome. By analysis of the DNAs of these mutants and wild-type cowpox virus, Pickup *et al.* (1986) identified and sequenced a gene that is essential for expression of the red pock phenotype. It codes for a protein resembling inhibitors of serine protease; the hemorrhagic phenotype may therefore be due to inhibition of serum proteases involved in blood coagulation.

The pathogenesis of the hemorrhagic pocks produced by rabbitpox and monkeypox virus is different from that of the red pocks produced by cowpox virus, the erythrocytes being located in the superficial tissues of the ulcerated pock. In rabbitpox virus the genes responsible for hemorrhage into the pock appear to be located at either end of the genome, since white pock mutants can be produced by deletions at either end of the genome (Lake and Cooper, 1980; Moyer *et al.*, 1980; see Fig. 2.6). The characteristics of the DNA of white pock mutants are described in Chapter 2; a possible way in which such mutants are generated during replication in Chapter 3.

Phenotype of Mixed Pocks. A white pock mutant may arise at any time during the development of a pock. After incubation for 3 days, "white" sectors can sometimes be seen in a "red" pock, although usually the whole pock becomes white. In membranes examined at 3 days about 1% of cowpox virus pocks are white, whereas only 0.1% of the slightly smaller number of pocks visible at 2 days are white.

The Skin Lesions of Smallpox

The epidermis, where the pustule develops, contains no blood or lymphatic vessels and its tough outer layer, the stratum corneum, is impermeable to viruses. The dead keratinized cells of the stratum corneum are continually being rubbed off and replaced from below, cell multiplication occurring in the stratum germinativum, which is separated from the dermis by a thin basement membrane. Of the three appendages of the skin—hair follicles, sweat glands, and sebaceous glands—only the last are destroyed by variola virus, and then only by variola major virus.

Early Changes in the Dermis. Bras (1952a) noted that the histological picture in the skin lesions in all clinical types of variola major was essentially the same. Since infection of the skin was caused by virus carried in the bloodstream, the earliest change was a dilatation of the capillaries in the papillary layer of the dermis, followed by swelling of the endothelial cells in the walls of these vessels, stasis of mononuclear cells within the lumen, and subsequently perivascular cuffing with lymphocytes, plasma cells, macrophages, and occasionally eosinophilic granulocytes (Plate 4-4A). Biopsies from cases of variola minor (MacCallum and Moody, 1921) showed that the affected papillae were edematous, with extravasation of erythrocytes and leukocytes, at a time when the overlying epidermis was unchanged or showed at most only an early vacuolation of the epidermal cells.

Epidermal Changes. All subsequent changes occurred in the epidermis. The cells of the Malpighian layer became swollen and vacuolated and underwent "ballooning" degeneration. This occurred in a sharply demarcated area and only a small number of cells separated the center of

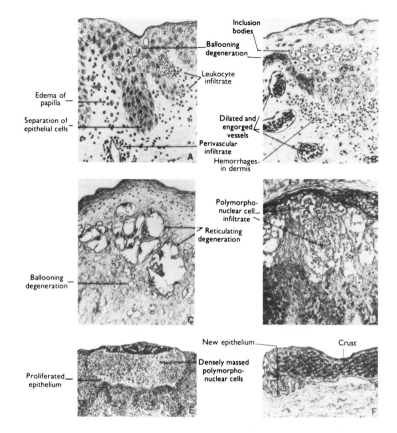

PLATE 4-4. *Stages in the development and evolution of the skin lesion. (From Michelson and Ikeda, 1927.) (A) The earliest change was edema of the dermis leading to the separation of epithelial cells of the papillae and lymphocytic infiltration in the dermis, especially around the small vessels. Ballooning degeneration was seen in a few cells in the lower Malpighian layer. (B) These changes progressed and the small vessels became dilated and engorged. Inclusion bodies were visible adjacent to cells showing ballooning degeneration. In early hemorrhagic-type smallpox, illustrated here, there was pronounced bleeding into the dermis. (C) As the pathological process progressed, the epithelial cells broke down by reticulating degeneration to produce a multilocular vesicle. (D) The vesicle formed by coalescence of the smaller cavities became infiltrated with polymorphonuclear leukocytes to produce a pustule, around which were cells containing inclusion bodies. (E) The fully developed pustule became packed with polymorphonuclear leukocytes and the epithelium on either side of the pustule proliferated. (F) Eventually the pustule became a crust, beneath which new epithelium grew in to repair the surface. Such lesions, in which the sebaceous glands were not involved, healed without leaving a pockmark.*

the lesions from the surrounding normal skin (Plate 4-4A and B). The degenerated cells were swollen, they stained faintly with acid dyes, and the characteristic B-type intracytoplasmic inclusion bodies (Guarnieri bodies) could be found in the cytoplasm.

The cells continued to increase in size, the cytoplasm became fainter, and the nucleus usually disappeared by lysis. Soon after this the cell membranes ruptured and the vacuoles coalesced to produce the early vesicle by what is called "reticulating" degeneration (Plate 4-4C). Because this coalescence occurred very quickly, a true papule was rarely seen; almost from the beginning the lesion was already vesicular.

Vesiculation.   The reticulating degeneration which produced the vesicle occurred exclusively in the middle and upper layers of the stratum spinosum; the basal cells were at first unaffected and the keratohyalin and horny layers showed no changes. Subsequently the cells of the lower stratum spinosum and the basal layer underwent a different kind of degeneration; the nuclei and cytoplasm became condensed, the cells became hyalinized, and the nuclei fragmented or lysed. Later these basal cells disappeared and the cavity of the vesicle (or pustule) was then immediately adjacent to the dermis.

The fully developed vesicle (Plate 4-4C) resembled a planoconvex lens with the following characteristics:

1. The roof, which was very thin over the summit of the vesicle, consisted of compressed cells of the stratum spinosum, keratohyalin layer, and horny layer.

2. The base consisted at first of cells of the stratum spinosum and basal layers, which showed a hyaline fibrinoid degeneration, becoming swollen, homogeneous, and refractile, losing their granular character, and staining more intensely with acid dyes. Later they lysed, so that the base of the vesicle was provided by the subjacent dermis.

3. Since a portion of the cytoskeleton persisted for a long period after the degeneration of the cells of the stratum spinosum, the cavity of the vesicle contained incomplete septa, creating a multiloculated appearance. Such loculation was never complete. Often there were heavier septa, which were made up of the coils of sweat glands traversing the cavity. Like the keratohyalin cells in the roof of the vesicle, the cells of sweat glands appeared resistant to the effects of variola virus.

4. Fluid accumulated inside the vesicle, with threads of fibrin and a few lymphocytes.

5. The cells immediately around the vesicle showed decreasing degrees of reticulating degeneration and the basal layers had often

proliferated, so that the wall around the vesicle was about twice the thickness of the unaffected epidermis. The rete pegs in the area adjacent to the vesicle were relatively deep.

The Pustule. Pustulation was due to the migration of polymorphonuclear cells from the subpapillary vessels into the vesicle, the dermis being relatively free of such cells (Plate 4-4D and E). This response was not due to secondary bacterial infection; numerous investigations showed that pustules containing abundant leukocytes were bacteria free, although occasionally secondary infection occurred.

Scabbing. With the development of an effective immune response, healing began. The contents of the pustule became desiccated and reepithelialization occurred between the cavity of the pustule and the underlying dermis. The pustular contents became a crust or scab (Plate 4-4F), which was subsequently shed; the newly formed epidermis had no rete pegs. In the absence of secondary infection, the dermis showed very few changes and in most parts of the body the lesions healed without scarring and therefore without producing pockmarks.

Scarring. The face bore the heaviest crop of lesions in most cases of smallpox and lesions usually appeared first on the face and evolved the most rapidly there. However, if the dermis was not involved such lesions should not have produced scars, yet scarring (pockmarks) were very much more common on the face than elsewhere (Jezek *et al.*, 1978). Bras (1952b) showed that this occurred because sebaceous glands are much larger and more numerous in the facial skin than elsewhere on the body. Although other skin appendages (hair follicles and sweat glands) were relatively unaffected by variola virus, cells of the sebaceous glands were highly susceptible. Degenerative changes began with cytoplasmic hyalinization accompanied by hyperchromatism of the nuclei, karyorrhexis, and cytolysis. This degeneration occurred simultaneously in several parts of the sebaceous gland, leading to extensive necrosis in the subepithelial layer of the skin. When healing occurred, the defect in the dermis was filled with granulation tissue, which subsequently shrank, leaving localized facial pockmarks.

Histopathology of Mucosal Lesions. Although showing a general resemblance to the lesions found in the skin, the pathological process in mucous membranes was modified by the nature of the tissue, in which there is no horny layer and the epithelial cells are in much looser relation with each other than in the skin. The earliest change in the mucous

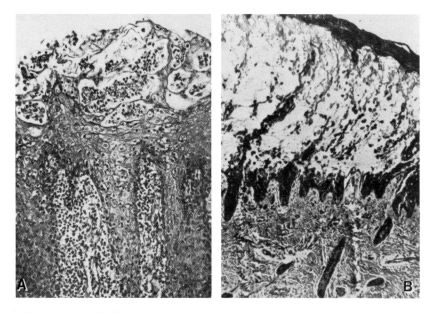

PLATE 4-5. *(A) Ulcer on oral mucous membrane in a case of smallpox. The pustule ulcerated early because there is no overlying horny layer. (B) Section of a vesicle produced in human skin by vaccinia virus in a child who developed progressive vaccinia. There are no inflammatory cells in the dermis. (A, from Michelson and Ikeda, 1927; B, from Keidan et al., 1953.)*

membranes appeared to be exudation into the epithelium, leading to separation of the cells, which underwent hyaline fibrinoid degeneration. Guarnieri bodies were numerous and the extensive necrosis of the epithelial cells, unrestrained by a horny layer, led to early ulceration (Plate 4-5). The base of the necrotic mass of cells was sharply outlined against the hyperemic tunica propria. Occasionally the tunica propria showed patchy necrosis unrelated to the superficial defects. Later there was increasing vascularization under the tunica propria and the tissue took on the appearance of granulation tissue, with numerous polymorphonuclear leukocytes in the demarcation zone beneath the necrotic epithelium. This combination produced a pseudomembrane, which could be easily detached. Because of the numerous bacteria on the mucous membrane of the pharynx, the necrotic lesions were usually found to harbor masses of bacteria of various kinds, but this was a secondary phenomenon.

The mucosal lesions, which appeared during the early papular stage of the rash, had usually healed, without scarring, by the end of the pustular stage.

Skin Lesions in Other Orthopoxvirus Infections

Although they have not been subjected to anything like the detailed histological studies made in human smallpox, essentially the same features are seen in the skin lesions in vaccinia in man (Howard and Perkins, 1905), and monkeypox in man (Stagles *et al.*, 1985) and monkey (Wenner *et al.*, 1969). However, Greene (1934a) noted that although there was an extensive rash in many cases of rabbitpox, with individual lesions passing through the stages of papule, pustule, and crust, the lesions occurred in the dermis and no epithelial changes were seen. In mousepox (see Chapter 9), the first histological evidence of the rash consists of localized areas of hyperplastic cells with dark-staining nuclei and occasionally with inclusion bodies. These areas of proliferation and edema increase in size until they become macroscopically visible as pale, slightly raised macules; large intracytoplasmic A-type inclusion bodies are present in most of the epithelial cells in the lesion. No vesiculation of pustulation occurs in the thin epidermis of the mouse, and within 24 hours the lesion becomes necrotic and then ulcerated, with an adhering scab.

Lesions of Internal Organs

Lymph Nodes. The lymph nodes are affected to a different extent in different orthopoxvirus infections. Bras (1952a) found few specific changes in the lymph nodes in smallpox, whereas in vaccinia the axillary nodes showed follicular hyperplasia with large germinal centers and a conspicuous proliferation of large pale cells among the lymphocytes of the cortical and paracortical regions—morphological evidence of an active immune response (W. E. D. Evans, unpublished observations, 1960, cited by Symmers, 1978). Generalized lymphadenopathy is a conspicuous feature of monkeypox in man (see Chapter 8), monkey (Wenner *et al.*, 1969), and chimpanzee (McConnell *et al.*, 1968), with lymphoid hyperplasia, medullary edema, dilatation of sinuses, and occasionally degeneration and necrosis of cells in the germinal centers. In mousepox, the regional lymph nodes draining the primary skin lesion are enlarged, with localized areas of necrosis in which there are many A-type inclusion bodies; later all lymph nodes are enlarged and hyperplastic, often with small foci of necrosis. In rabbitpox, Greene (1934b) found that the superficial lymph nodes were swollen and edematous and frequently showed focal areas of necrosis.

Spleen. In most generalized orthopoxvirus infections the spleen exhibits evidence of a developing immune response; however, in mousepox there is extensive viral replication in the spleen, producing large areas of necrosis (Allen *et al.*, 1981; Jacoby and Bhatt, 1987). In mice that recover from a severe attack, the spleen may be extensively scarred; in milder cases small plaques are produced on the serosal surface.

Liver. Few changes are seen in the liver in smallpox and monkeypox, but the acute stage of mousepox is characterized by great enlargement of the liver with extensive necrosis. Infection is initiated by blood-borne infection of the littoral cells of the liver parenchyma, from which virus spreads to contiguous parenchymal cells to set up numerous foci of necrosis throughout the liver. Viral titers in the liver are very high, but A-type inclusion bodies are rarely seen in hepatocytes.

Lungs. Few lesions are seen in the lungs except in cowpox virus infections of large felines, in which Marennikova *et al.* (1977) reported the occurrence of severe fibrinous bronchopneumonia in lions, cheetahs, and a black panther, all of which died. In rabbitpox, Greene (1934b) observed small areas of necrosis and consolidation in the lung and small subpleural nodules. In mousepox, antigen-positive foci were found in the mucosa of the upper respiratory tract even after subcutaneous inoculation of susceptible BALB/c mice; severe lesions of the bronchiolar epithelium and the alveoli occurred after intranasal inoculation (Jacoby and Bhatt, 1987).

Stomach and Intestines. Few lesions were seen in the intestinal tract except in mousepox, in acutely fatal cases of which the small intestine was often hyperemic and filled with blood. A careful histological survey showed that small necrotic foci with typical A-type inclusion bodies occurred in about 65% of acutely fatal cases (Greenwood *et al.*, 1936). Jacoby and Bhatt (1987) found that infection and necrosis of Peyer's patches preceded spread of viral antigen to the adjacent laminar propria and overlying enterocytes.

Brain. In man, postinfection encephalitis was a rare complication of both smallpox and vaccination (see Table 5-2), and one case has been reported after infection with cowpox virus (Verlinde, 1951). Although less common after vaccination than after smallpox, postvaccinial encephalitis was the more important complication, because it occurred after deliberate immunization of a healthy person and not as a rare complication of a disease with a case–fatality rate of about 25%. Apart from cases of concomitant disease due to other causes, two forms of encephalitis occurred after vaccination (de Vries, 1960). In infants under 2 years of age, the syndrome was termed encephalopathy, and was characterized by an incubation period of 6–10 days. In fatal cases there was a general hyperemia of the brain, widespread degenerative changes in ganglion cells, and occasionally perivascular hemorrhages. In persons over 2 years of age, postvaccinial encephalitis occurred after an incubation period of 11–15 days (Weber and Lange, 1961), and the pathological signs in fatal cases were those of an allergic response, characterized by acute perivascular demyelination, accompanied by destruction of the axis cylinders (Plate 4-6). The perivascular space contained many lymphocytes and the demyelinated areas contained lymphocytes and highly pleiomorphic microglia.

Postvaccinial encephalitis was virtually unknown after revaccination, and the reported incidence of cases in different countries ranged from 9 to 258 per million primary vaccinations (see Table 5-3). These differences may have been due in part to inconsistencies in reporting, but the strain of vaccinia virus used for vaccination also played a role. This was demonstrated when the Lister strain, which had long been used in the United Kingdom, was introduced in Austria, Belgium, the Netherlands, and Switzerland in the 1960s, and the incidence of postvaccinial encephalitis fell dramatically (Fenner et al., 1988).

After intracerebral inoculation in mice, most orthopoxviruses produce either no pathological effects (e.g., low to moderate doses of variola, camelpox, and taterapox viruses and many strains of vaccinia virus), generalized infection because of spread via the bloodstream (ectromelia virus), or infection of cells of the meninges and chorioid plexus (neurovaccinia virus, monkeypox virus). Raccoon poxvirus is unique in that even after footpad inoculation in suckling mice it can produce flaccid paralysis, with chronic multifocal meningoencephalomyelitis, as

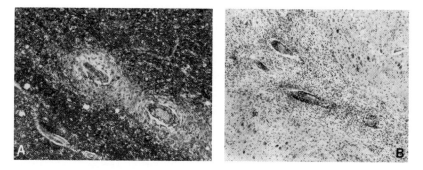

PLATE 4-6. *Lesions in the brain in postvaccinial encephalitis. (A) Perivascular demyelination; (B) perivascular cellular infiltration. (A, courtesy of Dr. R. T. Johnson; B, courtesy of Professor G. Bras.)*

well as small areas of necrosis in the lumbar plexus and some of the dorsal spinal ganglia (Thomas *et al.*, 1975).

VIRAL PERSISTENCE AND REACTIVATION

Persistence of Orthopoxviruses in Animals

The evidence that certain orthopoxviruses might persist for long periods after recovery from infection is difficult to evaluate; however, some examples of apparent persistence have been reported.

Mousepox. Several authors have suggested that ectromelia virus could produce a latent infection in mice, but most examples that have been quoted could be explained as plausibly by persistent infection in the population, rather than prolonged viral persistence in individual mice. However, there are a few examples of the persistence of ectromelia virus for at least several weeks in a sequestered site in healthy mice and also of persistent infection with shedding. For example, Fenner (1948c) noted the recurrence of foot swelling, and the isolation of virus from the swollen foot, in two mice that had been infected 2 and 7 months earlier and had not been subsequently exposed to infection. In other experiments Fenner (1948d) demonstrated the presence of ectromelia virus in the lungs and spleen of 2 mice that had recovered from infection acquired about 45 and 93 days earlier; 112 other mice that were tested more than 5 weeks after recovery gave negative results. There was no evidence that such mice shed virus. However, because mouse tissues or tumors are passaged in other mice, such persistent infections, especially if they occurred in animals that had just recovered from a mild or inapparent infection, could constitute a source for the dissemination of mousepox (see *Laboratory Animal Science,* 1981).

A more significant observation in relation to possible persistent infection with ectromelia virus was that reported by Gledhill (1962a,b), who demonstrated that mice infected by the oral route could sustain a chronic infection of the intestinal tract, with excretion of virus in the feces and scabs on the tail near the anus. However, he was unable to "activate" acute mousepox in such carriers, nor did they transmit infection to susceptible mice placed in the same cages.

Isolation of Orthopoxviruses from Normal Animals. A number of reports of the recovery of orthopoxviruses from the tissues of apparently normal animals have been published (Table 4-4). There are two problems in evaluating the significance of these findings, in terms of persistent infection. First, it is possible that most or all of the positive

TABLE 4-4

Recovery of Orthopoxviruses from Tissues of Healthy Animals Thought to Have Been Naturally Infected[a]

Example No.	Virus	Circumstances	Reference
1a	Cowpox (rat strain)	From lungs of 5 and kidneys of 14 out of 113 white rats in an interepidemic period	Marennikova (1979); Shelukhina et al. (1979)
1b	Cowpox (Turkmenia strain)	From kidneys of big gerbil (Rhombomys opimus) and yellow suslik (Citellus fulvus) captured in Turkmenia, USSR	Ladnyi et al. (1975); Marennikova et al. (1978b)
2	Monkeypox	From kidneys of several apparently healthy cynomolgus monkeys that had been exposed to infection in an outbreak in a laboratory colony	von Magnus et al. (1959)
3	Taterapox	From liver/spleen suspension of 1 of 95 wild Tatera kempi captured in Benin	Lourie et al. (1975)
4	Ectromelia	From brains of apparently normal mice in a colony enzootically infected with ectromelia virus	Schell (1964); Topciu et al. (1972)
5a	Vaccinia	Rabbitpox virus recovered from kidney of an apparently healthy Macacus rhesus	Alekseeva and Akopova (1966)
5b	Vaccinia	Vaccinia virus recovered from kidney of Cercopithecus ascanius killed in Zaire	Shelukhina et al. (1975)

[a] Excluding "whitepox" viruses (see Chapter 7).

TABLE 4-5

Recovery of Orthopoxviruses from Animals that Had Been Experimentally Infected Some Weeks Earlier
and Showed No Clinical Signs at Time of Recovery of Virus

Example No.	Virus	Circumstances	Reference
1	Ectromelia	From lungs and spleen of 2 out of 114 mice tested 5 weeks or more after recovery	Fenner (1948c)
2	Monkeypox	From kidneys and lungs of some hamsters up to 6 weeks after intracardiac injection	Shelukhina et al. (1979)
3	Cowpox	From kidneys and lungs of cotton rats and rats inoculated intranasally up to 6 weeks earlier	Shelukhina et al. (1979)
4	Cowpox (rat strain)	From several organs of white rats and Rattus norvegicus inoculated intranasally 4 weeks earlier	Shelukhina et al. (1979); Maiboroda (1982)
5	Cowpox (Turkmenia strain)	From kidneys and testes of big gerbils and yellow susliks inoculated 5 weeks earlier	Marennikova et al. (1978b)
6	Vaccinia	From spleen and testis of rabbits inoculated intradermally with neurovaccinia virus 114 and 133 days earlier	Olitsky and Long (1929)
7	Vaccinia	From brains of mice pretreated with cyclophosphamide 60 days after inoculation with vaccinia virus; by cocultivation only	Ginsberg and Johnson (1977)
8	Variola	From brains of mice that had been inoculated intracerebrally as infant mice up to 62 days earlier	Sarkar et al. (1959)

results were obtained with animals that were experiencing an inapparent infection at the time or were convalescing from infection (examples 1a, 1b, 2, 3, and 4 of Table 4-4). Secondly, the virus apparently isolated from a normal animal may have been a laboratory contaminant (examples 5a and 5b of Table 4-4).

Isolation of Orthopoxviruses after Recovery from Experimental Infection. Although most investigators have failed to demonstrate persistent infection with orthopoxviruses, there are several reports of the recovery of orthopoxviruses from the tissues of experimental animals that had been infected several weeks earlier and were apparently normal at the time of isolation of the viruses (Table 4-5). The results with mousepox have been described. The most systematic studies were those reported by Marennikova and her colleagues with monkeypox and cowpox viruses in hamsters and cotton rats (examples 2,3,and 4 of Table 4-5). The animals suffered inapparent infections and the relevant virus was recovered from some animals for up to 6 weeks after inoculation, when the experiment was terminated. Big gerbils and yellow susliks suffered severe disease with high mortality after inoculation with the Turkmenia strain of cowpox virus, but virus could be recovered from the kidneys and testes of animals that survived for 5 weeks (example 5).

Positive results were also reported with vaccinia virus in rabbits and mice (examples 6 and 7) and variola virus in mice (example 8).

Persistence of Variola Virus in Human Patients

The evidence against the persistence of variola virus in man, after recovery from infection, is strong but circumstantial; it is impossible to prove that cases of persistent infection never arose. In the whole history of smallpox, no case occurred which could be unequivocally traced back to infection acquired from a person who had recovered from the disease months or years earlier, nor has any suspicion of the recurrence of symptoms or of the excretion of variola virus ever arisen in a person who had recovered from smallpox and subsequently undergone immunosuppressive treatment. Nor has vaccinia virus, used until recently on a large scale in Europe and North America (where immunosuppression is practiced more widely than in the developing countries), ever shown any indication that it could cause a latent infection in man. Since alert clinicians and virologists have recognized the occurrence of latent infections with a wide variety of viruses, and their reactivation after immunosuppressive treatment, this negative evidence is important.

Epidemiological Significance

It is difficult to assess the epidemiological significance of the observations on apparent persistence of some orthopoxviruses in animals. Only in Gledhill's (1962a) experiments with ectromelia was there any evidence that virus shedding occurred but even then transmission to susceptible contact mice was not observed. In all other cases the virus apparently persisted for only a few weeks, and it was found in a sequestered site by laboratory tests. It is reasonable to suggest that the persistent carriage and shedding of virus is not a factor of epidemiological significance in orthopoxvirus infections in the way that is so important in some other viral infections, e.g., those due to arenaviruses and herpesviruses.

With smallpox, there is a further epidemiological observation of great significance. Over a period of some 50 years smallpox was progressively eliminated from every country in the world. In no instance was a "spontaneous" outbreak identified after smallpox had been eliminated from a region or country. Apart from laboratory-associated cases, all outbreaks of which the source was sought could be traced to the introduction of virus from known infected areas by known infected individuals.

The probable nonpersistence of orthopoxviruses after recovery from infection is also relevant to the searches for the natural reservoir animal(s) of monkeypox virus in Africa (see Chapter 8) and the claims that variola virus has been isolated from various healthy primates and rodents (see Chapter 7). Orthopoxviruses have only very rarely been recovered from wild animals, even when tests were carried out in areas and with species in which there was serological evidence that orthopoxvirus infection was widespread. It is likely that virus isolation in such circumstances usually depends on the chance use of tissues from an animal suffering or convalescing from infection, rather than from a long-term carrier. This appears to have happened with rodent strains of cowpox virus (examples 1a and 1b of Table 4-4). The chance of catching such an animal during an ecological survey in tropical forest areas, with their abundance and diversity of animals, is quite small, which may account for the fact that in spite of extensive efforts over many years, monkeypox virus has only once been recovered from a wild animal, a squirrel suffering from an acute infection (see Chapter 8). The same factor may account for the rarity of primary cases of human monkeypox; human monkeypox is a rare disease, in spite of the fact that inhabitants of rain forest areas in Zaire habitually eat meat of wild animals, several of which are known to be occasionally (monkeys) or frequently (squirrels) infected with monkeypox virus.

THE IMMUNE RESPONSE IN
ORTHOPOXVIRUS INFECTIONS

Variolation, the ancient practice whereby smallpox was transmitted to a susceptible person by the inoculation of material from smallpox scabs or vesicles, was based on observations that pockmarked persons never suffered from smallpox a second time. It provided the foundation on which the science of immunology was built (Needham, 1980). The next major landmark in immunology, as Pasteur (1881) recognized when he generalized the use of the term vaccination, was Jenner's substitution of an antigenically related nonvirulent agent obtained from pox infections of cows or milkmaids for the virus of smallpox.

Although immunology arose from observations of protection from infection or reinfection, the immune response also plays an important role in the process of recovery. In the following pages both aspects are reviewed. The data on the role of the immune response in protection against reinfection are taken largely from studies of vaccination and smallpox in man; information on the role of the immune response in recovery from orthopoxvirus infections from experiments with mousepox.

Protection against Reinfection

Second Attacks of Smallpox. It is difficult to obtain precise figures on the incidence of second attacks of smallpox. Dixon (1962), quoting from the observations of Barry (1889) in Sheffield, and Rao (1972), whose views were based on his own experience, supported by laboratory evidence, suggest that about 1 in 1000 pockmarked persons suffered a second attack of smallpox. Epidemiologists working in the field during the Intensified Smallpox Eradication Program believe that this figure is rather high. Although collectively they saw many thousands of cases of variola major, it was very rare indeed to find one in a pockmarked person.

Responses to Revaccination. Depending primarily on the interval since a previous vaccination, the responses obtained on revaccination with vaccinia virus range from a complete absence of reactivity, through an accelerated reaction, sometimes with vesiculation (which involves local viral replication), to a "major response," indistinguishable from that found in primary vaccination. Other things being equal, skin site plays a role in the severity of the response; revaccination on the ventral surface of the forearm more frequently produced positive reactions than revaccination over the deltoid muscle. Even more striking are the finger lesions sustained by workers in vaccine production laboratories, who often suffered from vaccinial whitlows in spite of the fact that they were revaccinated annually (Horgan and Haseeb, 1944). Such cases often exhibited enlargement of the epitrochlear and axillary lymph nodes and sometimes the reinfection progressed like a primary vaccinial reaction, with maximum vesiculation on the ninth or tenth day.

Heterologous Protection. Vaccination against smallpox was based on the observation that heterologous protection was effective. In general, vaccination within the previous 5 years protected persons exposed to smallpox against clinical disease, and statistically some protection was evident for over 30 years. In countries in which smallpox was endemic, estimation of the duration of protection was complicated by the fact that subclinical smallpox occurred in a substantial number of vaccinated individuals who were in close contact with cases of smallpox (Heiner *et al.*, 1971).

Experiments in laboratory animals reveal that infection with any one orthopoxvirus produces substantial protection against disease produced by any other orthopoxvirus (vaccinia virus against mousepox in mice: Fenner 1947a; ectromelia virus against rabbitpox in rabbits: Christensen *et al.*, 1967; vaccinia virus against monkeypox in monkeys: McConnell *et al.*,1964; vaccinia virus against cowpox in rabbits: Downie, 1939b; vaccinia virus against variola in monkeys: Horgan and Haseeb, 1939; taterapoxvirus against monkeypox in monkeys: Lourie *et al.*, 1975). Likewise, a prior attack of smallpox gave some protection against vaccination with vaccinia virus, which diminished with time and was greater after a severe attack of variola major than after a mild attack (Vichniakov, 1968).

Although other serological tests often provide a clue to the identity of a newly discovered poxvirus, cross-protection in a suitable animal is the ultimate biological test for the allocation of the virus to the appropriate genus.

Humoral Responses in Orthopoxvirus Infections

The genome of an orthopoxvirus codes for a very large number of different polypeptides, each of which is potentially antigenic, hence a large number of different antibodies are produced during orthopoxvirus infections. These may be short-lived IgM or persistent IgG. Many, but not all, of these antibodies are also elicited after the inoculation of large doses of inactive virions or viral antigens.

Methods for Measuring Antibodies to Orthopoxviruses. At various times almost every method that has been developed for detecting antibodies has been employed for titrating antibodies to orthopoxviruses. In the following paragraphs the methods more commonly used are listed, with comments about particular features of each, notably its sensitivity, the persistence of the antibodies detected by particular tests, the potential of the test for distinguishing between different species of orthopoxvirus, and the relation of antibodies detected by various tests to protection.

Complement-Fixation (CF) Test. Many poxvirus antigens react in CF tests, including an early antigen located on the surface of vaccinia-infected cells (Ueda *et al.*, 1972). However, only antibodies of certain classes and subclasses (in man: IgM and to a lesser extent IgG_1, IgG_2, and IgG_3) participate in CF reactions; in general, such antibodies are short-lived so that a positive CF reaction is an index of recent infection (in man, within 12 months; Wulff *et al.*, 1969). Because so many antigens are common to all orthopoxviruses, the CF reaction is useless for discriminating between different species of the genus, unless used with an antigen that is species specific.

Immunofluorescence Test. Like the CF test, immunofluorescence can be used to detect many different orthopoxvirus antigens, but the IgG antibodies involved in immunofluorescence reactions are long lived (Gispen *et al.*, 1974). Combined with serial absorption of antisera with suitable suspensions of virus-infected tissue, immunofluorescence can be used to recognize species-specific orthopoxvirus antibodies in sera of animals caught in the wild (Gispen *et al.*, 1976). However, it is relatively insensitive. Immunofluorescence has proved useful in the study of the sequence of intracellular events in orthopoxvirus infection (Ueda *et al.*,1972) and in studies of the pathogenesis of poxvirus infections in laboratory animals (Mims, 1964, 1966).

Radioimmunoassay. Radioimmunoassay, which is about a thousand times more sensitive than immunofluorescence, can be used for the detection of antigen–antibody reactions in tubes or plates, or it can be combined with gel precipitation and autoradiography to discriminate between antigens produced in orthopoxvirus infection (radioimmunoprecipitation, see below).

Hutchinson *et al.* (1977) developed a radioimmunoassay test for detecting species-specific orthopoxvirus antibodies in adsorbed sera, which because of its sensitivity could be applied to sera obtained during field surveys. Its suitability for such purposes was greatly enhanced by the use of staphylococcal protein A in place of anti-γ globulin (Richman *et al.*, 1982; Fenner and Nakano, 1988).

ELISA Method. If extracts of orthopoxvirus-infected cells are used as antigen, ELISA provides a cheap and sensitive screening test for serological surveys of populations of man, wild animals, and laboratory animals (e.g., in surveys for mousepox in laboratory mice, Collins *et al.*, 1981; Buller *et al.*, 1983). If used with a species-specific antigen or, in blocking tests, with suitable monoclonal antibodies, ELISA can be used for detecting orthopoxvirus-specific antibodies.

Neutralization test. Neutralization of infectivity is the traditional discriminative test in animal virology, and is used to distinguish between different species of *Alphavirus* and *Flavivirus,* for example. However, although titers are highest with the homologous virus (Downie and McCarthy, 1950; McNeill, 1968), the antigens which evoke neutralizing antibodies after infection with orthopoxviruses show a great deal of overlap. Neutralization tests are therefore useful for allocating poxviruses to the appropiate genus, but not for distinguishing between species.

Neutralization tests can be carried out in laboratory animals (rabbit skin: Parker, 1939; mouse brain: Bronson and Parker, 1941), or by looking for plaque or focus reduction in cultured cells (McNeill, 1968; Kitamura and Shinjo, 1972) or pock reduction on the chorioallantoic membrane (Keogh, 1936; McCarthy *et al.*, 1958a). In man, neutralizing antibody persists for many years after recovery from both smallpox and vaccination with vaccinia virus (Downie and McCarthy, 1958; McCarthy *et al.*, 1958a,b).

Traditionally, neutralization tests have been carried out by mixing serum with suspensions of virions obtained by disrupting infected cells. Since such virions lack the envelope antigens found in naturally released virions (see Chapter 3), such tests do not measure antibodies to envelope antigens, which are important in pathogenesis (see Table 4-6). Antibodies that neutralize the infectivity of eveloped virions can be assayed by the "anti-comet" test of Appleyard *et al.* (1971) (Plate 4-7). In essence, this test involves the use of a liquid overlay for cell monolayers infected with suitable concentrations of a strain of an orthopoxvirus, such as rabbitpox virus, that produces many enveloped virions. In the absence of antibody to the envelope, such virions migrate in the liquid overlay to produce comet-shaped areas of cell damage. Antibodies that neutralize the infectivity of envelope antigens, added after viral adsorption, prevent the formation of the "comets," but not the development of plaques. Antibodies to nonenveloped virions have no such effect, but prevent plaque production in orthodox neutralization tests, in which virus-serum mixtures are added to the monolayer.

TABLE 4-6

Comparison of the Protective Activity of Three Types of Antiserum Administered Passively to Rabbits before Challenge Infection with Rabbitpox Virus[a]

Antiserum			Response to challenge	
Against	Source	Neutralization titer	Fever	Death
Inactivated vaccinia virus	Horse	800,000[b]	12/12	5/12
	Sheep	500,000[b]	5/5	4/5
Live vaccinia virus	Sheep	150,000[c]	4/5	0/5
	Rabbit	19,000[c]	8/10	0/10
Live rabbitpox virus	Rabbit	32,000[c]	0/5	0/5
None	—	—	18/18	15/18

[a] From Boulter *et al.* (1971).

[b] Measured against nonenveloped virions; no neutralizing antibodies to enveloped virions.

[c] Measured against nonenveloped virions; neutralizing antibodies to enveloped virions also present.

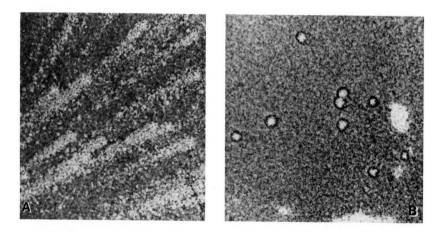

PLATE 4-7. *The "anti-comet" test of Appleyard* et al. *(1971). (A) Strains of virus which produce substantial numbers of extracellular enveloped virions produce "comet tails" from the initial plaques when suitable dilutions are inoculated in monolayers with a liquid overlay medium. (B) Addition to the overlay medium of antibody to viral envelopes or enveloped virions, or of antiserum from an animal that has been infected with vaccinia virus, prevents the production of the "comet tails" but not of the initial plaques. Antiserum to surface tubular elements or inactivated virions, although it neutralizes the infectivity of nonenveloped virions in a conventional neutralization test, does not inhibit "comet" formation. (From Appleyard* et al., *1971; courtesy of Dr. G. Appleyard.)*

Hemagglutination-Inhibition (HI) Test. All orthopoxviruses, but no other members of the family *Poxviridae,* produce a hemagglutinin which agglutinates cells from selected chickens (Nagler, 1942; Brown *et al.,* 1973). Ectromelia virus also agglutinates mouse cells (Burnet and Boake, 1946; Mills and Pratt, 1980). Because of its simplicity, the HI test has been widely used in serological surveys in man and in animals (e.g., in studies of the ecology of monkeypox virus, see Chapter 8). The hemagglutinin appears late in the course of viral synthesis as a new component of the plasma membrane of infected cells, where it may be recognized by hemadsorption tests (Driessen and Greenham, 1959; Blackman and Bubel, 1972). The hemagglutinin also occurs as a component of the viral envelope in virions that are released from cells (Payne and Norrby, 1976), but separately from infectious nonenveloped virions in extracts of infected cells. Hemagglutinins produced by different orthopoxviruses cross-react, although titers are usually higher with the homologous virus.

According to Downie and his colleagues (Downie and McCarthy, 1958; McCarthy *et al.,* 1958b), antibodies with HI activity persist for varying periods after recovery from orthopoxvirus infection, usually for only a few months, although somewhat longer than CF antibodies. However, using a somewhat different protocol for the preparation of vaccinia hemagglutinin, J. H. Nakano (personal communication, 1986) found that HI antibodies were more persistent, and were sometimes found in sera in which the neutralization tests were negative. Nonspecific inhibitors of orthopoxvirus hemagglutinin occur in some sera, especially if the specimens are old, have been improperly stored, or were collected at autopsy. They can usually be removed from human sera without loss of specific antibody by treatment with potassium iodate (Espmark and Magnusson, 1964).

Some strains and mutants of orthopoxviruses fail to produce a hemagglutinin or to promote the production of HI antibodies (Cassel, 1957; Fenner, 1958). Experiments with these viruses, and other evidence, show that HI antibodies are unrelated to those involved in neutralization reactions, or to protection against infection, although the presence of HI antibodies is evidence that the antigens that do evoke the production of protective antibodies have been produced.

Precipitation Tests. Much of the early work on antigens produced in orthopoxvirus-infected cells, involving both the time course of their production (e.g., Appleyard and Westwood, 1964a) and comparisons between strains and mutants of orthopoxviruses (e.g., Gispen, 1955; Rondle and Dumbell, 1962), utilized simple gel-precipitation tests. With the use of absorbed sera, reactions can be detected which differentiate variola, monkeypox, and vaccinia viruses (Gispen and Brand-Saathof, 1974; Esposito *et al.*, 1977b). However, the sensitivity of simple gel precipitation tests is low; they require highly potent sera and are not readily applicable to sera obtained from animals that have recovered from natural infections.

Both the sensitivity and the discriminative power of gel precipitation tests were greatly enhanced by two modifications: (1) electrophoresis of the viral antigens of antigen–antibody complexes in SDS–poly-acrylamide gels, and (2) radioisotopic labeling of the antigens. The radioimmunoprecipitation test was used by Ikuta *et al.* (1979) to demonstrate the presence of serologically related antigens among poxviruses of the same and different genera.

The Humoral Response in Relation to Pathogenesis.   Antibodies of many different specificities are generated during infection with a virus as complex as an orthopoxvirus. Most of these are probably irrelevant, as far as pathogenesis, recovery, and protection are concerned. The relevant antibodies belong to three classes: (1) antibodies that neutralize viral infectivity, of which there are two subclasses, directed, respectively, against nonenveloped and enveloped virions (for review, see Boulter and Appleyard, 1973), (2) those that, with complement, lyse virus-infected cells, and (3) antibodies that combine with circulating antigens to produce immune complexes.

It has long been believed that specific antibodies generated by the humoral immune response played important roles in both protection against orthopoxvirus infections and recovery from established infections. The protective effect of antibodies is most clearly demonstrated by passive immunization, their putative role in recovery was based mainly on temporal relationships observed during the course of established infections.

Passive Immunity. Passive immunization, either by the transmission of antibodies from mother to offspring, or by the inoculation of antisera, provides a means of studying the influence of antibodies on the disease process uncomplicated by cell-mediated immunity. The effectiveness and the limitations of passive immunization in generalized orthopox-virus infections are well illustrated in experiments on mousepox (Fenner, 1949a; Fig. 4-5). Immunity after recovery from ectromelia virus infection (not shown in Fig. 4-5) usually inhibited viral replication in the inoculated foot and always prevented generalization of the disease. Active immunization with vaccinia virus (Fig. 4-5B) had little effect on viral replication in the foot but greatly diminished generalization, although a transient rash sometimes occurred. The antibody titer rose

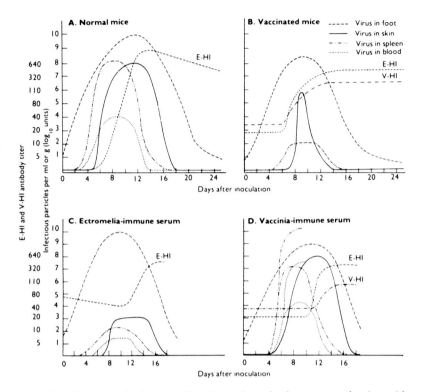

FIG. 4-5. *The spread of ectromelia virus through the organs of mice, either unprotected (A), or protected by active immunization with vaccinia virus (B), or by passive immunization with ectromelia-immune serum (C) or vaccinia-immune serum (D) obtained from convalescent mice. (From Fenner, 1949a.)*

and the relative titers against ectromelia and vaccinia hemagglutinins were reversed between 7 and 8 days after infection. Passive immunization was much less effective in modifying the course of the disease. Anti-vaccinial serum (Fig. 4-5D) had almost no effect and several mice died with acute hepatitis at the same time as the controls. Anti-ectromelia serum (Fig. 4-5C) was more effective but did not influence viral replication in the inoculated foot, and a low level of replication in the spleen and skin was found in most animals. In passively immunized mice the antibody level rose about 2 days later than it did in the vaccinated mice.

These experiments show that active immunization, whether heterologous or homologous, provided much greater protection than the administration of preformed antibodies, even when the antibody titers in the passively immunized mice were higher than those in the actively immunized animals. The reason for the difference is that active immunization with infectious virus provokes the complete range of cell-mediated and humoral immune responses; passive immunization provides only the antibodies present in the convalescent animals from which the sera were obtained. Further, passive immunization with the homologous antibody provided much greater protection than did heterologous antibody, even though the sera cross-reacted extensively in neutralization tests.

Experiments on passive immunization in rabbitpox (Boulter et al., 1961b) confirmed the protective value of potent homologous antiserum, which protected rabbits from death even when treatment was delayed until overt disease had developed. Subsequent experiments (Boulter et al., 1971; see Table 4-6) showed that antisera produced after infection with live virus had a much greater protective effect than antiserum produced by immunization with inactivated virus, even when the neutralizing titer, measured by orthodox neutralization tests, was much lower.

Thus antibodies as such do have an effect on viral replication and spread in generalized poxvirus infections, although as described below, cell-mediated immunity is more important.

Neutralizing Antibodies and Protective Immunity. From the point of view of protection against natural infection, the antibody that neutralizes enveloped virions is theoretically more important than the antibody that neutralizes nonenveloped virions (for review, see Boulter and Appleyard, 1973). Antibody which specifically neutralizes enveloped virions can be detected by the "anti-comet" test (see Plate 4-7). The differences between neutralizing antibodies to enveloped and nonenveloped virions are particularly important in relation to efforts to produce inactivated vaccines for protection against smallpox (see below). The relevant features of the experiments by Boulter, Appleyard, and their colleagues (Appleyard et al., 1971; Boulter et al., 1971; Turner and Squires, 1971) and later experiments by Payne (1980) can be summarized as follows:

1. Sera of rabbits which had recovered from infection with rabbitpox virus contained antibodies that neutralized enveloped virions; sera from rabbits inoculated with inactivated virions lacked such antibodies.

2. Sera from both groups of rabbits neutralized nonenveloped virions.

3. Cross-absorption of the two types of sera with concentrated enveloped or nonenveloped virions selectively removed neutralizing antibodies of the appropriate specificity.

4. Antibody to isolated envelopes was able to neutralize the infectivity of enveloped virions and to protect mice against the spread of infection.

5. Rabbits immunized with inactivated rabbitpox virus (i.e., with inactivated nonenveloped virions) had high levels of neutralizing antibodies to nonenveloped and none to enveloped virions, but showed only partial immunity on challenge inoculation, whereas rabbits immunized with enveloped virions were fully protected.

Cell-Mediated Immune Responses

Although observers had long been concerned with the frequent lack of correlation between circulating antibodies and recovery from infections, "cellular immunity," invoked in such situations to explain recovery from disease, had no precise meaning until the complex immunospecific responses of T cells were recognized during the 1970s, distinct and independent of the antibody-producing B cells. The T cell responses, other than those involved in modulation of the humoral response, constitute cell-mediated immunity. The mechanisms by which T lymphocytes exercise antiviral functions are complex; they may involve both the direct effects of T cells on cells with virus-modified surface membranes and the effects of their secreted products, which are called lymphokines and include, among others, γ-interferon.

The Delayed Hypersensitivity Reaction.   The classical method of measuring cell-mediated immune responses is the skin test for delayed hypersensitivity, an expression used to contrast the time course and nature of the reaction with that of "immediate" hypersensitivity, which comes on within minutes of exposure to the relevant antigen and is mediated by IgE (Gell *et al.*, 1975). Delayed hypersensitivity is a complex reaction involving three components: (1) an initial one in which antigen-sensitive T cells are sensitized, a procedure that may require the prior processing of antigens in macrophages; (2) a further component, consisting of antigen recognition and the proliferative response of T cells, with release of lymphokines; and finally (3) an inflammatory response which is amplified by chemotactic factors. Delayed hypersensitivity can be passively transferred by suspensions of lymphoid cells, but not by antiserum. Two subclasses of T cell may be involved in delayed hypersensitivity reactions: (1) cytotoxic T cells (T_c) are evoked by viral infection and have cytotoxic activity, reacting specifically with virus-induced antigens on cell membranes; (2) delayed hypersensitivity T cells (T_d) are evoked by antigen presented in a nonreplicating form, such as inactivated vaccines, as well as during viral infections; these T cells are not cytotoxic (Ada *et al.*, 1981).

Delayed hypersensitivity is recognized by the accelerated response to inoculation of the antigen(s) into the skin; its development during orthopoxvirus infection of man was recognized by Jenner (1798) and has been repeatedly demonstrated since then (e.g., Pincus and Flick, 1963). It used to be used as a method of assessing immunity to smallpox, but it is an indicator of resistance only when it is an indicator of the presence of cytotoxic rather than delayed hypersensitivity T cells (Ada *et al.*, 1981).

***In Vitro* Techniques for Analyzing T-Cell Function.**   There is no satisfactory *in vitro* assay of delayed hypersensitivity; the reaction involved is a particular kind of inflammatory response which must be tested in intact animals. However, in experimental systems there is a good assay for cytotoxic T cells. A known number of ^{51}Cr-labeled, virus-infected target cells are cultured together with varying numbers of lymphocytes obtained from the spleen or lymph nodes. The release of the radioactive label is an index of cytotoxic T cell activity; proof that it is due to T cells is provided by the absence of lysis if the lymphocyte preparation is treated with anti-theta serum and complement.

In vitro experiments with cells infected with ectromelia virus (Ada *et al.*, 1976; Jackson *et al.*, 1976) showed that some of the cell-surface changes relevant to T cell-mediated lysis occurred before viral DNA replication had begun—i.e., they were "early" synthetic functions coded for by the input DNA (see Chapter 3).

The mechanism of cytotoxic T cell lysis was shown by Zinkernagel and Althage (1977) to be a direct interaction between the appropriate T cells and cells infected with vaccinia virus. Recognition by T cells depended on the presence on the membranes of infected cells of both virus-specified antigens and compatible major histocompatibility gene products. The early lytic effect ensures that cells are lysed before progeny virions are assembled and thus accounts for the efficiency of cell-mediated immunity in promoting recovery in established infections with orthopoxviruses (see below).

Cell-Mediated Immunity in Relation to Pathogenesis.   Mice are particularly suitable for studying the role of various components of the immune response in orthopoxvirus infections, since so many genetically defined mouse lines are available. Mousepox, which proved so useful for studying other aspects of pathogenesis, has been used to provide information on the relative importance of humoral and cell-mediated immunity, and has been extensively exploited for this purpose (Blanden, 1970, 1971a,b; for reviews, see Blanden, 1974; Cole and Blanden, 1982).

Mechanisms controlling viral growth in the major visceral target organs (liver and spleen) become operative 4–6 days after primary infection by the natural route, which in the experimental studies was simulated by subcutaneous inoculation into the footpad. Cell-mediated immune responses occur soon after infection: virus-specific cytotoxic T cells are detectable 4 days after infection and reach peak levels in the spleen 1–2 days later, while delayed hypersensitivity is detectable by the footpad test 5–6 days after infection. In contrast, significant neutralizing antibody is not detectable in the circulation until the eighth day.

Mice pretreated with anti-thymocyte serum, which acts specifically on T lymphocytes, die from otherwise sublethal doses of virus, on account of uncontrolled viral growth in target organs. Such mice have impaired cell-mediated responses but their neutralizing antibody levels are normal, interferon levels in the spleen are elevated, and the innate resistance in target organs is unchanged.

Very large doses of interferon or immune serum transferred to preinfected recipients are relatively ineffective against the established infection in target organs, although high levels of interferon and high antibody titers can be demonstrated in the sera of the recipients. On the other hand, immune spleen cells harvested 6 days after active immunization of the donor transfer specific and highly efficient antiviral mechanisms which rapidly eliminate infection from the target organs of the recipients, in whose serum neither antibody nor interferon is detectable. The active cells in the immune population can be identified as cytotoxic T cells. Mononuclear phagocytes of immune T-cell recipients, labeled with tritiated thymidine before T-cell transfer, appear in foci of infection in the liver after T-cell transfer, and prior irradiation of immune T-cell recipients in a regimen designed to reduce blood monocyte levels significantly reduces the antiviral efficiency of the transferred cells.

Recognition of the diversity of T cells and the availability of monoclonal antibodies against various subclasses of T cell has made it possible to dissect the role of the subclasses with greater precision. *In vitro* experiments suggest that the generation of cytotoxic T cells ($CD8^+$) is strictly dependent on functions provided by helper T cells ($CD4^+$) (Wagner and Rollinghoff, 1978). However, Buller *et al.* (1987d) showed that genetically resistant C57BL/6 mice that had been depleted of $CD4^+$ cells by inoculation with an appropriate monoclonal antibody developed an optimum *in vivo* virus-specific cytotoxic T cell response and recovered from infection with ectromelia virus, which was lethal in nude mice ($CD4^-$, $CD8^-$). Experiments on depletion of $CD8^+$ cells, although not fully satisfactory, suggested that these cells played an essential part in the recovery process.

Taken together, these findings support the idea that blood-borne cytotoxic T cells with immunological specificity for virus-induced antigenic changes in infected cell surface membranes enter infectious foci and retard viral spread by lysing infected cells before the maturation and assembly of progeny virions. This T-cell activity attracts blood monocytes which contribute to the elimination of infection by phagocytosis and intracellular destruction of virus. Macrophage activation and locally produced interferon may increase the efficiency of virus control and elimination, but are less important than T cells.

Cytotoxic T cells also appear to be important in expressing the genetic resistance of C57BL mice to ectromelia virus infection (see Chapter 9). Using a highly sensitive assay system, O'Neill and Brenan (1987) showed that cytotoxic T cells could be detected in the popliteal lymph nodes of C57BL mice 3 days after inoculation of a large dose of ectromelia virus into the footpad, compared with 5 days in genetically susceptible BALB/c mice. The delay in the recruitment and proliferation of these cells could well account for the more rapid transmission of virus from the draining lymph node to the liver and spleen in susceptible strains of mice. Experiments with retrovirus immunosuppression (Buller *et al.*, 1987e) also support the view that the cytotoxic T-cell response is much more important than humoral immunity in protecting mice against mousepox. Nevertheless, the experiments illustrated in Fig. 4-5 and Table 4-6 showed that antisera from animals that have recovered from orthopoxvirus infection do have an effect on the progression of infection with the homologous virus, probably in controlling the extent of viremia.

Immunological Deficiency States in Man

The effects of immunological deficiency states in human subjects who have been vaccinated with vaccinia virus provides information on the relative importance of cell-mediated and humoral responses in determining recovery in orthopoxvirus infections. Fulginiti *et al.* (1968) described a number of cases of progressive vaccinia in immunologically deficient infants and children, and Kempe (1980) has summarized his extensive experience with these conditions, in relation to vaccination. In children with immunological defects in cell-mediated immunity, vaccinia virus replicated without restriction, resulting in a continuously progressive primary lesion, persistent viremia, and widespread secondary viral infection of many organs. This response was particularly severe in patients with thymic alymphoplasia. In patients with thymic dysplasia and partially or completely intact immunoglobulin-synthesizing capacity (Nezelof's syndrome) the progression of the primary disease was

Immunological condition	Immunological status	Response to vaccination
1. Normal, vaccinated	(+)CMI+ ; (+)Ab+	CMI+ ; Ab+ : no change
2. Normal, unvaccinated	(+)CMI− ; (+)Ab−	CMI+ ; Ab+ : primary vaccination
3. Thymic dysplasia	(−)CMI− ; (+)Ab−	CMI− ; Ab+ : progressive vaccinia ＼ VIG ↓ CMI− ; Ab+ : progressive vaccinia
4. Bruton's syndrome	(+)CMI− ; (−)Ab−	CMI+ ; Ab− : primary immunization or CMI− ; Ab− (CMI "overwhelmed") : progressive vaccinia ＼ VIG ↓ CMI+ ; Ab+ : CMI restored—recovery
5. Swiss syndrome	(−)CMI− ; (−)Ab−	CMI− ; Ab− : progressive vaccinia ＼ VIG ↓ CMI− ; Ab+ : progressive vaccinia
6. Acquired deficiencies (e.g., lymphoma)	(−)CMI− ; (−)Ab−	CMI− ; Ab− : progressive vaccinia ＼ VIG ↓ CMI+ ; Ab+ : CMI restored—recovery

FIG. 4-6. *The response of normal individuals and individuals with immunological defects to primary vaccination. Progressive vaccinia is associated with a defective cell-mediated immune response, but under some circumstances vaccinia immune globulin can be useful (see text). CMI, Cell-mediated immunity; Ab, antibody production; VIG, vaccinia immune globulin; +, positive immune response; −, absence of an immune response. (Based on Freed* et al., *1972.)*

sometimes slower and less persistent, but a fatal outcome was usual (Kumar *et al.*, 1977).

Figure 4-6 sets out various kinds of human immunological defect schematically and suggests the kinds of responses that occurred when such individuals were vaccinated. Cell-mediated immunity was clearly of major importance in controlling vaccinial infection, since individuals with defects in cell-mediated immunity but intact antibody production (category 3, Fig. 4-6) suffered from progressive vaccinia; those with defects in antibody production but a satisfactory capacity to mount a cell-mediated immune response (category 4) usually reacted normally to vaccination. Freed *et al.* (1972) suggest that in some circumstances, especially when the immunological defect was acquired (category 6), the administration of vaccinia-immune globulin (VIG) would allow cell-mediated immunity to recover sufficiently to control the vaccinial infection. Also, in some cases of Bruton's syndrome a partially deficient cell-mediated immune mechanism might have been "overwhelmed," but could be restored to effectiveness by the administration of vaccinia-immune globulin.

Delayed hypersensitivity reactions could not be evoked in patients with progressive vaccinia, nor could the peripheral blood lymphocytes of such patients be stimulated to mitosis by exposure to inactivated vaccinia virus (Fulginiti *et al.*, 1968). Although neutralizing antibody was sometimes present in the serum, its presence did not prevent the development of progressive vaccinia, if cell-mediated immunity was defective (Hansson *et al.*, 1966). Kempe (1980) records having seen 15 boys with very low levels of γ-globulin who had been routinely vaccinated in infancy without complications. All had histories of "enormous and hyperactive" delayed hypersensitivity responses to vaccinial antigens.

METHODS OF IMMUNIZATION AGAINST ORTHOPOXVIRUS INFECTIONS

Long before methods of protection against any other disease had been developed, sages and alchemists in India and China had developed a method of protecting against natural smallpox by introduction of variola virus by an unusual route, a practice that came to be called variolation. Several centuries later, vaccination was introduced and its efficacy established by the most relevant and stringent criterion; its ability to provide protection against smallpox. After smallpox had been eliminated from Europe and North America in the 1950s, the public in these countries did not readily tolerate the degree of illness—still less the occasional episodes of severe illness or even deaths—that were associated with standard smallpox vaccination. Two ways were sought to provide immunization against smallpox without the attendant risks of severe disease: the use of attenuated strains of vaccinia virus and the use of inactivated vaccines.

Apart from smallpox, protection against orthopoxvirus infections by immunization is justified in only a few situations. The most important are with laboratory animals—mousepox in laboratory mice and rabbitpox in laboratory rabbits. In Germany, cowpox virus infection of zoo and circus elephants is deemed sufficiently dangerous to these valuable animals to justify vaccination (see Chapter 6), and if economic considerations permitted, immunization of camels against camelpox would be justified.

In principle, four methods of immunization can be used for protection against orthopoxvirus infections: (1) homologous wild-type virus administered by an unusual route, (2) inoculation with a wild-type heterologous species of orthopoxvirus, (3) inoculation with attenuated live virus, either homologous or heterologous, and (4) inoculation with inactivated virus, either homologous or heterologous.

Homologous Wild-Type Virus Administered by an Unusual Route

This approach ("variolation") is not now used in any orthopoxvirus disease, although it was used in the past for smallpox, and also to protect young camels against camelpox before the onset of the wet season, when the disease was more severe (Leese, 1909). It is still used to protect lambs against scabby mouth, which is caused by a parapoxvirus.

Wild-Type of Heterologous Species of *Orthopoxvirus*

This, the Jennerian approach, has also been used in attempts to protect laboratory animals. Vaccinia virus has been extensively used to protect laboratory mice against infection with ectromelia virus (see Chapter 9), ectromelia virus can been used to protect rabbits against rabbitpox (Christensen *et al.*, 1967), and vaccinia virus to protect monkeys against monkeypox virus (McConnell *et al.*, 1964, 1968.

Attenuated Live-Virus Vaccines

Attenuation of Vaccinia Virus for Human Vaccination. After smallpox had been eliminated locally but before it was eradicated globally, efforts were made in the industrialized countries to reduce the pathogenicity of vaccinia virus used for vaccination of their citizens, so as to reduce the frequency of severe complications. The assessment of the efficacy of such vaccines in man presented obvious problems, since the real criterion for their value was the ability of the vaccine to protect against smallpox and the durability of this protection. It was clearly not possible to test new vaccines in this way, so two other criteria were used in their assessment: the neutralizing antibody response and protection against challenge inoculation of standard vaccine. Several attenuated strains of vaccinia virus were produced in different countries, but none of them was ever used on a large scale in a situation where there was a risk of exposure to smallpox. However, it is instructive to review experience with some of these vaccines (see Fenner *et al.*, 1988).

United States: CVI-78 and CVII Vaccines. These strains were produced by passaging the New York City Board of Health strain of vaccinia virus in minced chick cells (Rivers, 1931); the full passage history is summarized in Barker (1969). Rivers and Ward (1935) showed that the "second revived strain" (CVII) consistently produced less severe reactions in rabbits and humans than did the standard calf lymph vaccine. Primary vaccination produced only red papular lesions, without the development of pustules and with little constitutional disturbance (Rivers *et al.*, 1939). Using the strain CV1-78, Kempe *et al.* (1968) vaccinated 1009 children suffering from eczema, 326 by the multiple pressure method, and the rest by subcutaneous inoculation. Local reactions and temperature elevations were much milder than those seen in children vaccinated with standard vaccine and there were no serious complications. The antibody responses were comparable to those obtained after vaccination with calf lymph vaccine.

Tint (1973) summarized experiences with primary vaccination with the CV1-78 strain in 9000 subjects, 3500 of whom had eczema or other skin diseases, in the United States, England, and Japan. He suggested that vaccination with this attenuated strain on its own was probably not sufficient to provide protection against smallpox, but that its use in eczematous children as a preliminary to vaccination with standard vaccine would substantially lower the risks of eczema vaccinatum.

Because of increasing concern about morbidity and mortality associated with smallpox vaccination, the National Institute of Allergy and Infectious Diseases, National Institutes of Health (Bethesda, Maryland) sponsored a study of the reactogenicity and immunogenicity of four vaccines: calf lymph and egg vaccines made from the New York City Board of Health strain, egg vaccine made from CV1-78, and Lister sheep vaccine, administered at several dosages by percutaneous and subcutaneous routes (Galasso, 1970). The results were published in 1977 in six papers in the *Journal of Infectious Diseases* (Galasso *et al.*, 1977). Primary vaccination by the subcutaneous route, while accompanied by lower rates of fever, led to unsatisfactorily low antibody responses both initially and after standard percutaneous revaccination (Galasso *et al.*, 1977). The immunogenicity of the Rivers attenuated vaccine was too low for it to be used for vaccination against smallpox; furthermore, it was also found to be unsatisfactory for use as a priming inoculation, to be followed by standard calf lymph vaccine.

Germany: The MVA Strain. During the 1970s, workers in several other countries developed attenuated vaccines, since they believed then that vaccination procedures would have to be maintained for many years after eradication and that in these circumstances an attenuated vaccine would be necessary. In the Federal Republic of Germany, Stickl and his collaborators (Hochstein-Mintzel et al., 1975) produced a highly attenuated strain of vaccinia virus (MVA) by 572 serial passages of the Ankara strain of vaccinia virus in chick embryo fibroblast cells. In the process the M_r of the DNA was reduced by 9% and the strain was shown to have greatly reduced virulence for the chick embryo, for experimental animals, and for man (Mayr et al., 1978). Stickl et al. (1974) proposed that it should be routinely used as preimmunization for primary vaccinations, followed by conventional vaccine; and always used (for primary vaccination and revaccination) in individuals at special risk (e.g., with eczema or under immunosuppression). Its probable safety in immunosuppressed individuals was suggested by experiments in irradiated rabbits reported by Werner et al. (1980).

    Japan: The LC16m8 Strain. Hashizume (1975) developed several attenuated variants of the Lister strain of vaccinia virus by serial passage in rabbit kidney cells at 30°C and subsequent selection of a small pock from the chorioallantoic membrane. The variant most extensively studied, LC16m8, was much less pathogenic after intracerebral inoculation in monkeys than standard vaccine strains (Hashizume et al. 1973), but produced a satisfactory immune response (HI and neutralizing antibody) in humans as well as in vaccinated animals (Hashizume, 1975). A freeze-dried preparation of this vaccine produced in rabbit kidney cells was as stable as standard freeze-dried calf lymph vaccine. There were no severe complications in a field test in which more than 50,000 persons were vaccinated (Japan, Ministry of Health, 1975). The take rates in 10,000 of the vaccinees who were closely followed were not significantly different from those produced by the other vaccines tested, but the local and general reactions produced by LC16m8 vaccine were lower. Challenge vaccination of 138 of these vaccinees with the Lister strain 12 months later produced major reactions in 18.8%, a rate similar to that observed in 714 persons vaccinated with the standard Ikeda strain.

    This strain is licensed in Japan for primary vaccination against smallpox, and it has been proposed as a suitable strain for use in recombinant vaccinia virus vaccines (Morita et al., 1987). In an attempt to increase its capacity to multiply locally, Takahashi-Nishimaki et al. (1987) carried out homologous recombination with a selected fragment of Lister strain DNA. Some of the recombinants produced larger pocks and plaques, but retained the temperature sensitivity and low neurovirulence of the LC16m8 strain.

Attenuated Ectromelia Virus for Protection against Mousepox.   Immunization of mice against mousepox with vaccinia virus (i.e., vaccination with a heterologous wild-type orthopoxvirus) has given relatively satisfactory results (see Chapter 9), but more recently an alternative method has been developed, viz., attenuated ectromelia virus. Mahnel (1983) found that after 300 passages of ectromelia virus in chick embryo fibroblasts it was avirulent for mice, but seroconversion occurred after intradermal or intranasal inoculation, or administration in drinking water.

Inactivated Virus Vaccines

Ordinarily, inactivated viral vaccines are made with the virulent virus against protection is sought. However, variola virus was too dangerous to handle on a production scale, but an inactivated vaccinia virus vaccine seemed a feasible proposition, since the virus was easy to grow and purify. Many methods of inactivation were tested (Kaplan, 1969; Turner et al., 1970), including heat, formaldehyde, ultraviolet radiation, photodynamic inactivation, and gamma irradiation.

As described in Chapter 2, there are major antigenic differences between the surface antigens of enveloped and nonenveloped virions of vaccinia virus, both of which are infectious. Smallpox vaccine, however it was grown, was prepared in such a way that it consisted predominantly of nonenveloped virions. The infection provoked by vaccination with live virus vaccine led to the development of both enveloped and nonenveloped virions and the full range of humoral and cellular immune responses. In contrast, inactivated vaccine failed to provoke a humoral response to the envelope antigens (Appleyard et al., 1971; Turner and Squires, 1971; Payne, 1980), nor did it stimulate the production of cytotoxic T cells (Ada et al., 1981).

Experiments with inactivated vaccines highlighted the lack of correlation between levels of neutralizing antibody (measured against nonenveloped virions) and protection. Thus immunization of rabbits with a vaccinia virus "soluble antigen" gave a good antibody response and protection against intradermal challenge with vaccinia virus, but resistance to the more virulent rabbitpox virus was less than that induced by live vaccinia virus, despite the fact that the live virus induced much less antibody (Appleyard and Westwood, 1964b). Rabbits immunized by multiple intradermal injections followed by six intravenous injections of heat-inactivated vaccinia virus developed high titers of neutralizing antibody but even after this intensive course there was only partial protection against challenge infection with rabbitpox virus

(Madeley, 1968). Similarly, rabbits immunized with large doses of vaccinia virus inactivated by formaldehyde or ultraviolet irradiation developed extremely high titers of neutralizing antibody (tested against nonenveloped virions), but they remained susceptible to generalized rabbitpox infection, although protected from death (Boulter et al., 1971). The importance of antibody against enveloped virions was most clearly demonstrated by experiments on the passive transfer of resistance with antiserum (Table 4-6). Antisera against inactivated virus, with very high levels of neutralizing antibody to nonenveloped but none to enveloped virions, provided much weaker protection against challenge infection than apparently much lower titers of antibody induced by infectious virus, which, however, contained both kinds of neutralizing antibody.

It also appears that inactivated virus provokes a rather different kind of cell-mediated immune response from that found after infection, eliciting delayed hypersensitivity T cells but not cytotoxic T cells (Ada et al., 1981), perhaps because the surface antigens found in infected cells are not produced, or because the mode of inoculation results in too localized and immobile an antigenic mass. Indeed, the delayed hypersensitivity reaction itself was deficient in rabbits immunized with inactivated vaccine. For example, Turner et al. (1970) and Turner and Squires (1971) found that inactivated vaccines did not produce an obvious delayed hypersensitivity response, although the animals responded more rapidly than did the controls to challenge inoculation with live virus.

Thus inactivated vaccines suffered from two defects: they failed to elicit antibodies that neutralized enveloped virions and they failed to provoke the production of cytotoxic T cells. Nevertheless, in an effort to reduce the incidence and severity of postvaccinial encephalitis, formalin-inactivated vaccine ("vaccinia-antigen") was used in Germany during the late 1960s for a "priming" vaccination, followed by vaccination with standard vaccine. Subsequently, Marennikova and Macevic (1975) showed that preimmunization of rabbits with vaccine inactivated by ^{60}Co gamma-irradiation greatly enhanced their response to vaccination with active vaccine given 7–60 days later, both in the titer of antibody produced and in its rate of production. Preimmunization reduced the incidence of viremia in the rabbits 4–5 days after vaccination with live virus. Preliminary human trials on the use of this preparation as a priming antigen were carried out in eastern Europe in 1977, but by this time global eradication of smallpox was imminent, and no further trials were made.

NONSPECIFIC MECHANISMS INVOLVED
IN HOST DEFENSE

The efficacy of vaccination in protecting against smallpox and the increased susceptibility of individuals with certain immunological defects to severe complications after vaccination with vaccinia virus illustrate clearly the great importance of the immune response in orthopoxvirus infections. There are nevertheless a number of defense mechanisms against viral infections whose activity is not specific in an immunological sense. Most of these are ill understood and it is difficult to evaluate their importance.

Body Temperature

Determination of the ceiling temperature of viral replication is a useful laboratory method of distinguishing between certain orthopoxviruses (see Table 1-2) and between variola major and certain strains of variola minor virus (see Table 7-2). In animal models, body temperature has a dramatic effect on the severity of the leporipoxvirus disease, myxomatosis (Marshall, 1959), and mice housed at 2°C are about 100 times more susceptible to mousepox than those maintained at 20°C (Roberts, 1964). There is no evidence that raised body temperature affected the progress of variola major; severe cases (flat-type and haemorrhagic-type smallpox) were often associated with higher temperatures than found in ordinary-type smallpox. However, Dumbell and Wells (1982), comparing variola major and alastrim viruses, found that many fewer virions of alastrim virus (which had the lower ceiling temperature) than of variola major virus were released from infected cells when the temperature was raised. Perhaps the decreased dissemination of virus acted in concert with developing immunity to reduce the severity of variola minor.

Nutrition

Almost any severe nutritional deficiency will interfere with the activity of phagocytes, and the integrity of the skin and mucous membranes is impaired in many types of nutritional deficiency (for review; see Scrimshaw et al., 1968). The numbers of circulating B cells, immunoglobulin levels, and antibody responses are generally normal in cases of moderate to severe malnutrition, but cell-mediated immunity is consistently impaired, whether measured by cutaneous delayed hypersensitivity tests or by the numbers of circulating T cells (Chandra, 1979). The number of null cells, i.e., cells without the surface characteristics of

T or B cells, which suppress the activity of other lymphocytes, was relatively increased in cases of malnutrition. The proportion of such cells was substantially increased in patients with smallpox and there seemed to be a correlation between the height of the null cell count and the prognosis (Jackson *et al.*, 1977).

Little information is available about the effect of malnutrition on smallpox, although it seems clear that its effects were not as dramatic as those seen in measles in young children in many African countries. The mortality of variola major in unvaccinated infants was so high that it was difficult to determine whether nutritional deficiency was important. However, WHO epidemiologists in Somalia noticed that variola minor was much more severe in malnourished than in well-nourished infants (Z. Jezek, personal communication, 1982).

The occurrence of blindness after smallpox is said usually to have been associated with secondary bacterial infection or nutritional deficiencies.

Age

Among unvaccinated persons, smallpox produced its highest mortality in the very young and the aged, and its lowest in the 5- to 20-year age group. A similar response was observed in mousepox (Fenner, 1949c), in which both suckling and aged mice (over a year old) were much more susceptible than 6-week-old mice. Viral invasion of the internal organs occurred much earlier in suckling mice, and lethal changes were produced before the immune response developed. However, invasion of the liver and spleen occurred at about the same time in 6- to 8-week-old and 56-week-old mice, but replication usually progressed inexorably to produce a lethal outcome in the older mice, but was controlled in the young adult mice. There is no obvious explanation for these age-related effects except perhaps that the mechanisms of specific and nonspecific resistance function less effectively at the extremes of life.

Hormonal Effects

Pregnancy has a very pronounced effect on the severity of orthopoxvirus infections. In variola major, for example, pregnant women were much more likely than any other group of people to suffer from hemorrhagic-type smallpox, which was almost always lethal. Pregnant women have elevated levels of 17-dihydroxycorticosteroids, which have an antiinflammatory effect, depress the immune response, and inhibit interferon production. Studies in rabbits infected with vaccinia virus showed that cortisone diminished the local inflammatory reaction and increased viral titers in the blood and internal organs (Bugbee *et al.*, 1960), and vaccination can produce severe effects in humans receiving corticosteroid therapy.

Rao *et al.* (1968) found that cortisone converted experimental smallpox in monkeys from a nonlethal into a lethal disease. Viremia was greatly enhanced in intensity and persisted for a longer time, the internal organs contained much more virus than in control animals, and there were numerous hemorrhages in the lungs and in the mucous membrane of the gastroinestinal tract.

Interferon

Vaccinia virus was the first virus shown to be sensitive to interferon in an intact animal, when Isaacs and Westwood (1959) showed that interferon prepared in rabbit cells protected rabbits completely against intradermal infection with a large dose of vaccinia virus, when given a day before the virus was administered, and against a smaller dose when both were administered intradermally on the same day. However, Blanden (1970, 1971a) showed that passively administered interferon had no effect on recovery from mousepox. Interferon seems unlikely to have played a role in determining differential host responses in smallpox, but it may have been important in determining the differences in severity of inoculation smallpox and the "natural" disease.

Interferon and Inoculation Smallpox. Inoculation smallpox (variolation) is much milder than "natural" smallpox (see Chapter 7). Following the demonstration that vaccinia scabs contained interferon, Wheelock (1964) suggested that the presence of interferon in scab material that was used for variolation might have so interfered with the replication of the inoculated virus that the consequent disease was milder than smallpox acquired by the inhalation of virus contained in oropharyngeal secretions. This is unlikely to be the complete explanation; intradermal inoculation was associated with a shorter incubation period and the immune response would have been differently stimulated, but interferon in the inoculum and possibly the local production of interferon induced by inactivated virus in the inoculum may have played a role.

CHAPTER 5

Vaccinia Virus: The Tool for Smallpox Eradication

Vaccinia virus is the type species for the genus *Orthopoxvirus,* on which the vast bulk of the experimental work with members of that genus and, indeed, with viruses of the family *Poxviridae* has been carried out (see Chapters 2 and 3). It was also used as a live virus vaccine more extensively, and for a much longer period of time, than any other immunizing agent.

In their report on the classification of poxviruses, carried out at the request of the Virus Subcommittee of the International Nomenclature Committee of the International Association of Microbiological Societies, Fenner and Burnet (1957) suggested the name *Poxvirus officinale* for vaccinia virus. Although this name has never been widely used, it emphasizes that man has played a key role in producing the various strains of vaccinia virus that are currently recognized. The origin and natural host(s) of vaccinia virus remain matters of speculation.

VACCINIA VIRUS AS A SEPARATE SPECIES

Vaccinia virus can be differentiated from other species of *Orthopoxvirus* by biological tests and by restriction endonuclease analysis of its DNA.

Because of its long history of use as a vaccine and in the laboratory, there are a large number of strains of vaccinia virus that differ somewhat from each other in their properties. Nevertheless, because of the constellation of biological properties that they exhibit (see Table 1-2) and their characteristic DNA maps (see Fig. 1-1), all strains clearly belong to a distinct species of *Orthopoxvirus*. Biological characteristics that are useful for species diagnosis are the broad host range (shared only with cowpox and monkeypox viruses), the rapid growth of large pocks on the chorioallantoic membrane, and the high ceiling temperature for growth on the chorioallantoic membrane.

Vaccinia virus used for vaccine production has at various times been thought to have been derived from either cowpox virus or variola virus (see below). Table 5-1 summarizes some of the differences between these species in terms of their biological characteristics. In addition to those listed, vaccinia virus grows more rapidly on the chorioallantoic membrane (and in cultured cells) than other orthopoxviruses, so that pocks produced by vaccinia virus are about 2 mm in diameter after 2 days, at a time when pocks of other species are very much smaller.

Restriction endonuclease maps of strains of vaccinia virus from different sources resemble each other much more closely than they

TABLE 5-1

Biological Characteristics of Vaccinia, Cowpox, and Variola Viruses[a]

	Virus		
Characteristic	Variola	Vaccinia	Cowpox
Pock on chorioallantoic membrane	Small opaque white	Strains vary; large opaque white or ulcerated	Large, bright red
Ceiling temperature	37.5–38.5°C	41°C	40°C
Skin lesion after intradermal inoculation of rabbit	Nil or small nodule, nontransmissible	Strains vary; indurated, sometimes hemorrhagic nodule	Large, indurated, hemorrhagic nodule

[a] Unequivocal species diagnosis can be made by the examination of electropherograms produced after restriction endonuclease digestion of viral DNAs.

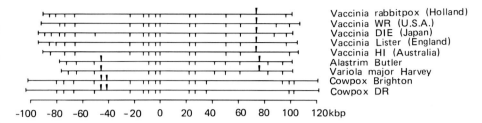

Vaccinia rabbitpox (Holland)
Vaccinia WR (U.S.A.)
Vaccinia DIE (Japan)
Vaccinia Lister (England)
Vaccinia HI (Australia)
Alastrim Butler
Variola major Harvey
Cowpox Brighton
Cowpox DR

FIG. 5-1. *Comparison of the cleavage sites after digestion with* XhoI *and* SmaI *of DNA from several strains of vaccinia virus and DNA from the two species of virus from which it has been suspected to have been derived, cowpox and variola viruses. (From Mackett and Archard, 1979.)*

resemble those of other species of orthopoxvirus (Fig. 5-1; see also Fig. 1-1). However, it is not necessary to map the DNA to differentiate vaccinia virus from other orthopoxviruses; satisfactory comparisons can be made by comparing electropherograms of fragments produced after digestion with *Hin*dIII or *Pst*I (see Plate 5-7).

HISTORY OF SMALLPOX VACCINE

The novelty of Jenner's discovery and promotion of the use of a related virus, derived from another species of animal, for protection against smallpox (Jenner, 1798) is apparent when it is realized that it was not until the 1880s that Louis Pasteur introduced the next immunizing agents (including "fixed" rabies virus), and not until the late 1930s that the third live virus vaccine was developed for human use—17D yellow fever vaccine (Theiler and Smith, 1937). Over the years the method of production of smallpox vaccine evolved from serial passage in humans to a quality-controlled freeze-dried vaccine, but until the completion of the smallpox eradication campaign in 1980 smallpox vaccine production retained features that would not have been tolerated in a newly developed vaccine, notably the unavoidable bacterial contamination of vaccine lymph prepared in animal skin.

Jenner's Discovery

Although others had inoculated children with material from cowpox lesions to protect them from smallpox before 1796 (Dixon, 1962; Baxby, 1981), Jenner's demonstration of its efficacy by challenge inoculation with variolous material, his recommendation for maintenance of the virus by serial passage in humans, his recognition of "spurious"

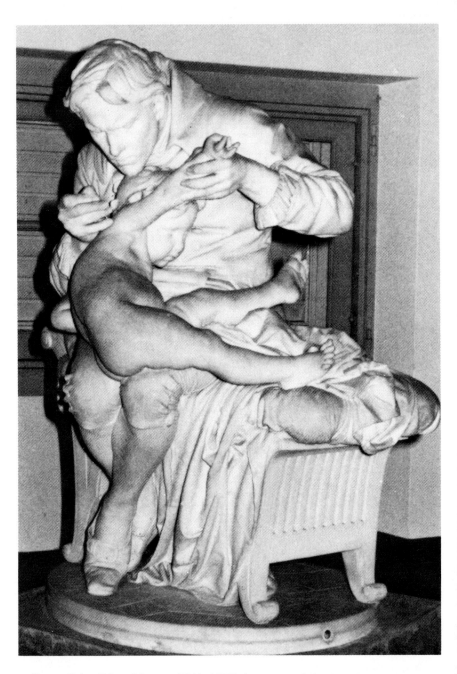

PLATE 5-1. *Edward Jenner (1749–1823) demonstrated that inoculation of cowpox virus ("the vaccine") would protect against smallpox and popularized its use. (Sculpture by Giulo Monteverde, 1878, in the Palazzo Bianco, Genoa. Photograph courtesy of Dr. S. Kato.)*

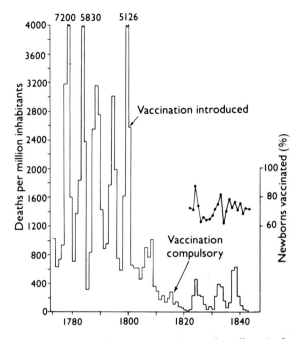

FIG. 5-2. *The effect of vaccination on the incidence of smallpox in Sweden between 1722 and 1843, showing from 1820 onward the proportion of newborn children who were vaccinated in infancy. (From Fenner et al., 1988.)*

cowpox, and his vigorous promotion of the practice justify full credit for the discovery of vaccination being given to him (Plate 5-1).

The discovery was timely, for smallpox was then a major scourge in every country of Europe and one of Europe's most feared exports to North America. Within 5 years, Jenner's *Inquiry* was translated into most European languages, "vaccination institutes" had been established in many countries, and the vaccine had been transported to every continent. The most celebrated distribution was the Balmis–Salvany voyage to New Spain, the Philippines, and Macao, commissioned by King Carlos V of Spain (Smith, 1974). Although there were many bitter attacks on Jenner, and much opposition to vaccination, the ameliorative effects of vaccination were too dramatic to ignore (Fenner *et al.,* 1988; Fig. 5-2).

Production in the Prebacteriological Era

Jenner's discovery antedated the recognition of viruses by a century. Threads covered with dried vaccine were sent from England to physicians in every country in Europe and indigenous sources were found, cases of cowpox and horsepox yielding material that was then carried for several passages in children, who supplied virus for individual physicians to vaccinate others. In some parts of Europe rewards were offered for the discovery of cases of cows with cowpox, in order to augment supplies of vaccine.

Serial passage in calves was used as a source of virus in Italy from 1840 and the practice was taken up in France in 1864, stimulated by the fear of the transmission of vaccinal syphilis from human sources of vaccine. Calves were not used in England until 1881 and arm-to-arm vaccination was popular there until it was banned in 1898.

Production Methods during the Nineteenth Century

Calves were inoculated in multiple sites from which material was reaped when the lesions were judged to be "ripe" and the "pulp" was ground in a mortar before being suspended in diluent as vaccine "lymph." Glycerine was introduced as the diluent in Italy at an early date and its use spread, especially after Copeman (1892) had demonstrated that at low temperature it was bactericidal but not virucidal. Glycerine was still used for the preparation of liquid vaccine in the 1970s, in the few places where this product was still manufactured.

By the end of the nineteenth century there was a multitude of small "backyard" factories in most countries in Europe, each supplying vaccine to a small clientele. In the Netherlands, for example, there were some 15 "parcs vaccinogène" in 1875, and Hime (1896) complained that "The country [England] is flooded with cheap stuff 'made in Germany' and elsewhere, of unknown nature and origin. It is cheap and therefore sells."

It is clear that with such a history "vaccine virus" included many different strains, with properties that differed according to their origin and their passage history. Yet all that have been adequately tested can be classified as belonging to the species "vaccinia virus" by both biological characterization and restriction endonuclease mapping of their DNA.

Development of Freeze-Dried Vaccine

Glycerinated lymph was reasonably stable at refrigerator temperatures and the eradication of smallpox from Europe was accomplished with liquid vaccine. But it was quite unsatisfactory in tropical countries, where there was an inadequate public health infrastructure and few refrigeration facilities. From Jenner's day air-dried vaccine had been used for despatch of virus to distant places. Otten (1927), in Java, improved on this by drying vaccine *in vacuo,* but the material was often heavily contaminated and it was difficult to reconstitute. The solution to the problem, embarked upon independently in France, the United States, and England, was the introduction of freeze-drying. The method that was finally adopted worldwide was based on the work of Collier (1954, 1955).

Vaccine Production in the Intensified Smallpox Eradication Program

Global eradication of smallpox was adopted as a goal by the World Health Organization in 1959, and increased amounts of vaccine were supplied to some of the endemic countries of Asia. However, no funding was provided and little was achieved until the Intensified Smallpox Eradication Program was initiated in 1967 (see Chapter 11). One of the first actions of the Intensified Program was to survey smallpox vaccine production facilities throughout the world (Fenner *et al.*, 1988). In 1967 information was obtained from 67 producers in 45 countries. Most of these producers were using calves, a few sheep, and some water buffalo for vaccine production; three were producing vaccine in cultured cells or developing eggs. The major changes since the nineteenth century were that the animals were kept in more hygienic conditions and high titer seed vaccine was applied by scarification (Plate 5-2), rather than by inoculation.

Seventeen of the 67 producers were then using Lister strain vaccine, 5 the New York City Board of Health strain, and at least 10 other strains of vaccinia virus were used in various laboratories. Tests of the vaccine produced showed that less than half of the samples came up to the standards of potency and stability of freeze-dried vaccine that had been laid down by the World Health Organization in 1959.

A program was therefore established by the World Health Organization for upgrading vaccine production and ensuring that WHO standards for vaccine quality were met for all vaccine used in the Intensified Smallpox Eradication Program. By 1973 most vaccine producers, in developing as well as developed countries, were producing satisfactory vaccine. Many had changed from the strain previously used to the Lister strain, because of its lower encephalitogenic potential and the fact that it was easy to titrate on the chorioallantoic membrane.

Vaccination Techniques

Scarification. Jenner inoculated human subjects with vaccine virus by the same method then in use in England for variolation, viz., by a light scratch of the skin through a drop of vaccine, which introduced

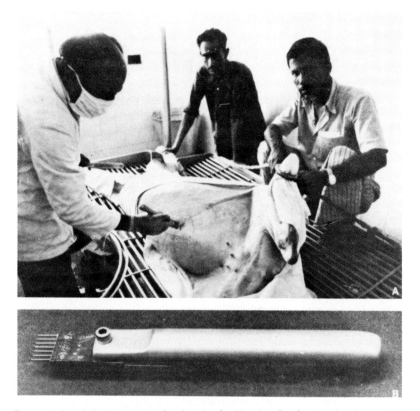

PLATE 5-2. *(A) Vaccine production in the Vaccine Production Institute, Dhaka, Bangladesh, during the Intensified Smallpox Eradication Program, using a scarifier developed by the Wyeth Laboratories. (B) Enlarged view of the scarifier. The only changes since the mid-nineteenth century were the more hygienic maintenance of the calves and the use of a scarifier. (From Fenner et al., 1988, courtesy World Health Organization.)*

virus into the epidermal cells. With variations in the depth of the "insertion" and the number of sites (from one to four, usually over the deltoid muscle) this method remained in vogue for the next 150 years. In 1927 Leake introduced a "multiple pressure" method that introduced vaccine into the epidermis with less chance of subdermal inoculation, and this method was widely adopted. In India a special instrument called a "rotary lancet," which produced severe local damage to the skin, was widely used.

Jet Injection. As the worldwide campaign for smallpox eradication got under way tests were begun for a more "up-to-date" method of

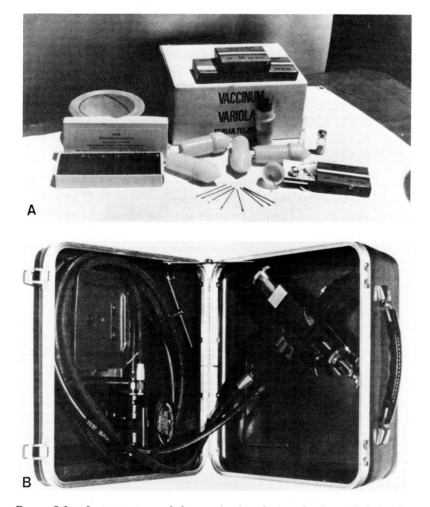

PLATE 5-3. *Instruments used for vaccination during the Intensified Smallpox Eradication Program. (A) Bifurcated needle and plastic container. The top of the container screwed off so that it could be packed with needles. The holes in the bottom allow excess water to be removed after sterilization by boiling. Sterile needles are delivered one at a time through the hole in the conical lid, by shaking. (B) Ped-o-Jet jet injector, assembled in case for transport. (Courtesy World Health Organization.)*

delivery of vaccine on a mass scale and the jet injector (Plate 5-3), introduced in the 1950s for mass immunization by subcutaneous inoculation, was adapted for intradermal delivery of smallpox vaccine. Jet injectors were used extensively in the eradication campaigns in West Africa and Brazil, but elsewhere the need to assemble large numbers of vaccinees in one place, and difficulties of maintenance of the instruments, militated against their use.

Bifurcated Needle. The major advance in vaccination technique during the global eradicaton campaign was the development of the bifurcated needle, which carried the requisite quantity of vaccine between its prongs, to be introduced into the dermis by pricking the skin some 15 times with the needle, at right angles to the skin; "multiple puncture" rather than "multiple pressure." The technique could be learned rapidly even by illiterate persons, and it could be applied to small numbers of vaccinees in rural areas. An ingenious container was developed for carrying and dispensing sterile needles for each day's use and boiling them after each day's work (see Plate 5-3). The bifurcated needle used about half as much vaccine as was needed for scarification and it was a much more reliable technique, even when used by newly recruited and inexperienced vaccinators.

Interpretation of the Results of Vaccination

In susceptible individuals, smallpox vaccination produces a typical Jennerian pustule at the inoculation site and frequently swelling and tenderness of the draining lymph node. At the height of the reaction there is usually slight fever and the subject may feel miserable for a few days. A feature of smallpox vaccination, which among the variety of agents now used for immunization against infectious diseases is shared only with BCG vaccine, is that successful vaccination produces a characteristic skin reaction which can be readily observed and which usually leaves a permanent and characteristic scar. This has both immediate and long-term consequences. Observation of the nature of the cutaneous lesion after recent vaccination or revaccination enables the vaccinator to decide whether the virus has multiplied and the patient thus rendered immune to smallpox. In the longer term, the vacinnation status of an individual or a population can be determined, with considerable accuracy, by visual examination for vaccination scars, rather than requiring a serological survey. For these reasons, special attention was devoted to the reactions in the skin after both primary vaccination and revaccination.

Reactions to Primary Vaccination. A typical Jennerian pustule was termed a "major reaction," and constituted evidence that the vaccinee would be protected against smallpox. The course of the reaction is illustrated in Plate 5-4A. A papule appears at the vaccination site on the third day after vaccination and within 2 or 3 days this becomes vesicular,

DAY 0	DAY 3	DAY 7	DAY 10	DAY 14

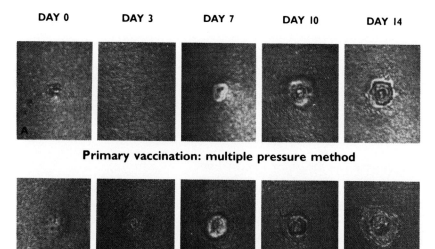

Primary vaccination: multiple pressure method

Late revaccination: multiple pressure method

PLATE 5-4. *Skin reactions after primary vaccination and late revaccination (several years after primary vaccination) by the multiple pressure method. (From Fenner et al., 1988.)*

to constitute the umbilicated and loculated "Jennerian vesicle." The vesicle soon becomes pustular due mainly to the entry of polymorphonuclear cells, the migration of which is stimulated by the viral infection itself, and the surrounding area becomes erythematous and indurated to a much greater extent than is found in the skin lesions of smallpox. The area of erythema reaches a maximum between the eighth and the twelvth day, and at this time the draining lymph nodes are enlarged and tender and the patient often sustains a mild fever and may feel miserable. The pustule dries from the center out to become a dry brown or black scab which falls off about 3 weeks after vaccination, to leave a typical pitted scar.

Observation of the pustule on the seventh day confirms whether vaccination has been successful. The reaction to vaccinia virus can be readily distinguished from reactions due to bacterial infection, both by its time course and its characteristic appearance.

Revaccination. The response to revaccination depended on the potency of the vaccine and the interval since primary vaccination. Except when a long interval had elapsed, it always progressed more rapidly than the lesion produced by primary vaccination (Plate 5-4B). Interpretation of the results of revaccination was sometimes difficult, in terms of evaluating their significance in relation to protection against smallpox. Sometimes there was no reaction at all, a result that was usually due to the use of vaccine of low potency, and which was impossible to interpret correctly (WHO Expert Committee on Smallpox, 1964). If it did occur, the reaction to revaccination could be maximal at any time between the second and the eighth day.

Complications of Smallpox Vaccination

Compared with variolation, which it replaced, vaccination produced a much milder reaction, both local (Plate 5-5) and general; nor could it cause smallpox in unvaccinated contacts. However, smallpox vaccination involved infection of man with a virus that had not been deliberately attenuated. Not only did a successful vaccination always produce a local lesion and some constitutional disturbance, but as Sir Graham Wilson has said, "Smallpox vaccine has probably been followed by more complications and been responsible for more deaths than any other vaccine" (Wilson, 1967; see also Chapter 12).

The complications fell into two groups; those affecting the skin (see

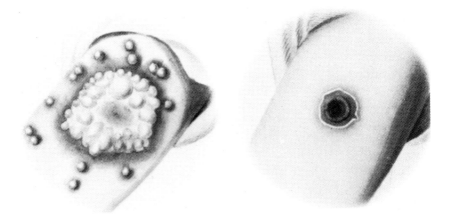

PLATE 5-5. *Local lesions 14 days after variolation (left) and vaccination (right). The local lesion was much more severe after variolation, there were always satellite pustules, and there was often a generalized rash. The degree of erythema around the vaccination lesion recalls vaccinia virus rather than cowpox virus. Engravings made by G. Kirtland (1806) from paintings by R. Gold (1802).*

Plate 12-1) and those affecting the central nervous system. The first category includes two types of disease, neither of which should have been allowed to occur, because the predisposing conditions were contraindications to vaccination. The most common was eczema vaccinatum, which concsisted of extensive skin lesions in eczematous subjects. This also occurred in eczematous family contacts of vaccinated subjects. The other response (progressive vaccinia) was found only in subjects with a serious impairment of their cell-mediated immune system of either genetic or acquired origin (immunosuppressive drugs, lymphoreticular malignancy).

The most important group of complications were diseases of the central nervous system, which included encephalopathy, a rare but serious disease in infants, and postvaccinial encephalitis, a syndrome in older persons associated with demyelination, with a case–fatality rate of about 30%. It was impossible to predict the occurrence of either of these diseases, but statistically their frequency was correlated with the virulence of the virus. Postinfection encephalomyelitis also occurred in smallpox, with an incidence of about 0.2% in variola major and 0.05% in variola minor. The incidence after vaccination has varied greatly at different places and times (Table 5-2). The complications of vaccination with vaccinia virus remain a concern for scientists and public health officials interested in immunization based on the use of vaccinia virus as a vector for genes of other infectious agents; for this reason they are discussed in somewhat greater detail in Chapter 12.

TABLE 5-2

Postinfection Encephalitis Associated with Smallpox and Primary Vaccination[a,b]

Infection	Cases per million	Reference
Variola major	2000	Rao (1972)
Variola minor	500	Marsden (1936)
Primary vaccination		
Holland (1924–1943)	258	van den Berg (1946); Stuart (1947)
Bavaria (1945–1953)	121	Herrlich (1954)
United States (1968)	9	Lane et al. (1970)
England and Wales (1951–1970)	10	Cited by Fenner et al. (1988)

[a] In persons over 2 years of age.
[b] Postvaccinial encephalitis is virtually unknown after revaccination.

VARIANTS OF VACCINIA VIRUS

Differences between Strains Used for Vaccines

When the Intensified Smallpox Eradication Campaign began in 1967 it was found that there were 67 establishments, in various countries of the world, producing smallpox vaccine (Fenner *et al.*, 1988). This was already a great reduction on the situation in Europe during the nineteenth century, when many physicians maintained their own strains by arm-to-arm vaccination, with occasional recourse to "retrovaccination" (i.e., back bassage in cows) or new material from naturally infected cows, and when there were numerous small private vaccine-producing factories in many countries.

Strains of vaccinia virus in use for the production of smallpox vaccine as late as the 1960s varied considerably in their virulence for man, as determined by the severity of the local skin lesion and associated constitutional symptoms. Some strains, e.g., Tashkent, were too severe in their effects for use in mass vaccination campaigns; others, e.g., certain tissue culture-derived strains (Rivers *et al.*, 1939), produced too mild a reaction to promote the level of immunity required in smallpox-endemic countries (Galasso *et al.*, 1977). There was also a correlation between the frequency of the one unpredictable serious complication of vaccination, postvaccinial encephalomyelitis, and the strain of virus employed (see Table 5-2).

Neurovaccinia and Dermal Vaccinia Strains

Two terms occur in older works on vaccinia virus that need some explanation: "dermal vaccinia" and "neurovaccinia." Early workers usually maintained vaccinia virus by passage through calves, sheep, or rabbits, the animals being inoculated by scarification (see van Rooyen and Rhodes, 1948). When the skin lesions reached a sufficient size the infected skin area was scraped, the material thus obtained being called "dermal vaccinia" or "dermovaccine." Virus passaged in this way usually produces white pocks on the chorioallantoic membrane and erythematous, nonulcerating nodules when inoculated intradermally in rabbits. However, some strains produce mixtures of white and grayish pocks. In the period 1920–1940 a great deal of experimental work was carried out with vaccinia virus that had been maintained by intracerebral inoculation of rabbits, sometimes with occasional testicular passage (Levaditi *et al.*, 1922, 1938). Such strains produced encephalitis in rabbits, and they were called "neurovaccinia" virus. This early work suffered from a failure to use cloning techniques and to realize that the many different stocks of vaccinia virus maintained by serial passage in different laboratories had undergone different kinds of selective pressure before any deliberate experimental procedures were begun. One important aspect of neurovaccinia virus was that some strains could establish an enzootic disease in rabbit colonies, often with substantial mortality (Greene, 1933); these were often called "rabbitpox virus." It is characteristic of such strains that they produce ulcerated, slightly hemorrhagic pocks on the chorioallantoic membrane (see Plate 5-6), large indurated lesions with a purple center after intradermal inoculation in rabbits, and are more virulent than dermal vaccine strains. Restriction endonuclease digestion reveal that both neurovaccinia (rabbitpox) and dermovaccinia strains have very similar DNA maps (Wittek *et al.*, 1977).

Differences between Laboratory Strains of Vaccinia Virus

Two systematic studies have been made of the biological characteristics of laboratory strains of vaccinia virus (Fenner, 1958; Ghendon and Chernos, 1964). Both workers found that some of the strains tested produced pocks of different morphology when grown on the chorioallantoic membrane. A variety of other differences in biological behavior were found, involving production of hemagglutinin, heat resistance of the virion, and pathogenicity in rabbits and mice (Fenner, 1958; Table 5-3), and plaque morphology and virulence for monkeys (Ghendon and Chernos, 1964). Of the strains examined by Fenner, using only a limited number of biological characteristics, only two (Gillard and Connaught) were identical. The traditional divisions of strains into "dermovaccinia" and "neurovaccinia" broadly differentiated viruses of lower and higher virulence; in particular the occurrence of hemorrhagic pocks on the chorioallantoic membrane was correlated with large indurated skin lesions with a purple center after the intradermal inoculation of rabbits. White pock mutants of the "neurovaccinia" strains produced small pink nodules in the rabbit skin.

Some strains or mutants of vaccinia virus fail to produce hemagglutinin and do not provoke the production of hemagglutinin-inhibiting antibodies in infected animals. Rabbitpox virus (Utrecht strain) is one example (Fenner, 1958). Another HA$^-$ variant (IHD-W) produces a nonglycosylated form of the 89K polypeptide, the glycosylated form of which Payne (1979) identified as the hemagglutinin. Infection of rabbits with this mutant did not evoke the production of hemagglutinin-inhibiting antibodies, suggesting that glycosylation must produce important conformational changes in the secondary structure.

TABLE 5-3

Some Biological Properties of Several Different Laboratory Strains of Vaccinia Virus[a]

Strain	Pocks on chorioallantoic membrane	Hemagglutinin production	Heat resistance of infectivity	Mouse virulence[b]	Rabbit virulence[b]	Skin lesions in rabbit
Gillard	Opaque white	+	High	−	−	Small pink nodule
Connaught	Opaque white	+	High	−	−	Small pink nodule
Mill Hill	Opaque white	+	High	+	−	Small pink nodule
Lederle-7N	Opaque white	+	Low	−	−	Small pink nodule
Nelson	Opaque white	+	Moderate	++	−	—
Williamsport	Opaque white	+	High	++	++	Small pink nodule
Pasteur	Opaque white hemorrhagic center	+	High	+	++	Large nodule with purple center
IHD	Pale with hemorrhagic center	+	High	+++	+++	Large nodule with purple center
Rabbitpox Utrecht	Pale with hemorrhagic center	−	High	+++	+++	Large nodule with purple center
Rabbitpox Rockefeller	Pale with hemorrhagic center	+	High	+++	−	Large nodule with purple center

[a] From Fenner (1958).
[b] After intracerebral inoculation.

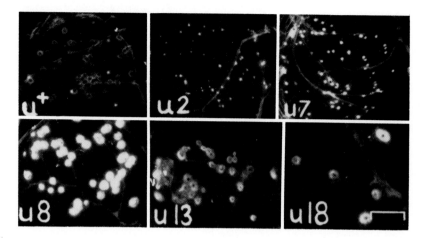

PLATE 5-6. *Pocks produced by white pock mutants of rabbitpox virus, showing their diverse appearance. u⁺, Wild type; u2, etc., various white pock mutants. Bar = 1 cm. (From Gemmell and Fenner, 1960.)*

Variation within One Strain of Vaccinia Virus

Several investigators have shown that uncloned stocks of vaccinia virus are in fact mixtures of genetically dissimilar virions. The most obvious examples of genetically mixed stock cultures were revealed by examination of the pocks produced by various commercial strains of vaccinia virus. Fenner (1958) found that the vaccine produced by Lederle (Led O) produced a mixture of grayish and white pocks, which could be easily separated by cloning, and Ghendon and Chernos (1964) found 2 obviously mixed strains of vaccinia virus among the 10 that they examined. Using other methods of study, stocks of vaccinia virus which apper to be homogeneous when the pock character is examined may be shown to be mixed, either in the plaques produced on certain kinds of cells (Ghendon and Chernos, 1964) or by heterogeneity in the patterns produced after digestion with restriction enzymes (Wittek *et al.*, 1978a). The latter investigators showed that whereas an uncloned stock of a standard vaccine strain produced heterogeneous end fragments when cleaved with selected restriction enzymes, this heterogeneity could be eliminated by cloning the virus.

Strains of vaccinia virus which produce hemorrhagic pocks yield, on cloning, a substantial proportion of white nonulcerated pocks (up to 1%; Fenner, 1958). Gemmell and Fenner (1960) recovered 18 white pock mutants of rabbitpox virus, all of which were different. The pock morphology of many of these mutants was distinctive (Plate 5-6).

Recombination between certain pairs of white pock mutants was used to produce a primitive "genetic map" of rabbitpox virus (Gemmell and Cairns, 1959; Gemmell and Fenner, 1960). Subsequently Fenner and Sambrook (1966) found that 16 of a group of 34 white pock mutants were host cell restricted, and that a further 16 mutants recovered by Sambrook *et al.* (1966) on the basis of their host cell restriction all produced white pocks. Certain of these mutants were subsequently used by Lake and Cooper (1980) and Moyer and his colleagues for the study of the genome changes responsible for the white pock and host cell-restricted phenotypes (see Chapter 2).

SOURCES OF VACCINIA VIRUS STRAINS USED FOR SMALLPOX VACCINE

The virus originally used by Jenner (1798) to protect human subjects against smallpox was obtained from a milkmaid infected with "cowpox." As well as demonstrating that it provided protection with much less constitutional disturbance than did the practice of variolation, and without the risk of transmitting smallpox to others, Jenner suggested that supplies could be maintained by serial arm-to-arm passage in children. This method was long used as the source of virus for vaccination, especially in England (Dudgeon, 1963), but periodically new strains were obtained from cows or from lesions of "horsepox" (Fleming, 1880a,b; see Chapter 6). It is impossible to know what viral species were involved, except that variola virus can be excluded, although throughout the nineteenth and early twentieth centuries it was widely believed that "variolae vaccinae" (as Jenner called it) had been derived from smallpox virus.

It was not until 1938–1939 that Downie (1939a) showed that cowpox virus differed in many biological properties from the material then being used for smallpox vaccination (vaccinia virus), a finding that found support from studies of DNA mapping (see Figs. 5-1 and 6-1). Although the studies are far from complete, no strain of virus used for vaccine production during the last 20 years has been characterized as cowpox virus; all have had the biological characteristics of vaccinia virus. However, the Evans strain produced very hemorrhagic pocks on the chorioallantoic membrane, with very little infiltration (K. R. Dumbell and K. McCarthy, unpublished observations), but unfortunately it has not been possible to examine its DNA. The "Chambon-St. Yves Menard strain," which Dr. F. Fasquelle (personal communication to Dr. M. Soekawa, January 25, 1972) claimed "possessed the characteristic properties of cowpox according to Downie," is said by Dr. R. Netter (personal communication, 1982), to have "never exhibited any cowpox aspect" in tests carried out in his laboratory during the previous 20 years.

If at least some strains of smallpox vaccine virus were indeed derived from cowpox virus, when were such strains replaced by vaccinia virus? We can only speculate, but there are some hints. The first depends on the very prominent regular-shaped eosinophilic A-type inclusion bodies produced by cowpox virus but not by vaccinia virus. If cowpox virus was used for vaccine production at the time when pathologists first described poxvirus inclusion bodies (Guarnieri, 1892; Prowazek, 1905), they would have recognized the A-type inclusions. However, the only type of inclusion body described in tissues infected with variola virus and "vaccine virus" is the irregular B-type Guarnieri body. The second clue depends on the fact that cowpox virus does not occur naturally in North America. Vaccine strains were repeatedly sent to the United States from England, and the strain long used for vaccination in that country (New York City Board of Health strain) is said to be a lineal derivative of a strain first used for vaccine production in 1876, but imported from England in 1856 (Berg and Stevens, 1971). This strain is classical vaccinia virus, not cowpox virus, which suggests that vaccinia virus was being used for vaccination at least as early as 1856.

THE ORIGIN OF VACCINIA VIRUS

Isolation from Natural Sources

Strains of vaccinia virus have been isolated from skin lesions of domestic animals on many occasions (see Table 1-4), usually associated with contact with vaccinated humans. However, some isolations of vaccinia virus have been made in circumstances in which there had been no known contact with persons who had been recently vaccinated, as in some outbreaks of rabbitpox (Jansen, 1946; Christensen et al., 1967) and, more recently, buffalopox in India (see below). Some cases of "cowpox" in cattle (Dekking, 1964; El Dahaby et al., 1966; Maltseva et al., 1966; Lum et al., 1967; Topciu et al., 1976) have been due to vaccinia virus acquired from vaccinated humans, as were other outbreaks of buffalopox in both India (Baxby and Hill, 1971; Lal and Singh, 1977) and Egypt (Tantawi et al., 1977; Iwad et al., 1981) and possibly some cases of "camelpox" (Krupenko, 1972). During periods when vaccination of humans against smallpox was being vigorously pursued there were clearly numerous opportunities for infection to be transferred from recently vaccinated persons to various domestic animals, with subsequent spread in herds either by milkers acting as vectors, or by some other route.

Apart from speculations about horsepox (Baxby, 1981) and reports on buffalopox and rabbitpox (see below), the only suggestions that vaccinia virus might occur as a natural infection of animals were the recovery of strain MK-10-73 from the kidney of a sala monkey (*Cercopithecus ascanius*) shot in Zaire (Shelukhina *et al.*, 1975), at the time "whitepox" viruses were being recovered from organs of other wild animals (see Chapter 7), and of strain 65.3993 in cynomolgus kidney cell culture (Gispen and Kapsenberg, 1966). MK-10-73 is a strain of vaccinia virus; it seems unlikely that it occurred as a natural infection of a wild animal in Zaire. In a personal communication (1983), Dr. J. G. Kapsenberg noted that strain 65.3993 was a strain of vaccinia virus present in vesicle fluid together with herpesvirus, of unknown origin; she thought that it was unlikely to have originated from the kidney of a cynomolgus monkey.

Two other strains of vaccinia virus, "Lenny" (Bourke and Dumbell, 1972) and "Radebe," were recovered from persons with a severe papulovesicular rash, who had had no known exposure to smallpox vaccine and had died shortly after admission to hospital in Nigeria and South Africa, respectively. They differed from standard vaccinia virus strains in some biological characteristics, giving rise to the suggestion that they were recombinants between vaccinia virus and variola virus, or that they may have been circulating in nature independently of vaccination. Restriction endonuclease digestion of their DNA (Carra and Dumbell, 1987) showed that each resembled the locally used vaccine strains, Wyeth and South African smallpox vaccine, respectively.

Buffalopox. Buffaloes are used for milk production in many tropical countries and pock lesions on the teats have been reported from several of these countries: India, Bangladesh, Pakistan, Indonesia, USSR, and Egypt (for review, see Lal and Singh, 1977), most commonly in India (Singh and Singh, 1967; Ghosh *et al.*, 1977) and Egypt (Tantawi *et al.*, 1977; Iwad *et al.*, 1981). Occasionally skin lesions occur on other sites, such as the perineum, the medial aspects of the thighs, the ears, and the eyes. About 20% of untreated cases suffer from mastitis. Calves that suckle from infected teats may get severe lesions of the tongue and lips which may be fatal because of interference with feeding.

Epidemiologically, the disease appears to be spread by milkers and infection of their hands is common. In one outbreak, for example, 70% of the milkers had lesions on the hands and forearms, even though most of them had been vaccinated (Ghosh *et al.*, 1977). Markets and fairs at which animals are exchanged and traded may serve to spread infection. Most observers agree that the infection did not spread from buffaloes to cows that were in close contact with them. Some outbreaks were ascribed to infection of the buffaloes by milkers who had been recently vaccinated (e.g., in USSR: Ganiev and Ferzaliev, 1964) and routine vaccination against smallpox has been practiced in all countries where the disease has been reported, so that there have always been opportunities for buffaloes to be infected from vaccinated humans.

Although long suspected to be a poxvirus infection (Sharma, 1934), the causative agent was first isolated and characterized by Singh and Singh (1967) and has been further studied by Kataria and Singh (1970), Baxby and Hill (1971), Lal and Singh (1973), Mathew (1976), Sehgal *et al.* (1977), and Tantawi *et al.* (1977). Most strains that have been examined cannot be differentiated from vaccinia virus, although several workers have reported that vaccinia virus and buffalopox virus, which cross-react strongly in neutralization and complement-fixation tests, show higher titers with homologous than heterologus antisera.

The best supported claim for the designation of buffalopox virus as a separate species is that made by Baxby and Hill (1971). After examining four Indian strains by a variety of methods, they suggested that three of them could not be differentiated from vaccinia virus, but a fourth ("virus A") had some distinctive biological characteristics. Besides the serological differences described by Lal and Singh (1973), this virus had a lower ceiling temperature than vaccinia virus (38.5°C compared with 41°C), produced only a small nodule after intradermal inoculation in rabbits, and produced much smaller plaques in RK13 cells than vaccinia virus or the other three strains of "buffalopox virus." However, restriction mapping shows that virus A is vaccinia virus (M. Richardson and K. R. Dumbell, unpublished results, 1987).

With the cessation of routine vaccination it was expected that buffalopox would disappear, but in July 1985 advice was received from India that outbreaks were still occurring in several districts, and that some severe infections were occurring in persons who milked the buffaloes. Further enquiries led to the availability of scab specimens from 13 outbreaks occurring in Maharashtra state between December 1985 and May 1986, from which 9 orthopoxviruses were isolated (M. Richardson and K. R. Dumbell, unpublished results, 1987). One isolate had a biotype fully consistent with that of vaccinia. The other eight resembled the virus A of Baxby and Hill (1971), in that they had a lower ceiling temperature and produced smaller plaques than did typical vaccinia virus. Two were tested in rabbits and produced only a small nodule after intradermal inoculation. Anti-vaccinia rabbit serum had a significantly higher neutralization titer against vaccinia virus than against the isolates from the buffaloes. DNA from each of the isolates and from virus A had a *Hin*dIII digestion profile characteristic of vaccinia virus. All the Maharashtra isolates (except the one with the vaccinia biotype) gave identical fragment profiles in *Pst* 1 digests and this profile differed from

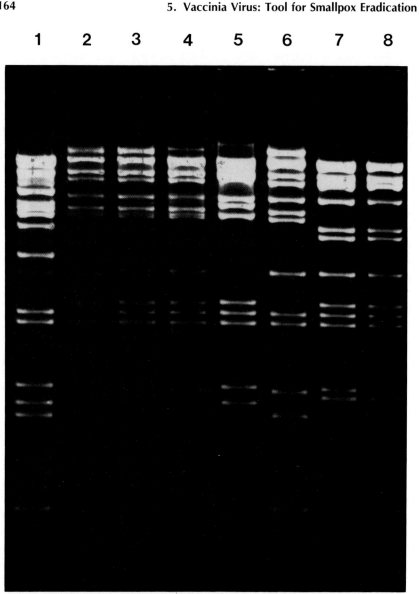

PLATE 5-7. *Electropherograms of DNA fragments obtained by digestion with Pst1 of samples of buffalopox virus and vaccinia virus genomic DNA. Origins of DNA: 1, 2, 3, 4, vaccinia virus strains: Dairen, King Institute, Patwadangar, USSR; 5, buffalopox virus strain A (Baxby and Hill, 1971); 6, buffalopox virus Maharashtra 1985 (vaccinia virus biotype); 7, 8, buffalopox virus Maharashtra 1985 strains B3 and B4 (buffalopox virus biotype.) (M. Richardson and K. R. Dumbell, unpublished observations, 1987.)*

that of virus A and from those of three smallpox vaccine strains previously used in India (Plate 5-7). Thus a number of recent buffalopox isolates from different parts of Maharashtra state shared common characteristics and the total picture was sufficiently individual to justify designation as a separate viral entity, which may appropriately carry the name buffalopox virus. This point is highlighted by the one isolate which was fully typical of vaccinia virus. Buffalopox virus is so closely related to vaccinia virus that it must be regarded as a subspecies of vaccinia virus rather than as a separate species and it seems most likely that smallpox vaccine was the original source of these outbreaks. Nevertheless the recent isolates have a biotype similar to that of the original buffalopox virus characterized by Baxby and Hill (1971) and the virus appears to be capable of sustained transmission some years after the general use of smallpox vaccine has been stopped. The natural history of buffalopox virus has not been fully investigated, but it may well be maintained in a manner similar to that in which cowpox virus is maintained in Europe (see Chapter 6).

Rabbitpox. Aspects of rabbitpox of interest to veterinarians have been reviewed in Fenner (1988). The association of "rabbitpox" with vaccinia virus is clearer than that of buffalopox. The name was first used to designate a pox disease caused in a colony of laboratory rabbits in New York in 1932–1934 which was known to have been caused by a strain of vaccinia virus, and which subsequently was called "rabbitpox virus" (Greene, 1933; Rosahn and Hu, 1935). In 1941 a similar virus caused outbreaks in laboratory rabbits in Utrecht, which killed most rabbits before skin lesions developed (Jansen, 1946, 1962); in this case there was no clear association with smallpox vaccination or vaccinia virus. Subsequently the Utrecht strain of rabbitpox virus was used extensively for genetic studies on orthopoxviruses (see above), and for investigation of the pathogenesis of generalized poxvirus infections (see Chapter 4). Its biological characters (Fenner, 1958) and its genome map (Wittek *et al.*, 1977) clearly identify the Utrecht strain of rabbitpox virus as vaccinia virus.

Rabbitpox appears to be due to the infection of laboratory rabbits with "neurovaccinia" variants of vaccinia virus. Two types of situation have been described: "latent" infection within the colony, with few or no symptoms until the animals were stressed in some way, and acute epizootics of severe disease.

Verlinde and Wensinck (1951) described a situation in which apparently healthy rabbits died after being inoculated intracerebrally with a variety of indifferent materials, and neurovaccinia virus was recovered from their brains. In fact the rabbits were not completely healthy, for subsequently an epizootic occurred in the stock rabbits, with symptoms of acute respiratory infection with conjunctivitis, keratitis, and orchitis and a case–fatality rate of 60%. Other investigators who had been working with neurovaccinia virus have described unnoticed infections among their laboratory rabbits which led either to immunity (Duran-Reynals, 1931; Levaditi and Sanchis-Bayani, 1927) or to clinical rabbitpox after animals were subjected to various nonspecific stimuli (Nicolau and Kopciowska, 1929; Levaditi et al., 1931).

However, attention has usually been directed to rabbitpox by epizootics of severe, acute disease. Greene (1934a,b, 1935) described a devastating outbreak in a breeding colony of rabbits in the Rockefeller Institute, which he likened to smallpox in man. Other scientists had been working with neurovaccinia virus in rabbits in an adjacent room prior to the outbreak. The virus recovered from the outbreak, designated rabbitpox-Rockefeller Institute, was regarded by Fenner (1958) as being very similar in its biological properties to neurovaccinia virus as developed by workers in the Pasteur Institute in the 1920s to 1930s.

The other major outbreak, from which the strain of "rabbitpox virus" (Utrecht) was derived, was described by Jansen (1941, 1946). It began among rabbits bought from a dealer a few days after they were introduced into the laboratory colony, and spread among the stock rabbits. The disease was usually lethal, death occurring before there was time for the development of a rash. The virus that caused this outbreak caused similar highly lethal epizootics when it "escaped" in the Institut Pasteur in Paris (Wang, 1947).

Another outbreak of rapidly lethal rabbitpox, without obvious skin lesions, but with a few papules on the lips and tongue ("pockless rabbitpox") occurred in the New York University Medical Center in 1965 (Christensen et al., 1967); the source was not discovered. A third epizootic in United States occurred shortly after this in the rabbit colony of the Bowman-Gray School of Medicine, and followed injection of some rabbits with "inactivated" rabbitpox virus, Utrecht strain, causing a very high mortality (Christensen et al., 1967).

In most of these outbreaks, spread appeared to occur by the respiratory route, and experiments by Bedson and Duckworth (1963) confirmed that infection occurred readily by this route. Rabbits infected by contact were not infectious for other rabbits until the second day of illness, which was usually 5 days after infection. Actual contact was not necessary; transmission could occur across the width of a room (Westwood et al., 1966). Air sampling revealed the presence of rabbitpox virus in the air of rooms housing infected rabbits (Thomas, 1970).

The clinical features of rabbitpox have been described by several authors (Greene, 1934a; Jansen, 1941, 1946; Bedson and Duckworth, 1963; Westwood et al., 1966). The Rockefeller Institute strain was somewhat less virulent than the Utrecht strain; more rabbits survived and deaths occurred later. Longer survival allowed time for pocks to appear on the skin and mucous membranes; these signs were almost always present in rabbits infected with the Rockefeller Institute strain (Greene, 1934a), but rabbits dying of hyperacute infection with the Utrecht strain showed no obvious skin lesions, although titrations showed that virus was present in the skin (Westwood et al., 1966).

Animals appeared normal during the incubation period of 4–6 days, but then the temperature rose, the animals became listless and lost weight, and profuse discharges occurred from the nose and eyes. The popliteal and inguinal lymph nodes were enlarged, and severe orchitis occurred in male rabbits. Skin lesions usually appeared about 5 days after infection, initially as an erythematous or macular rash only readily visible on shaved skin, but later developing into papules that varied in size from a few millimeters to about a centimeter in diameter. Papules occurred all over the body and on the mucous membranes of the oral and nasal cavities. Sometimes the skin lesions were hemorrhagic. Eye lesions varied from mild blepharitis to purulent conjunctivitis. Death usually occurred between the seventh and tenth day, or the animals recovered. The disease was particularly severe in pregnant does, which usually aborted.

As well as the severe infections with obvious clinical signs, Greene (1934a) noted that abortive infections occurred in some animals, and it is apparent from the descriptions of "latent" infections (see above) that sometimes strains of vaccinia virus that caused few clinical symptoms could become established enzootically in colonies of rabbits.

The Origin of Vaccinia Virus Used for Smallpox Vaccine

All strains of smallpox vaccine in use during the last 30 years, except possibly the Evans strain (see above), consist of vaccinia virus and not cowpox virus. The origin of these strains has long been disputed, and no certain answer is possible. Five hypotheses have been advanced:

1. Vaccinia virus was derived from variola virus by transfer and adaptation by passage in cows.
2. By continuous passage in human skin through variolation in the eighteenth and early nineteenth centuries, variola virus became attenuated and altered to become vaccinia virus (Razzell, 1977a).
3. Vaccinia is a hybrid between cowpox and variola viruses, occurring

in the early years of the nineteenth century when cowpox virus was used to vaccinate patients in smallpox hospitals (Bedson and Dumbell, 1964b).

4. Vaccinia virus is a "fossil" and represents the maintenance in the laboratory of a virus of a domestic or wild animal that has otherwise become extinct (Baxby, 1981).

5. Vaccinia virus was derived from cowpox virus by repeated passage on the skins of cows, sheep, and other animals.

Possible Derivation from Variola Virus by Passage in Cows. In his published paper, although not in earlier manuscripts, Jenner used the term "variolae vaccinae" (smallpox of the cow). Crookshank (1889) and Creighton (1889) castigated Jenner for what they saw as the adoption of a misleading name for the purpose of promoting his vaccine, but its use in just the title of the published version suggests that this "may have been a last-minute decision" (Baxby, 1985). The notion that cowpox was really smallpox in bovines was popular among scientists in the late nineteenth century and even later, and led to the frequent introduction of variola virus from patients into the inoculum used for the preparation of smallpox vaccine. In consequence, several strains in use for vaccine production in the 1970s are said to have been "derived" from variola virus (Wokatsch, 1972). It is impossible to determine what happened in these cases. Variola virus has a very narrow host range and rarely produces even a transient nodule in cows (see Chapter 7). On the other hand, since orthopoxviruses are notoriously resistant to environmental conditions, laboratory contamination was an ever-present risk when such manipulations were carried out (as they usually were) in institutions where vaccine was in use. Indeed, Kelsch et al. (1909) showed that in such circumstances cows scarified with sterile glycerine often developed a few vaccinia pustules.

Eventually, the careful experiments of Herrlich et al. (1963), carried out over several years in premises in which neither vaccinia nor cowpox virus had ever been used, were entirely negative—these workers could not achieve the "transformation" of variola virus into vaccinia virus. This conclusion is supported by the DNA maps of the two viruses; there is no way in which a variola virus DNA molecule could be readily converted into the quite different DNA molecule of vaccinia virus (see Fig. 5-1).

Serial Passage of Variola Virus in Humans, by Variolation. In an attempt to "debunk" Jenner, Razzell (1977a) has postulated that vaccinia virus was derived from variola virus by continued intradermal passage in humans, using as a virological analogy the production of an attenuated "cold" variant of vaccinia virus by Kirn and Braunwald (1964).

However, all strains of variola virus have a much lower ceiling temperature than vaccinia virus. As with the suggested derivation of vaccinia virus from cowpox virus, DNA mapping of variola and vaccinia viruses negates this hypothesis.

Hybridization between Cowpox and Variola Viruses. In the early days of vaccination there would have been ample opportunities for recombination to occur between cowpox virus and variola virus (Baxby, 1981). Recombination between these viruses occurs readily in the laboratory (Bedson and Dumbell, 1964b); variants showing a wide variety of biological characteristics occur (see Table 6-3), although none of those examined resembled vaccinia virus in all characteristics. A final decision on the likelihood that vaccinia virus originated in this way would be assisted by restriction endonuclease mapping of the DNA of selected recombinants; so far this has not been done. One argument against such an origin is that all strains of vaccinia that have been examined have very similar DNA maps and very similar biological characteristics—such homogeneity would be unlikely to occur in recombinants between variola and cowpox viruses produced at different times and places.

Vaccinia Virus Is a "Fossil" Virus. Apart from the recent observations on buffalopox in India, no animal reservoir has been discovered. Considering its broad host range, it would be reasonable to postulate that as with the other two viruses that have a broad host range, cowpox and monkeypox viruses, there would be several alternative "natural hosts," Baxby's (1981) suggestion that vaccinia virus may have been derived from horsepox, a disease that is now extinct, does not solve the problem, since it now seems more likely that both horsepox and cowpox were due to the infection of horses and cows, respectively, with "cowpox" virus, or even vaccinia virus, perhaps from a wild rodent source (see Chapter 6).

Derivation from Cowpox Virus by Mutation. In some ways the most appealing suggestion is that vaccinia virus is the result of adaptation of cowpox virus to growth in the human and bovine skin. Cowpox virus has the largest genome of all orthopoxviruses and deletion mutants occur with high frequency. Many such mutants resemble dermal vaccinia virus in two related properties: the type of pock produced on the chorioallantoic membrane and the type of lesion produced in the rabbit skin. Most retain the capacity to produce A-type inclusion bodies, but Amano and Tagaya (1981) have reported the recovery of mutants of cowpox virus that failed to produce A-type inclusion bodies, although they behaved like wild-type cowpox virus in pock production and pathogenicity for the rabbit skin. Further, they and Patel *et al.* (1986)

have shown that there is a close relationship between the 160K protein of cowpox virus A-type inclusion bodies and the 92K LS antigen of vaccinia virus. Clearly, mutants of cowpox virus could be found that resembled vaccinia virus in all three characteristics.

However, there are problems with this explanation too. If vaccinia virus was derived from deletion–transposition mutations in cowpox virus, what is the origin of the rabbitpox strains, that produce somewhat hemorrhagic pocks and large lesions in the rabbit skin, but have the characteristic DNA map of vaccinia virus? And why did the many strains of cowpox/horsepox in use in different parts of Europe during the first half of the nineteenth century all give rise to viruses with DNA maps as similar to each other, and as different from cowpox virus, as those illustrated in Fig. 5-1, unless, as Baxby (1981) suggests, vaccinia virus once occurred as a natural infection but is now extinct? The origin of vaccinia virus remains an unsolved mystery.

VACCINIA VIRUS AS A VECTOR FOR OTHER ANTIGENS

Because of the local lesion that smallpox vaccination always produced and the rare but serious complications, one of the great advantages of the global eradication of smallpox was that it obviated the need for vaccination of anyone except persons experimenting with orthopox-viruses. It is ironic that just as every country in the world had agreed to the WHO recommendation that routine vaccination against smallpox should be discontinued, scientists involved in molecular biological studies of vaccinia virus should propose its use as a vector of other antigens (Fenner, 1985).

The rationale is simple. Vaccinia virus is the most stable of all vaccines and it can be administered satisfactorily even under the most difficult field conditions. Potent vaccine was produced for the smallpox eradication campaign in many developing countries. The use of restriction enzymes made it possible to clone fragments of the vaccinia genome, and experiments in 1980 (Sam and Dumbell, 1981) showed that fragment rescue was easy to achieve, with the production of viable virus containing the rescued fragment of DNA. The way was therefore clear for the fragment of DNA that was rescued to be that coding for any desired polypeptide; for example, the surface antigen of the hepadnavirus, hepatitis B. When such foreign DNA is properly engineered into the vaccinia virus genome, efficient expression can be achieved in the infected cell. Further development of this technique is described in Chapter 12.

CHAPTER 6

Cowpox Virus

From the point of view of their natural history, orthopoxviruses fall into two groups (see Table 1-3); those with a wide host range in laboratory animals and thus potentially in nature as well, and those with a narrow host range. Of the species that have been adequately studied, the first group comprises vaccinia, cowpox, and monkeypox viruses. Various aspects of the biology of vaccinia virus, which is the "model" poxvirus, are elaborated in Chapters 2, 3, and 5. In this chapter and Chapter 8 we describe the biology of cowpox and monkeypox viruses and the diseases they cause.

Cowpox entered medical history with the publication of Jenner's "Inquiry" in 1798. It was known to country people in Europe long before that time as a sporadic disease of cows that could infect milkers, and was confused with a more common parapoxvirus infection of cow's teats which also affected milkers and is now known as "milker's nodules." After Jenner, claims were made that others had protected humans against smallpox by inoculation of cowpox virus (Dixon, 1962): Fewster (in 1765), Jesty (in 1774), and Nash (in 1781) in England, Bose (in 1769) in Germany, Plett (in 1791) in Denmark, and Jensen (in 1791) in Holstein. But as Baxby (1981) points out, none of these claims came to light until after Jenner's publication in 1798, and they made no impact on medical opinion. It was Jenner who tested the efficacy of cowpox inoculation by challenge with variola virus, devised a method of maintaining the virus in the absence of lesions in cows, and brought it to public attention.

Even when the naming of chickenpox and the "great pox" are excluded, diseases due to poxviruses illustrate well the problems of naming diseases on the basis of clinical signs and of designating viruses by the names of particular host animals. On the one hand, the term "cowpox" has been used to describe skin lesions on the teats of cows which may be produced by four different viruses (see Table 6-5): two orthopoxviruses, cowpox virus and vaccinia virus, milker's nodule virus (a parapoxvirus), and bovine alphaherpesvirus 2. On the other hand, viruses that we now regard as belonging to the cowpox virus species, when recovered from different animals, were designated as "elephantpox," "carnivorepox," "ratpox," etc., viruses. When we speak of cowpox virus in this book, we use the term for the orthopoxvirus identified by its biological characteristics as a distinctive species by Downie (1939a), a view subsequently confirmed by DNA mapping (Mackett and Archard, 1979; see Chapter 1). Association with a particular host can then be indicated by appropriate adjectives: bovine cowpox virus, rodent cowpox virus, etc.

PROPERTIES OF COWPOX VIRUS

For many years the cause of all cases of "classical cowpox" in cows was regarded as what Jenner (1798) had called the virus of "variolae vaccinae" (smallpox of cows), which came to be more generally known as vaccine or vaccinia virus (Blaxall, 1930). Downie (1939a) then showed that cowpox virus as obtained from infected cows or milkers was different from vaccinia virus (as then available in the laboratory) in a number of biological characteristics: the hemorrhagic pocks produced on the chorioallantoic membrane, the indurated, dark purple, hemorrhagic lesion in rabbit skin, and the presence of eosinophilic A-type inclusion bodies in the cytoplasm of infected cells. Other differences were found (Gispen 1955; Rondle and Dumbell, 1962), and Mackett and Archard (1979) then showed that vaccinia and cowpox viruses have quite different DNA maps.

Behavior in Laboratory Animals

Like vaccinia virus, cowpox virus produces lesions in all laboratory animals in which it has been tested, except that vaccinia virus, but not cowpox virus, produces a lesion after inoculation into feather follicles of chickens (Mayr, 1966; Mahnel, 1974). The most characteristic lesions for diagnosis are those in the rabbit skin and on the chorioallantoic membrane, but from the point of view of its natural history it is also necessary to review the behavior of cowpox virus in experimentally infected rodents.

Cowpox Virus Infection of Rabbits. Downie's (1939a) description has been confirmed by other investigators. Twenty-four hours after intradermal inoculation of a large dose of cowpox virus in rabbits a small papule develops at the site of inoculation and enlarges until it reaches a diameter of 2–3 cm by the fourth or fifth day. Initially bright red, the lesion, which is sharply circumscribed and markedly elevated, becomes dark purple in color. After reaching a maximum size about the sixth or seventh day, the central area becomes depressed, and slight vesiculation sometimes occurs. Many rabbits develop a generalized rash about the seventh day. Healing begins by the tenth or eleventh day. The lesions that develop after scarification are also bright red initially and then become darker red and hemorrhagic.

Pocks Produced by Cowpox Virus on the Chorioallantoic Membrane. When inoculated at suitable dilutions, single pocks develop as discrete, raised, opaque spots with a central bright red, hemorrhagic area, which is much more prominent than the hemorrhagic center found in pocks produced by monkeypox or neurovaccinia viruses (see Plate 4-2). Histological examination shows that there are many A-type and B-type inclusion bodies in the hyperplastic ectodermal cells and in cells in the mesoderm, especially in the capillary endothelium. The bright red central area is associated with an accumulation of erthrocytes which lie beneath the ectodermal layer, rather than in the surface tissue layer of the chorioallantoic membrane, as in pocks produced by monkeypox and neurovaccinia viruses (see Plate 4-3). Secondary pocks may develop by the third day, but are rarer than with vaccinia virus. The ceiling temperature for cowpox virus on the chorioallantoic membrane is 40°C (Bedson and Dumbell, 1961).

All strains of cowpox virus produce white pock mutants (see below).

Cowpox Virus Infection of Rodents. A few experiments have been carried out in mice with the Brighton strain of cowpox virus, and in rats with cowpox virus recovered from white rats (see Table 6-4).

Experiments in Mice. Subrahmanyan (1968) showed that between the second and third week of life there a sharp change in the susceptibility of mice to inoculation with cowpox virus by any route (footpad, intracerebral, intranasal, or intravenous). Both age groups were equally susceptible to infection by the footpad route, but viral growth was more rapid in the younger mice, higher titers were reached in the liver and spleen, and foci of infection could be detected in the skin by immunofluorescent staining, although no rash developed. However, when large doses of cowpox virus were inoculated intravenously in adult mice, skin lesions developed on the tail, feet, ears, snout, and anogenital region, initially as bright red, swollen spots which ulcerated and healed with scabbing in mice which survived (Mims, 1968). A characteristic feature of fatal infections was the transformation of most lymph nodes into "hemal" nodes, which contained large numbers of erthrocytes in their sinuses, apparently because of leakage from small vessels in the skin in which the virus replicated (Wallnerova and Mims, 1970).

Intracerebral injection of mice with orthopoxviruses produces meningitis and ependymitis (Mims, 1960), which is lethal with some strains of cowpox virus but not with others (Fenner, 1958; Baxby, 1975).

Experiments in Rats. Attention was directed to the infection of rats with cowpox virus by an outbreak of pox disease in carnivores in the Moscow zoo in 1973–1974, which was traced to infection of the white rats which constituted part of their food supply (Marennikova *et al.*, 1978a; Marennikova and Shelukhina, 1976a; Table 6-4). Intranasal inoculation of white rats with authentic cowpox virus and with strains derived from the rat colony produced similar lesions, two clinical syndromes being seen: a severe disease with pulmonary signs, which was usually fatal, and a milder disease in which a red, papular rash developed on hairless areas of the body (Marennikova *et al.*, 1978a). Subsequently, Maiboroda (1982) showed that wild rats (*Rattus norvegicus*) were highly susceptible to the rat strain of cowpox virus, given by scarification or intranasal inoculation or after natural infection by placing them in cages which had housed rats that had died of the disease. Virus was excreted in the urine and feces for up to 5 weeks after infection.

Growth in Cell Cultures

Cowpox virus grows in many cell lines and plaque counting provides an alternative to pock counting for assay of viral infectivity (Porterfield and Allison, 1960). Baxby and Rondle (1967) found that rabbit kidney cells (RK13) were as sensitive to cowpox virus as the rabbit skin, both RK13 cells and rabbit skin being considerably more sensitive than inoculation on the chorioallantoic membrane or in chick embryo fibroblasts (Dumbell *et al.*, 1957). In most cell lines, cowpox virus replicates more slowly than vaccinia virus and produces smaller plaques.

Inclusion Bodies Produced by Cowpox Virus Infection

The types of inclusion body produced by orthopoxviruses are described and illustrated in Chapter 4. Cowpox virus produces both B-type inclusion bodies, which are the site of viral replication, and the great majority of strains of cowpox virus also produce highly eosinophilic, A-type inclusion bodies. With most strains of cowpox virus these contain large numbers of mature virions (V^+ strains; see Plate 4-1B). With some strains of cowpox virus the A-type inclusion bodies are devoid of virions (V^-; Plate 4-1C), or rarely, the virions are clustered around the periphery (V^i; Plate 4-1D).

Distinctive Antigens and Polypeptides

In early experiments with gel precipitation tests, Gispen (1955) showed that a major band present in extracts of cells infected with vaccinia virus was absent from the cowpox virus material. However, Rondle and Dumbell (1962) showed that antibody to this antigen was present in cowpox antisera and that the antigen was present in cowpox tissue extracts, but in a nondiffusible form. These observations have been explained by the demonstration by Patel et al. (1986) that the abundant 94K LS antigen of vaccinia virus cross-reacts serologically with the 160K protein of the A-type inclusion bodies of cowpox virus.

Rondle and Dumbell (1962) also showed that an antigen "d" was produced by orthopoxviruses that produced red pocks on the chorioallantoic membrane, but not by white pock mutants of cowpox virus. Using hybrid viruses (Bedson and Dumbell, 1964b), production of the d antigen and pathogenicity for the rabbit skin were shown to be closely linked (Rondle and Dumbell, 1982; see Table 6-3).

Maltseva and Marennikova (1976) used a two-step gel precipitation test and a hyperimmune vaccinia antiserum to compare the soluble antigens of several orthopoxviruses. They showed that bovine cowpox virus and the cowpox viruses recovered from the Moscow Zoo carnivores, from an elephant, and from an okapi (see Table 6.4) were indistinguishable from each other by this test but could be distinguished from vaccinia, variola, and monkeypox viruses.

Studies using one-dimensional gels showed general similarities between the structural polypeptides of various cowpox virus isolates (Turner and Baxby, 1979; Arita and Tagaya, 1980). A difference in the 37K polypeptide between bovine cowpox virus strains and the Moscow rat–carnivore cowpox virus and the elephant cowpox virus (Turner and Baxby, 1979) may be a reflection of the unique polypeptide that Arita and Tagaya (1980) found to be present in the Brighton strain but absent in three Dutch isolates of bovine cowpox virus.

Analysis of the intracellular polypeptides by Harper et al. (1979) showed a distinctive 177K "late" polypeptide in cells infected with cowpox virus, the Moscow rat–carnivore cowpox virus, and the Whipsnade cheetah cowpox virus, but not in any of 20 other orthopoxviruses belonging to 3 other species: vaccinia, variola, and monkeypox viruses.

Radioimmunoprecipitation tests further increased the capacity to recognize species-specific and cross-reactive polypeptides. This method was exploited by Ikuta *et al.* (1979), who showed that there were several polypeptides common to cowpox and vaccinia viruses and a few common to these two orthopoxviruses and a *Leporipoxvirus*, Shope fibroma virus. Kitamoto *et al.* (1984) reported the production of several monoclonal antibodies that showed cross-reactivity between these three viruses. Other monoclonal antibodies reacted with extracts of both vaccinia virus- and cowpox virus-infected cells, and one reacted specifically with an antigen found in the A-type inclusion bodies of cowpox virus.

Other proteins that differ from those of vaccinia occur in the envelope (Payne, 1986). The major glycoprotein of cowpox virus (76K, probably equivalent to the 89K hemagglutinin of vaccinia virus envelopes) and a 44K glycoprotein did not comigrate with any vaccinia virus glycoproteins, whereas the 42K and 20K–23K glycoproteins did. The envelopes of both species contain a 37K nonglycosylated protein.

Size and Structure of Cowpox Virus DNA

Restriction maps have been constructed by Mackett (1981) of the DNA from five strains of classical cowpox virus derived from cows or humans and the cowpox viruses recovered from elephant (Baxby and Ghaboosi, 1977), cheetah (Baxby *et al.*, 1982), and the Moscow Zoo outbreak (Marennikova *et al.*, 1977, 1978a). *Hind*III and *Sma*I cleavage sites of these strains, two strains of variola virus, and two strains of vaccinia virus, are shown in Figure 6-1A.

The molecular masses (M_r) of the viral DNAs were estimated by summing those of the restriction fragments; M_r values of the DNAs for the eight strains of cowpox virus were between 141 and 145 million, except for the Whipsnade cheetah isolate, which is considerably smaller (124 million). The genome of cowpox virus is thus substantially larger than that of any other *Orthopoxvirus* (vaccinia virus: M_r 119–124 million).

An analysis of similarities and differences between the restriction maps of these eight strains of cowpox virus and, for comparison, two strains of variola virus and two strains of vaccinia virus is shown in Figure 6-1B. The four classical cowpox strains from Britain are very similar to each other in both *Hind*III and *Sma*I cleavage sites. The other four cowpox viruses fall into two groups: Whipsnade (cheetah) and

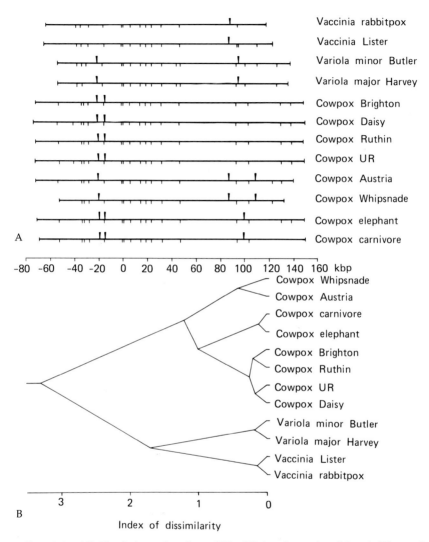

FIG. 6-1. *(A) Physical map locations of* HindIII *(I at bottom) and* SmaI *(▼ at top) restriction fragments of eight strains of cowpox virus (including elephant, Moscow Zoo carnivore, and Whipsnade cheetah isolates), two strains of variola virus, and two strains of vaccinia virus (data from Mackett, 1981). (B) Dendrogram showing dissimilarities between these orthopoxviruses, using centroid sorting strategy (Gibbs and Fenner, 1984).*

Austria have almost identical restriction sites, although the Whipsnade virus DNA has deletions at each end, and the zoo/circus animal isolates from Moscow, USSR, and Germany are similar to each other. All these cowpox virus DNAs cluster together compared with the DNAs of variola and vaccinia viruses, and in broader comparisons they differ from all other orthopoxvirus DNAs (Esposito and Knight, 1985; see Fig. 1-1). Patterns obtained by digestion with *Xho*I (Mackett, 1981) gave comparable results. Thus there is good agreement between the distinctive biological characteristics shown by cowpox viruses, whatever their source, and restriction maps of their DNA.

Range of Variation in Biotype and in DNA

Strains from Different Animals and Areas. Strains of cowpox virus have been isolated from a wide range of animals in various parts of Europe and the USSR (see Table 6-4). All share two biological characteristics which, occurring together, differentiate them from all other orthopoxviruses: the bright red, hemorrhagic pocks on the chorioallantoic membrane and the presence of large A-type inclusion bodies in infected cells. However, they differ in certain other biological attributes (Table 6-1): the size of the pock, the ceiling temperature, pathogenicity for rats, and the composition of certain virion proteins. However, all these isolates have DNA maps more like that of classical cowpox virus than any other species of orthopoxvirus (see Figs. 1-1 and 6-1), and they are best regarded as strains of cowpox virus that have become somewhat different from each other because of geographic isolation.

Strains Recovered from Cows and Humans in Britain and Holland. It might also be expected that differences would occur between isolates recovered from "classical" sources (cows and humans) at different times within geographically more restricted areas. To examine this question, Baxby (1975) tested 10 strains of cowpox virus obtained from infected cows or humans in Britain, and 8 from similar sources in Holland, all of which had been recovered between 1937 and 1971. Four tests were used, as follows: (1) the location of virions in relation to the A-type inclusion body, (2) heat resistance, (3) mouse virulence, and (4) level of hemagglutinin production (Table 6-2). Several strains were tested after 7–10 further passages on the chorioallantoic membrane; this did not alter their behavior. By these tests, Baxby distinguished 8 groups among the 17 strains, 2 of which (C and F, Table 6-2) contained strains derived from both Britain and Holland. Differences between strains included occlusion or absence of virions from the A-type inclusion bodies, production or virtual absence of hemagglutinin in extracts of infected cells, heat

TABLE 6-1

Biological Characteristics of Strains of Cowpox Virus Recovered from Various Animals and Various Geographic Areas

Characteristic	Classical cowpox virus (England)[a]	Elephant cowpox virus (Germany)	Rat–carnivore cowpox virus (Moscow, USSR)	Rodent cowpox virus (Turkmenia, USSR)[b]
Pocks on chorioallantoic membrane	Large, hemorrhagic	Small, hemorrhagic	Small, hemorrhagic	Small, hemorrhagic[b]
Ceiling temperature	40°C	39°C	38.6°C	38.6°C
Rabbit skin lesion	Indurated, hemorrhagic, moderate size	Indurated, hemorrhagic, small	Indurated, hemorrhagic, large	Indurated, hemorrhagic, large
A-type inclusion bodies	+	+	+	+
Pathogenicity for rats	+	. .	+++	+
Differentiation by absorption-neutralization[c]	−	+	−	. . .
Differentiation by virion polypeptide composition[d]	+	−	−	. . .
Differentiation by intracellular polypeptides[e]	−	. . .	−	. . .

[a] Viruses recovered from Whipsnade cheetahs and domestic cats (Table 6-4) were similar.

[b] After further passages on the chorioallantoic membrane the pocks were more like those of classical cowpox (Marennikova *et al.*, 1978b).

[c] Baxby (1982b). +, difference; −, no difference; . . . , not done.

[d] Turner and Baxby (1979). +, difference; −, no difference; . . . , not done.

[e] Harper *et al.* (1979).

TABLE 6-2

Some Biological Properties of Strains of Cowpox Recovered from Bovines and Humans in Britain and Holland[a]

Group	Strain	Designation	Source[b]	Type of inclusion[c]	Log yield virus/HA	Heat resistance (56°C × 20 min)[d]	Mouse virulence (% mortality)[e]
A	1	Ruthin	Britain	V+	5.2	3.0	20
B	2	Pritchard	Britain—cow	V+	5.2	3.3	25
	3	188	Britain—cow	V+	5.2	3.0	0
	4	Poole	Britain	V+	5.1	2.9	0
	5	Verlinde	Holland	V+	5.1	2.9	0
	6	Brandsen	Holland	V+	5.1	3.0	0
	7	Wiersema	Holland	V+	5.5	2.9	0
C	8	Dorchester	Britain	V+	5.5	3.2	55
	9	53	Holland	V+	5.2	3.0	65
D	10	Carmarthen	Britain	V+	5.4	5.6	15
	11	Maund	Britain	V+	5.5	5.8	25
E	12	60	Holland	V+	8.0	1.9	20
	13	De Haagen	Holland	V+	8.0	2.0	25
F	14	Brighton	Britain	V−	5.5	2.0	90
	15	Tyson	Britain—cow	V−	5.6	1.9	100
	16	Middlesborough	Britain	V−	5.3	2.2	95
G	17	Friesland	Holland—cow	V+	5.0	3.3	90
H	18	61	Holland	V^i	8.0	3.0	20

[a] From Baxby (1975).
[b] From humans infected by contact with cows, except where indicated.
[c] V+, Virions throughout the inclusion body; V−, no virions associated with inclusion body; V^i, virions only on the surface of the inclusion body.
[d] Fall in titer in $\log_{10}$ units.
[e] Young adult mice inoculated by intracerebral route with 10^5 pfu of virus.

resistance of infectivity, and mouse virulence. It is clear that there is a good deal of variability in certain biological characteristics between different strains of cowpox virus recovered from lesions in cows and man, but all produced bright red hemorrhagic pocks on the chorioallantoic membrane and A-type inclusion bodies in infected cells.

White Pock Mutants

Because wild-type cowpox virus produces such bright red hemorrhagic pocks on the chorioallantoic membrane, it was with this species of virus that the first observations were made of pock mutants of an animal virus. Working independently, Downie and Haddock (1952) and van Tongeren (1952) observed that among the bright red pocks produced by cowpox virus there were always a few white pocks (about 1% of all pocks). Careful subculture from one of these white pocks yielded a mutant virus which produced only white pocks on subsequent passages. The white pock mutant had the same virulence for the chick embryo as did the wild-type virus, but was less virulent for mice than wild-type cowpox virus and produced a smaller local lesion after intradermal inoculation in rabbits. White pock mutants can also be isolated in rabbit skin; K. R. Dumbell (unpublished observations, 1955) recovered such a mutant from a nonhemorrhagic nodule in a rabbit that had been inoculated intradermally with limit dilutions of a specimen of Brighton cowpox virus that had never been passaged on the chorioallantoic membrane.

White pocks are marked by heavy infiltration with leukocytes, and A-type inclusion bodies are less common than in pocks produced by wild-type virus, and they are restricted to the ectoderm (Allsop *et al.*, 1958). The yield of virus from white pocks was about 10-fold less than that from red pocks, but there was no difference in the growth in cell cultures (Allsop, 1958). This difference could be explained either by a nonproductive infection of leukocytes or interference with extraction from heavily infiltrated chorioallantoic membranes.

Stability of Mutant Phenotype. The mutant retained its white pock character for at least 20 serial passages on the chorioallantoic membrane (A. W. Downie and K. R. Dumbell, unpublished observations), although wild-type virus rapidly became dominant during serial passage of a mixed inoculum. Apparent reversion to wild type was produced by passage in mouse brain (D. J. Bauer, personal communication, 1958), or by passage in mouse, rabbit, human, cow, and chick embryo fibroblasts (Allsop *et al.*, 1958). None of these was a true reversion; red lesions were produced only on the first passage on the chorioallantoic membrane and thereafter all pocks were white (K. R. Dumbell, unpublished observations, 1958). The white pock mutant could be maintained more easily than wild-type virus in arginine-deficient cell cultures (Williamson and Mackett, 1982).

The complete absence of at least one antigen in white pock mutants of cowpox virus (Rondle and Dumbell, 1962; Amano *et al.*, 1979) suggested a substantial deletion at the right end, and this was substantiated by Archard and Mackett (1979), who showed the loss of over 20 kbp of unique DNA sequence from the right-hand end of the genome. Further study by Archard *et al.* (1984) and by Pickup *et al.* (1984) showed that the DNA structure of these mutants was a deletion–transposition like that found with rabbitpox virus. There was substantial variation in the amount of the transposed sequence (5–50 kbp), but little variation in the left margin of the deleted right terminus. The critical gene was subsequently identified and sequenced (Pickup *et al.*, 1986), and shown to code for a protein resembling inhibitors of serine protease.

Genetic Analysis of White Pock Mutants.   Early attempts to exploit the white pock mutants of cowpox virus for studies of genetic recombination were frustrated by the failure to obtain wild-type virus from pairwise crosses of many separately derived and phenotypically distinct white pock mutants of the Brighton strain of cowpox virus (K. R. Dumbell, unpublished observations, 1960; R. Greenland and F. Fenner, unpublished results, 1960). This result is now explained by the discovery that all white pock mutants of that particular strain of cowpox virus (but not of some other strains, C. K. Sam and K. R. Dumbell, unpublished observations, 1980) involve deletions and transpositions of sequence from the left-hand end of the genome. White pock mutants of species of orthopoxvirus which do recombine, like rabbitpox virus, show deletions at one or the other end of the genome, often accompanied by transpositions from the opposite end; mutants with deletions at different ends recombine to yield red pock-producing progeny (Lake and Cooper, 1980).

Hybridization of Cowpox Virus with Other Orthopoxviruses.   Frustrated by the failure to obtain recombination between any of some 50 different white pock mutants of cowpox virus, R. Greenland (unpublished experiments, 1960) readily obtained red pock-producing progeny from mixed infections with cowpox and rabbitpox white pock mutants. In other experiments Woodroofe and Fenner (1960) demonstrated recombination between a white pock mutant of cowpox virus and

TABLE 6-3

Biological Characteristics of Cowpox and Variola Major Viruses and of Several Hybrid Clones Derived from Them[a]

Virus	Pock type[b]	A-type inclusion bodies	Diffusible LS antigen	"d" antigen[c]	Ceiling temperature (°C)	TTP sensitivity[d]	Plaque type	Plaques appear (day)	Skin lesions in rabbit[e]
Cowpox	RU	+	0	+	40	−	Trabeculated	2	+++
Variola major	WO	0	+	−	38.5	+	Rimmed	4	0
Hybrid viruses:									
VC2	IU	+	0	+	40	−	Trabeculated	2	+++
VC5	IU	0	0	+	38.5	−	Trabeculated	3	0
VC6	WO	+	0	−	40	−	Trabeculated	3	+
VC7	IU	+	0	+	40	+	Trabeculated	3	+++
VC8	WU	0	+	−	38.5	−	Rimmed	4	+
VC10	WU	+	0	−	39.5	−	Trabeculated	2	+
VC12	WU	+	0	−	40	−	Rimmed	3	+
VC13	IU	+	0	+	39	−	Trabeculated	2	+++
VC14	IU	+	0	+	40	−	Trabeculated	3	+++
VC16	WU	0	0	−	40	−	Trabeculated	2	+

[a] Based on Bedson and Dumbell (1964b).

[b] R, Red; W, white; I, intermediate; U, ulcerated; O, nonulcerated.

[c] Presence of "d" antigen (Rondle and Dumbell, 1982).

[d] Sensitivity of viral thymidine kinase to inhibition by thymidine triphosphate (Bedson, 1982).

[e] +++, Large papule with hemorrhage and necrosis; +, small pink papule; 0, insignificant lesion.

wild-type rabbitpox virus, and between wild-type cowpox virus and ectromelia virus.

The most comprehensive study of hybridization involving cowpox virus was that reported by Bedson and Dumbell (1964b). Using a combination of nongenetic reactivation (Fenner *et al.*, 1959) for cowpox virus and incubation at a temperature above the ceiling temperature for variola virus (Bedson and Dumbell, 1961), they obtained many hybrids between cowpox virus and variola major virus. The characteristics of some of these hybrids are summarized in Table 6-3. They showed a wide variety of combinations of properties, some of the recombinants being like those of one or the other of the parental species, others being intermediate. Each of the seven markers examined by Bedson and Dumbell (1964b), as well as sensitivity of the viral thymidine kinase to inhibition by thymidine triphosphate, studied later by Bedson (1982), and the presence of the *d* antigen (Rondle and Dumbell, 1982), was capable of segregating independently. This suggests that if enough hybrids were tested, it would be possible to obtain one that resembled vaccinia virus in all of these biological characteristics. It would be of considerable interest to determine whether the restriction map of such a hybrid would resemble that of vaccinia virus.

NATURALLY OCCURRING INFECTIONS WITH COWPOX VIRUS

Classical cowpox virus (as well, sometimes, as vaccinia virus; see Chapter 5) has been implicated as a sporadic cause of pustular lesions on the teats of cows and the hands of milkers ever since its recognition as a distinctive species of *Orthopoxvirus* in 1939. Since about the 1970s viruses with many of the biological characteristics of cowpox virus have been recovered in various parts of Europe and the USSR, from domestic cats, rats, and zoo and circus animals of several kinds, from animals suffering from localized pustular lesions or diseases with a generalized rash or pneumonia, as well as from apparently normal rodents (Table 6-4).

Cowpox Virus Infection in Cattle

Cowpox virus derives its name from its occurrence in milking cows and humans who become infected during milking operations. Bovine cowpox is not a common disease (Gibbs *et al.*, 1973) and apparently never was, even in Jenner's time (Crookshank, 1889). The occurrence of lesions of "spurious cowpox" on cow's teats (see below) gave rise to much confusion when these lesions were used as a source of vaccine.

TABLE 6-4

Animals from Which Cowpox Virus Has Been Recovered

| Animal | Disease | | Place | Reference |
	Form	Severity		
Man	Lesions on hands	Mild	England	Davies *et al.* (1938)
Cow	Lesions on teats	Mild	Holland	Dekking (1964)
Domestic cat	Multiple skin lesions	Mild	Britain	Bennett *et al.* (1986)
White rat	Pulmonary Generalized rash	Severe	Moscow Zoo	Marennikova *et al.* (1978a)
Norway rat	Generalized rash	Mild	Moscow	Maiboroda (1982)
Lion	Pulmonary	Severe		
Cheetah	Pulmonary	Severe		
Black panther	Pulmonary Generalized rash	Severe mild		
Ocelot	Generalized rash	Severe	Moscow Zoo	Marennikova *et al.* (1977)
Jaguar	Generalized rash	Mild		
Puma	Generalized rash	Mild		
Anteater	Hemorrhagic rash	Severe		
Far Eastern cat	Generalized rash	Mild		
Cheetah	Pulmonary Generalized rash	Severe	Whipsnade Zoo	Baxby *et al.* (1982)
Elephant	Generalized rash	Severe	Germany (zoos and circuses)	Gehring *et al.* (1972); Baxby and Ghaboosi (1977); Pilaski *et al.* (1982)
Rhinoceros	Generalized rash	Moderate	Germany (zoos)	Pilaski *et al.* (1982)
Okapi	Generalized rash	Moderate	Copenhagen Zoo Rotterdam Zoo	Basse *et al.* (1964) Zwart *et al.* (1971)
Giant gerbil Yellow suslik	Normal animals captured in wild	—	Turkmenia (USSR)	Marennikova *et al.* (1978b)

TABLE 6-5
Differential Diagnosis of Viral Lesions of the Bovine Teat[a]

	Bovine herpes mammillitis	Pseudocowpox	Cowpox[b]
Causative agent	*Alphaherpesvirus*	*Parapoxvirus*	*Orthopoxvirus*
Geographic incidence in Britain	Common, particularly in western areas	Extremely common	Rare
Seasonal incidence in Britain	July to December	None, but outbreaks more common in autumn and spring	None
Characteristic clinical appearance	Extensive edema and vesication giving rise to extensive ulcers which later scab. Lesions are essentially ulcerative	No vesication; papules progress to small scabs that are shed, leaving the pathognomonic circinate lesions. Lesions are essentially proliferative	Moderate edema. Vesicles soon form pustules which subsequently rupture and scab or remain as large ulcers. Lesions are essentially proliferative
Course of infection in the individual animal	3 to 4 weeks	6 weeks	3 to 4 weeks
Immunity	Probably life long	3 to 4 months, though some cows may be chronically infected	Probably life long
Human infection	Not recognized	Localized skin lesions (milker's nodule)	May produce severe localized pustular lesions

[a] From Gibbs *et al.* (1970).

[b] Bovine vaccinia mammillitis (caused by vaccinia virus and originating from vaccinated milkers) is clinically indistinguishable from cowpox caused by cowpox virus.

After obtaining material for the vaccination of Phipps from the hand of Sarah Nelmes, Jenner had to wait 2 years before he could find another case of human or bovine cowpox. Ceely (1842), who provided one of the best and most detailed early descriptions of cowpox in cattle, noted that "The disease is occasionally epizootic . . . more commonly sporadic or nearly solitary. It may be seen sometimes at several contiguous farms, at other times one or two farms entirely escape its visitation. Many years may elapse before it recurs at a given farm or vicinity, although all the animals may have been changed in the meantime."

Clinical Features. What is called "cowpox," both in popular usage and in scientific papers, includes four different viral infections which can be more accurately designated cowpox, bovine vaccinia mammillitis, pseudocowpox, and bovine herpes mammillitis (Gibbs *et al.*, 1970). The distinctive clinical features of these infections in cattle are shown in Table 6-5; cowpox and bovine vaccinia mammillitis are clinically indis-

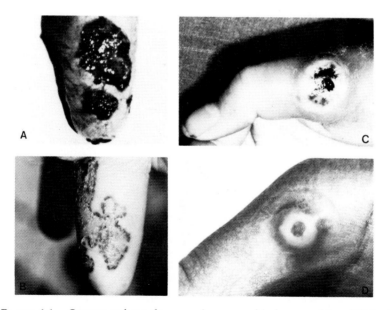

PLATE 6-1. *Cowpox and pseudocowpox in cows and in humans. (A and B) Lesions on teats of cows. (A) Cowpox, 7 days after onset of symptoms; (B) pseudocowpox. (C and D) Lesions caused by these viruses on the hands. (C) Cowpox; (D) milker's nodule (pseudocowpox virus). (A and B, courtesy of D. E. P. J. Gibbs; C, courtesy of Dr. A. D. McNae; D, courtesy of Dr. J. Nagington.)*

tinguishable. The lesions produced in cattle by the two orthopoxviruses appear to be restricted to cows and calves and to occur only on the teats and udder (Plate 6-1A and B) or on the muzzle of calves sucking affected cows. Pseudocowpox, cowpox, and bovine vaccinia mammillitis can produce lesions in humans who milk affected cows (Plate 6-1C and D); outbreaks of each of these diseases in cattle often come to attention because of human cases.

The production of lesions on cows' teats by viruses other than cowpox virus, notably the parapoxvirus that causes milker's nodules, was of considerable historical importance, since transfer of material from such lesions to man was one of the reasons for the failure of vaccine inoculations. In his second pamphlet, Jenner (1799) recognized the problem and coined the term "spurious cowpox" to describe lesions of the teats caused by agents other than an orthopoxvirus, and he was later to use this idea to explain all vaccination failures. Baxby (1981) devotes an interesting chapter to a discussion of "true" and "spurious" cowpox, seen in a historical context.

Geographic Distribution.   As far as is known, bovine cowpox caused by cowpox virus is known only in Europe, including parts of Poland. Outbreaks of bovine vaccinia mammillitis, in cows, milch buffaloes, and/or persons engaged in milking, could of course have occurred anywhere where smallpox vaccination was practised. From published evidence it appears that cases of "cowpox" in Egypt (El Dahaby et al., 1966), the USSR (Maltseva et al., 1966), South America (Lum et al., 1967), and some cases in Holland (Dekking, 1964) were in reality bovine vaccinia mammillitis. Other outbreaks, in North America (Nakano, 1986) and probably in Japan (Soekawa et al., 1964a,b) were caused by the parapoxvirus that produces milker's nodules in man. Mathew (1976) reported obtaining "cowpox virus" from outbreaks in cattle byres in the suburbs of Bombay which produced hemorrhagic pocks on the chorio-allantoic membrane, in contrast to the white flattened pocks produced by vaccinia and "buffalopox" viruses. This isolate has not been further investigated.

Epidemiology.   In view of its rarity and limited geographical distribution, is "cowpox" enzootic in cattle? This question has been evaluated by Baxby (1982a) and Gibbs and his colleagues (1973; Gibbs and Osborne, 1974) by comparing the natural history of pseudocowpox (caused by a parapoxvirus) and classical cowpox in cattle in England (Table 6-6). The conclusion is inescapable; pseudocowpox is a viral infection which is enzootic in cattle, but cattle are merely incidental hosts of cowpox virus.

TABLE 6-6

Comparison of Epidemiological Features of Cowpox and Pseudocowpox in Cattle

	Cowpox	Pseudocowpox
Causative agent		
Genus	*Orthopoxvirus*	*Parapoxvirus*
Species	cowpox (vaccinia[a])	pseudocowpox
Host range	Broad	Narrow
Lesions in man		
Nature	Pustule	Milker's nodule
Frequency	Rare	Common
Geographic distribution	Western Europe	Worldwide
Frequency in cattle in England	Rare[b]	Common[c]

[a] Not further considered here.

[b] Gibbs *et al.* (1973); Baxby (1977a); Baxby and Osborne (1979).

[c] Gibbs and Osborne (1974).

Human Infections with Cowpox Virus

Cowpox has long been recognized as an occupational disease to which milkers in various countries in Europe were subject, although it was always rare. However, Baxby (1977a) has pointed out that cows were directly implicated as a source of cowpox virus in only 3 out of 16 virologically confirmed cases in humans in England between 1969 and 1981 (Table 6-7). No source of infection could be discovered for the other 13 cases, and only 4 of the cases occurred in farm workers.

Clinical Features.   Downie (1965) has described the lesions found in humans infected with cowpox virus (Plate 6-1C) as follows. One or more lesions usually appear on the hands—the thumbs, the first interdigital cleft, and the forefinger being especially liable to attack. Scratches or abrasions of the skin may determine localization of the lesions elsewhere on the hands, forearms, or face. The lesions resemble those of primary vaccination, passing through stages of vesicle and pustule before a scab forms. Local edema is usually more pronounced than in vaccination and there is lymphangitis, lymphadenitis, and often fever for a few days. Baxby (1977a) noted that cowpox in children was sometimes rather severe. However, although there are sometimes multiple primary lesions, a generalized rash has not been reported, although one case of postcowpox encephalitis has been described (Verlinde, 1951).

TABLE 6-7

Features of 16 Virologically Confirmed Cases of Infection of Humans with Cowpox Virus in Britain[a]

Outbreak		Contact with infected cows	Human cases		
Place	Year		Farm worker[b]	Age	Lesions
Dorchester	1969	+	+	Adult	Hand
Winchester	1969	−	−	Adult	Hand
Middlesborough	1971	−	−	8 years	Chin
Exeter	1971	+	+	Adult	Hand
Burnley	1974	−	−	14 years	Hand, chin
Penrith	1974	−	−	Adult	Hand
Scarborough	1975	−	−	6 years	Face
Lincoln	1975	−	−	17 years	Hand
Bristol	1976	−	−	17 years	Face
Taunton	1976	+	+	Adult	Hand
Leeds	1978	−	−	Adult	Hand
Newcastle	1978	−	−	Adult	Hand
Shrewsbury	1978	−	−	11 years	Hand
Taunton	1978	−	−	Adult	Hand
Stoke	1979	−	+	Adult	?
Norwich	1981	−	−	9 years	Hand

[a] From Baxby (1977a) and Dr. D. Baxby, personal communication (1983).
[b] Occupations of the other patients were diverse.

Cowpox Virus Infection in Wild Rodents

Susliks and Gerbils in the USSR.   A clue to the natural reservoir host(s) of cowpox virus emerged as a result of a virological study of wild rodents in Turkmenia (Ladnyi *et al.*, 1975; Marennikova *et al.*, 1978b). In 1973–1974, independently of the outbreak of poxvirus infection in the Moscow Zoo (see Table 6-4), a survey of wild rodents was undertaken in desert areas of the Turkmenian Soviet Socialist Republic. Wild rodents were trapped, serum taken for serological studies, and organs removed for virus isolation in a laboratory in which no work with poxviruses had ever been carried out (Nebit-Dag Anti-Plague Laboratory). Orthopoxvirus antibodies were detected in four species of rodents (Table 6-8), being relatively common in great gerbils and large-toothed susliks (Plate 6-2). Two strains of cowpox virus were recovered from the kidney and spleen, respectively, of 2 different great gerbils, out of 1102 tested; and 1 strain from the kidney of a large-toothed suslik, out of 173 tested.

TABLE 6-8

Evidence of Cowpox Virus Infection in White Rats in Moscow and in Wild Rodents in Turkmenia

Species	Serological test			Virus isolation		
	HI	Precipitin	Neutralization	Number	Organs	Clinical condition
Domesticated rats[a]						
White rat (zoo)	12/31	10/24	· · ·	· · ·	Lungs and kidney	Sick
White rat (breeding colony)	33/100	18/100	· · ·	4/100	Lungs	Apparently healthy
Wild rodents[b]						
Great gerbil (Rhombomys opimus)	57/306	· · ·[c]	43/258	2/1102	Kidneys, spleen	Apparently healthy
Large-toothed (yellow) suslik (Citellus fulvus)	25/163	· · ·	9/103	1/173	Kidneys	Apparently healthy
Midday gerbil (Meriones meridianus)	2/35	· · ·	2/35	0/133	—	—
Meriones erythrourus	1/32	· · ·	1/32	0/184	—	—

[a] From Marennikova et al. (1978a).
[b] From Landyi et al. (1975).
[c] · · · , Not done.

PLATE 6-2.   *Reservoir hosts of cowpox virus in Turkmenia. (A) Yellow suslik* (Citellus fulvus); *(B) great gerbil* (Rhombomys opimus). *(Courtesy Dr. I. D. Ladnyi.)*

Subsequent examination of these strains showed that they were identical to the isolates from the Moscow Zoo outbreak in all properties examined, although Moscow and Turkmenia are some 2000 km apart and in quite different climatic zones.

Inoculation of captured great gerbils and yellow susliks with the Turkmenia strain of cowpox virus showed that both species were highly susceptible, many deaths occurring in animals inoculated by a variety of routes (Marennikova *et al.*, 1978b). Virus was recovered from the kidneys and testes of a few animals of each species as long as 5 weeks after their recovery.

Wild Rodents in Britain.   The only attempt to identify the natural reservoir of cowpox virus in England had negative results. After the occurrence of a cowpox virus infection in cheetahs in the Whipsnade Zoo in 1977, Baxby *et al.* (1982) examined 138 samples collected from 90 mammals and 12 birds captured in the parklike zoo, with negative results for evidence of orthopoxvirus infection. Kaplan *et al.* (1980) have provided a possible clue to the reservoir hosts of cowpox virus in Britain, with their demonstration of orthopoxvirus antibodies in 2 out of 28 woodmice and 7 out of 20 short-tailed voles trapped in southern England and Wales. However, these animals might equally well have been infected with ectromelia virus or some other orthopoxvirus; proper evaluation of these results awaits testing of such positive sera by an efficient method for orthopoxvirus species-specific serological diagnosis, or the recovery of virus.

Cowpox Virus Infection of Domestic Cats

In 1978 Thomsett *et al.* described a case of poxvirus infection in a domestic cat in England, from which cowpox virus was recovered. Since then a single case has been described in Austria (Schonbauer *et al.*, 1982) and many more cases have been recognized in various parts of Britain. In the series of 30 cases reported by Bennett *et al.* (1986), 53% had a history of a recent primary skin lesion on the face, neck, or foreleg, or in 1 case on the scrotum. Multiple secondary skin lesions occurred on most parts of the body in 29 of the 30 animals, the interval between appearance of primary and secondary lesions being 4–16 days (mean, 11 days). There were usually no other clinical signs, or vague symptoms only.

Cases were recorded throughout England, but only two cases have been reported from Scotland and none from Wales or Ireland. They occurred most frequently in the period August–November. The mode of infection was not determined; 17 of 25 cats whose origins and habits were investigated were from a rural environment and 19 of 25 were known to hunt rodents.

Cowpox Virus Infections in Unusual Hosts

Cowpox Virus Infection in Zoo and Circus Animals. *Zoo Animals.* Cowpox virus has been recovered from sporadic infectons in many different species of animals in zoos (see Table 6-4), and causes a serious disease in several species of zoo animal, with considerable morbidity and mortality. The most dramatic outbreaks occurred in the Moscow Zoo in 1973 and 1974 (Marennikova *et al.*, 1977), in a separate air-conditioned building in which six different species of the family *Felidae*, bears belonging to two species, and two giant anteaters were housed. First the lions contracted a severe pulmonary disease, which was rapidly fatal, and infection then occurred in several other animals in the building, but not in the bears. A second outbreak occurred in the same building 11 months later. Signs were predominantly dermal (a generalized rash), pulmonary (pneumonia), or a combination. Cowpox virus was recovered from 18 of the 19 animals examined, although the pocks were slightly smaller and the ceiling temperature lower than for classical cowpox virus (Baxby *et al.*, 1979b).

Colony of White Rats. The origin of the virus that infected the large felines appears to have been the white rats which were used as food for the carnivores. Other zoo animals were infected either by repeated exposures to infected rats, or from other diseased zoo animals, by airborne infection. Marennikova *et al.* (1978a) reported that prior to the outbreak in carnivores there had been epizootics of a poxvirus infection in colonies of white rats bred as food for them. A virus that was virtually identical to that recovered from the zoo animals was isolated from organs of affected rats (see Table 6-8). The case–fatality rate in the rats was over 30%, pulmonary, dermal, and mixed forms of the disease occurring. It was thought that the white rats may have been infected from wild Norway rats which were in contact with them. Several other references to "ratpox" have appeared in the Russian literature (see Maiboroda, 1982) and what is probably the same disease came to the attention of scientists in the United States when tissues from rats flown in the Russian satellite Cosmos 1129, and ground-based control rats, were examined by Kraft *et al.* (1982) as part of a USA/USSR Science and Applications Agreement. The first recovery of cowpox virus from rats in the USSR appears to have been made by Krikun (1974).

Maiboroda (1982) studied the rat strain of cowpox virus in wild Norway rats, which could be infected by contact (as could mice, Maiboroda *et al.*, 1980) and developed scabbed lesions on the skin of the ears and tail. Virus could be recovered from the urine and feces between 10 and 35 days after infection. Kulikova and Lobanova (1980) noted that the course of natural infection was benign in white rats housed under good conditions, but these animals suffered a severe disease with high mortality if they were unduly stressed.

Cowpox Virus Infection In Elephants. Infections among circus elephants have been particularly common in Germany (see Table 6-4), where these animals are now vaccinated with the attenuated MVA strain of vaccinia virus (H. Mahnel, personal communication, 1983).

Horsepox. "Horsepox" is a tantalizing disease for a modern virologist who is interested in the history of Jenner's vaccine. Jenner confused the situation by suggesting that "variolae vaccinae" in cows originated from "grease" of horses; a disease syndrome that may be caused by several different agents. There is no doubt that horsepox, a generalized pox disease of horses (Plate 6-3B) which sometimes caused the lesions in the fetlocks described as "grease" (Plate 6-3A), did occur in Europe during the nineteenth century. However, grease is more commonly caused by other microbial agents, notably *Dermatophilis congolensis* (Gillespie and Timoney, 1981). Horsepox lesions were sometimes used

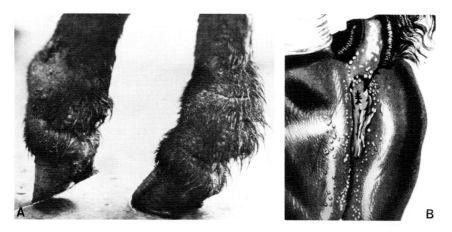

PLATE 6-3. *(A) Grease, a lesion of the fetlocks caused by a variety of agents; (B) horsepox. Illustration of a case investigated by Professor Peuch of Toulouse, which occurred during an outbreak of horsepox in Toulouse in 1880. (A, courtesy Dr. M. Soekawa; B, from Crookshank, 1889.)*

as a source of virus for human "vaccination" ("equination") in Europe. Chaveau (quoted by Crookshank, 1889) and Fleming (1880a,b) believed that horsepox was due to infection of horses with the virus that caused cowpox. It seems likely that horses, like cows, were incidental hosts for an orthopoxvirus. Also, they may have been infected by human transfer of the virus from cows, and vice cersa. It is difficult to understand why occasional cases do not occur at the present time, since there are still many horses in rural areas of Europe.

Geographic Distribution and Possible Reservoir Hosts

Cowpox virus appears to occur naturally among some unknown reservoir hosts—probably rodents of several species—in Europe and parts of the USSR. Strains recovered from different geographic areas differ somewhat in their biological characteristics, but all produce pocks with a hemorrhagic center, large A-type inclusion bodies in infected cells, and large indurated lesions with a hemorrhagic center in rabbit skin. Restriction maps differ according to the geographic origin of the strain, but all resemble each other and classical cowpox more than any other orthopoxviruses.

Cowpox virus has a very wide host range, causing lesions in almost every species that has been tested except chickens inoculated by the feather follicle. Sometimes there is only a local lesion, sometimes a severe disease in which the signs may be pneumonia or a generalized rash.

Sporadic infections have now been recognized in a wide range of animals, both domestic (cow, man, cat, laboratory rat) and wild zoo animals. Elephants and large carnivores, in particular, suffer a severe disease. Virus has been recovered from enzootic situations in colonies of laboratory rats in the USSR, and from wild susliks and gerbils in Turkmenia in the USSR.

CHAPTER 7

Variola Virus

By the end of the eighteenth century the epidemiology of smallpox was sufficiently well understood for Haygarth (1785) to publish his "Rules for Preventing the Smallpox," based on 6 years of successful operation of them in Chester. Although a few years later Jenner and his contemporaries used the word "virus" when referring to the infectious agents of smallpox and cowpox, they had no idea of the nature of the "infectious principle" but used the word in the general sense common at the time (Wilkinson, 1979). When Guarnieri (1892) described inclusion bodies in cells from smallpox lesions, he thought that they were protozoa and regarded them as the causal organisms of the disease. Before this, Buist (1886) had described minute particles in variolous lymph. Negri (1906) and Paschen (1906) showed that both infectivity and these minute particles were filterable, and believed that they were the infectious agent. Gradually it became clear that smallpox was indeed caused by a virus whose virion was large enough to be just visible with the light microscope (see Plate 1-1).

Although studies of variola virus were carried out in monkeys early in the history of virology (Zuelzer, 1874; Brinckerhof and Tyzzer, 1906), vaccine virus (vaccinia or cowpox virus) was much more widely used by laboratory scientists. This trend has been maintained, and more recent laboratory studies of variola virus have been restricted to a small number of laboratories with maximum containment facilities. Apart from improving diagnostic methods, most of this work was concerned with differentiating variola virus from other orthopoxviruses and with attempts to find laboratory correlates of its virulence for humans.

VARIOLA VIRUS IN ITS HUMAN HOST

The pathogenesis and pathology of smallpox are described in Chapter 4 and a brief description of the clinical features of smallpox is given in Chapter 11. More detailed accounts can be found in Fenner *et al.* (1988).

Variola Major and Variola Minor Viruses

The most important property of variola virus, in relation to its effects on the human host, is its virulence. Observations on outbreaks of smallpox in the United States and South Africa at the end of the nineteenth century and in the United States and England during the early twentieth century led to the recognition that, regardless of the vaccination status of the community involved, some epidemics of smallpox were associated with a high mortality and others with a low mortality. Although subsequent studies showed that strains of variola virus which covered a wide spectrum of virulence for man (case–fatality rates in unvaccinated individuals ranging from less that 1% to about 40%) occurred in different parts of the world, for practical purposes only two clinicoepidemiological varieties of smallpox were recognized: variola major (case–fatality rates 5 to 40%), and variola minor (case–fatality rates 0.1 to 2%). Laboratory studies of strains of variola minor virus recovered from Africa and the United States (or countries with variola minor originally derived from the United States) revealed that these two groups of variola minor viruses differ from one another in several characteristics (Dumbell and Huq, 1986); it is convenient to distinguish between African and "American" strains of variola minor virus by calling the American and American-derived strains "alastrim" virus. Alastrim virus, but not African strains of variola minor, can be distinguished from variola major virus by laboratory tests, but this is believed to represent geographic variation rather than being correlated with virulence (see below).

THE PROPERTIES OF VARIOLA VIRUS

The virion of variola virus has the same size and shape as that of all other orthopoxviruses (see Chapter 2). Enveloped and nonenveloped forms are seen in material submitted for laboratory diagnosis, the former being uncommon and found in vesicle fluid but not in scabs (Nakano, 1978). In serological tests, variola virus cross-reacts with all other orthopoxviruses (see Chapter 2), but can be distinguished from vaccinia and monkeypox viruses by tests with absorbed sera (Gispen and Brand-Saathof, 1974; Esposito et al., 1977a,b; see Chapters 4 and 8).

The biological characteristics of variola virus are compared with those of other orthopoxviruses in Table 1-2, and the restriction map of the DNA of variola virus is compared with the maps of DNAs of other species of orthopoxviruses in Fig. 1-1.

Thermal Stability

Like other orthopoxviruses, variola virus is relatively stable, its inactivation kinetics being similar to those of vaccinia virus. The infectivity of alastrim virus is much more thermolabile than that of variola major virus (Bedson and Dumbell, 1961). Assessment of the viability of variola major virus in scabs at different temperatures and humidities suggested that the virus would not survive for more than a few months at ambient temperature (MacCallum and McDonald, 1957; Huq, 1976). However, viable virus has been extracted from smallpox scabs preserved for over a year at cool temperatures, and Wolff and Croon (1968) have reported the survival of alastrim virus in scabs for as long as 13 years.

Pathogenicity for Laboratory Animals

Experimental observations with variola virus demonstrated that, unlike vaccinia, cowpox, and monkeypox viruses, each of which has a wide host range, few experimental animals could be infected with variola virus, except under unusual conditions, although several species of primates are susceptible. The most valuable experimental animal for studies of variola virus was the developing chick embryo.

Effects on the Chick Embryo. Although Torres and Teixeira (1935) had cultured variola virus on the chorioallantoic membrane, the first use of this method for diagnosis was suggested by Buddingh (1938), who noted that variola virus did, and varicella virus did not, grow on the chorioallantoic membrane. Using dilute suspensions, North et al. (1944)

and Downie and Dumbell (1947) showed that variola virus produced opaque white pocks on the chorioallantoic membrane which could readily be distinguished from those of cowpox and vaccinia viruses (see Plate 4-2), at that time the only other orthopoxviruses known to infect man. The particle:infectivity ratio was estimated at about 12:1 (K. R. Dumbell, A. W. Downie, and R. C. Valentine, unpublished observations, 1957), which is about the same as that of vaccinia virus (Dumbell *et al.*, 1957). Individual lesions appear by 48 hours and are most characteristic after incubation at 35–36°C for 3 days. They are small, white, domed, and not ulcerated. As the incubator temperature is raised the lesions become fewer and smaller, and variola virus does not produce pocks at temperatures higher than 38.5°C (Bedson and Dumbell, 1961). Variola virus inoculated on the chorioallantoic membrane spreads to the embryo; large doses are fatal and high titers of virus are found in the liver (Helbert, 1957). Cultivation on the chorioallantoic membrane became the routine method for growing variola virus from scabs or vesicle fluid, and remained the preferred method throughout the Intensified Smallpox Eradication Program.

The discovery of human monkeypox in 1970 (see Chapter 8) presented a new challenge in diagnostic work, for the lesions that monkeypox virus produces on the chorioallantoic membrane are readily distinguishable from those of variola virus only at strictly controlled temperatures of incubation. When eggs are incubated at 36–38°C, the pocks of monkeypox virus are white and nonulcerated and are difficult to distinguish from those of variola virus. Further tests were therefore required to make a reliable identification, and a more extended profile of the marker characteristics of variola virus was required. This became even more necessary when camelpox and taterapox viruses had to be considered, although these viruses never presented a diagnostic problem with material from human cases. The most useful biological characteristics for this purpose were growth in the rabbit skin and the ceiling temperature of growth on the chorioallantoic membrane (see Table 1-2).

Ceiling Temperature on the Chorioallantoic Membrane. The ceiling temperature for variola virus on the chorioallantoic membrane (see Chapter 4) readily distinguishes variola virus from monkeypox and camelpox viruses as well as from vaccinia and cowpox viruses. However, variola virus itself occurred in two varieties that can be distinguished by their ceiling temperature: alastrim virus (variola minor virus of American origin) and all other strains of variola virus (see below).

Virulence for the Chick Embryo. Although North *et al.* (1944) found that after passage on the chorioallantoic membrane variola virus often killed the embryo, in general variola virus is less virulent for the chick embryo than cowpox, monkeypox, or vaccinia viruses (see Table 1-2). This property proved to be a useful marker to distinguish alastrim virus from other varieties of variola virus.

Stability of Characteristics. The behavior of variola virus on the chorio-allantoic membrane is remarkably stable. Nelson (1939, 1943) maintained a strain of variola virus for 200 passages without changes in either pock morphology or virulence, and K. R. Dumbell (unpublished observations) passaged 2 strains of variola virus, 1 for 80 passages, the other for 2 series of 50 passages, at limit dilution and at maximum concentration. None of the three series resulted in any detectable change in pock character, in ceiling temperature or in mean survival time of the chick embryos. Limited success attended attempts to increase the ceiling temperature of variola major virus by repeated passage on the chorioallantoic membrane using small increments of incubation temperature (Dumbell et al., 1967). One adapted strain produced pocks with a low content of infectious virus at 39°C and replicated to a somewhat higher titer than unadapted virus when grown at 38°C, and growth of the adapted strain was significantly less than that of the unadapted virus at 35°C. Its virulence for the chick embryo was also much reduced.

Rabbit. Paul (1915) observed that inoculation of variolous material on the cornea of the rabbit produced keratitis, with Guarnieri bodies in the affected epithelial cells; this was the basis of the diagnostic test for smallpox which bears his name. Intradermal inoculation of rabbits with variola virus usually produces slight erythema and a transient papule (Bedson and Dumbell, 1964a,b), but some strains of variola virus may produce hemorrhagic, necrotic lesions (Gispen and Brand-Saathof, 1972; Dumbell and Kapsenberg, 1982). Primary cell cultures from rabbits are readily susceptible to infection with variola virus (Baltazard et al., 1958) and after passage in such cells variola virus produces superficial lesions in the scarified skin of rabbits, increases in titer, and becomes capable of sustained experimental transmission (Dumbell and Bedson, 1966). However, in general virulence for the rabbit skin was a stable character which was useful for differentiating variola virus from monkeypox virus.

Variola virus has been carried through several successive passages in rabbits by inoculation in the testis. Nelson (1943) noted that there were no pathological signs of infection, but 1 strain was maintained for 11 passages and probably could have been maintained indefinitely; another strain failed to survive beyond the third passage. Other workers have reported passage for a few generations with diminishing titer, the virus eventually being lost (see Downie and Dumbell, 1956).

Mouse.   Variola virus is much less pathogenic for mice than most other species of orthopoxvirus (see Table 1-2). Adult mice are not seriously affected by the inoculation of variola virus intranasally (Nelson, 1939) or intracerebrally (Brown *et al.*, 1980). However, variola virus may be passed serially in suckling mice, which may be killed by adequate inocula administered intraperitoneally (Mayr and Herrlich, 1960) or intracerebrally (Brown *et al.*, 1960). Inoculation of variola virus into the footpads of baby mice produces local infection and runting; in contrast, monkeypox virus produces generalized infection and is always lethal (Nakano, 1978). Intracerebral inoculation of variola virus may be fatal to baby mice up to the age of 10 days, but susceptibility is rapidly lost after this age (Marennikova and Kaptsova, 1965). The dose response of the harmonic mean survival time of suckling mice has been used to quantitate the effects of variola virus inoculated intracerebrally (Bauer and Sadler, 1960; Bauer *et al.*, 1962). F. Huq and K. R. Dumbell (unpublished observations) used this method to show that strains of variola virus from Asia, Africa, and South America had different levels of virulence for suckling mice (see Table 7-4).

Primates.   It had been demonstrated during the latter part of the nineteenth century that monkeys (probably rhesus) could be infected with variola virus (Zuelzer, 1874; Copeman, 1894) and later studies (Hahon, 1961) showed that many species of monkey and apes were susceptible. Noble (1970) found that three species of New World monkeys that he tested were insusceptible to alastrim virus, although they reacted serologically but without symptoms to experimental infection with variola major virus.

Experimental Transmission of Smallpox between Primates. Noble and Rich (1969) showed that serial infection could be maintained for as many as six successive passages in cynomolgus monkeys placed in contact with other monkeys during the period in which they had a rash, but transmission then failed (see Fig. 4-1). Among chimpanzees, Kalter *et al.* (1979) observed that two animals situated in cages near an inoculated chimpanzee contracted smallpox, one suffering a severe illness. One other animal in the group escaped infection. Thus primates of several species are susceptible to variola virus, get a rash when infected, and can transmit the disease to other primates in contact with them. However, in cynomolgus monkeys, the only primate in which adequate studies were performed, the infection persisted with some difficulty and then died out.

Reported Infections of Primates in Nature. Arita and Henderson (1968) reviewed the published accounts of supposed smallpox in primates as

TABLE 7-1

Episodes of Presumed or Proved Naturally Occurring "Pox" Infection in Nonhuman Primates[a]

Country	Year	Species	Author
France	1767	?	Barrier, quoted by Schmidt (1870)
Panama	1841	?	Anderson (1861)
France	1842	?	Rayer, quoted by Schmidt (1870)
Trinidad	1858	?	Furlong, quoted by Schmidt (1870)
Brazil	1922	*Mycetes seniculus* *Cebus capucinus*	Bleyer (1922)
India	1936	*Macaca mulatta*	M. A. Rahman (quoted by Arita and Henderson, 1968)
Indonesia	1949	Orangutan[b]	Gispen (1949)
India	1966	*Macaca mulatta*[c]	Mack and Noble (1970)

[a] Based on Arita and Henderson (1968).
[b] Variola virus isolated.
[c] Serological and epidemiological evidence suggesting infection with variola virus.

well as other naturally occurring epidemics of "pox" infections among monkey populations. Only eight such episodes are known and only four of these occurred during the present century (Table 7-1). In only two instances was laboratory confirmation available; in each of these there had been close association between the primate concerned and cases of human smallpox. All other episodes must be regarded with caution as far as their significance as evidence of an animal reservoir of variola virus is concerned. They may have been instances of the infection of primates from human cases of smallpox or they may have not been due to variola virus or indeed a poxvirus at all.

Cows.   During the nineteenth century the cow was often inoculated with variola virus in efforts to obtain new strains of "variolae vaccinae" for human vaccination. However, these experiments were usually conducted in premises in which vaccine had been prepared, and in such circumstances Kelsch *et al.* (1909) showed that scarification of cows with sterile glycerine resulted in the appearance of a few vaccinial lesions. The most recent experiments in cows are those reported by Herrlich *et al.* (1963), who inoculated very large doses of variola major virus into each of 10 calves. Only one reacted with small papules, from which variola virus was recovered, but could not be further passaged in cows. Only the animal that reacted was found to be immune when subsequently challenged with vaccinia virus.

Growth in Cultured Cells

In contrast to its limited host range in laboratory animals, variola virus will grow and produce a cytopathic effect in cultured cells derived from many species (Hahon, 1958; Pirsch *et al.*, 1963), including, as well as mammals, chick embryos (Mika and Pirsch, 1960; Bedson and Dumbell, 1964b), and even a continuous cell line, Th.I, derived from the heart of the turtle (Wells, 1967). However, it grows best in cells from humans and other primates, in which it produces characteristic "hyperplastic" foci, first noted by Vieuchange *et al.* (1958), and later made the basis for infectivity assays (Pirsch and Purlson, 1962; Kitamura, 1968). This appearance (Plate 7-1) is not due to proliferation of the infected cells, but to their aggregation. Because of the low cytopathogenicity of variola virus, infected cells remain in the monolayer and are pushed together by the growing noninfected cells around them. In diploid human cells the virus is more cytopathic, the cells becoming shrunken and rounded, with the production of contracted round clumps (Marennikova *et al.*, 1964; Baxby, 1974; Dumbell and Wells, 1982).

Variola virus can be readily adapted to grow in primary rabbit kidney cells (Baltazard *et al.*, 1958; Dumbell and Bedson, 1966), but not in RK13 cells (Gispen, 1970). However, in RK13 cells grown in medium containing rabbit serum instead of calf serum, hyperplastic foci appeared and

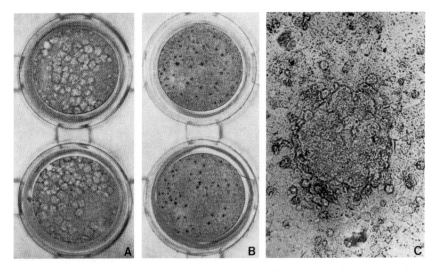

PLATE 7-1. *(A) Lytic plaques produced in HeLa cell monolayers by vaccinia virus. (B) Hyperplastic foci produced by variola virus. (C) Enlarged view of a hyperplastic focus in FL cells. (A and B, from Kitamura and Tanaka, 1973; C, from Ono and Kato, 1968.)*

continued to enlarge; after 3 or 4 days the centers of the foci broke down, forming small plaques with a dense rim. A day or two later cytolysis slowly extended through the cell sheet. The cytolytic effects were perpetuated on passage, moderate titers of hemagglutinin could be demonstrated, and the virus appeared to have become adapted to indefinite serial passage in these cells. Not every batch of rabbit serum was satisfactory; the operative constituent in the rabbit serum was not identified (K. R. Dumbell, unpublished observation). Some strains of variola virus from southern Africa produced cytolytic effects and hemagglutinin even on first passage in RK13 cells (Dumbell and Kapsenberg, 1982).

Variola virus will replicate and produce a cytopathic effect in continuous line pig embryo kidney (PEK) cells (Marennikova et al., 1971), a test which has been used to differentiate it from monkeypox virus. However, Y. Ueda and K. R. Dumbell (unpublished observations, 1974) found that different strains of monkeypox virus varied in their capacity to grow and produce a cytopathic effect in PEK cells; some hardly grew at all, others moderately well, but none grew as well as variola or vaccinia viruses.

Hemagglutinin Production and Hemadsorption. Hemadsorption is readily demonstrated in human embryonic cell cultures infected by variola major or alastrim viruses. However, at 40°C no hemadsorption is demonstrable in cells infected by alastrim virus that has been grown on the chorioallantoic membrane (Wells, 1967; Dumbell and Wells, 1982), but if the virus is adapted by three to six passages in human embryonic cells it will show hemadsorption even at 40°C (Gurvich and Marennikova, 1964). A strain of variola virus passaged 73 times on the chorioallantoic membrane had changed its behavior in human embryonic cell cultures so that it produced a syncytium which rapidly extended through the cell sheet, but no hemadsorption could be demonstrated. Attempts to repeat this observation with a different strain of variola virus showed no change in the effects seen in human embryonic cell cultures during 40 passages on the chorioallantoic membrane.

When isolates of variola virus grown on the chorioallantoic membrane were tested in HEp2 cells, some produced hemagglutinin and hemadsorption while others did not (Dumbell and Huq, 1975, 1986). However, all strains replicated well and on the second and subsequent passages moderate titers of hemagglutinin were produced. A variola virus strain thus adapted to be hemagglutinin positive in HEp2 cells was passed once on the chorioallantoic membrane and this chorioallantoic membrane stock was again hemagglutinin negative in HEp2 cells (Huq and

Herd, cited by Dumbell and Huq, 1986). Using HEp2 cells, Dumbell and Huq (1986) found hemadsorption useful for distinguishing between variola virus strains from different geographic areas (see below).

Thymidine Kinase Activity. All orthopoxviruses produce a virus-coded thymidine kinase in infected cells (see Chapter 2). Bedson (1982) reported that there were functional differences between the thymidine kinase of variola virus and that of other orthopoxviruses, in that only the variola virus enzyme was sensitive to inhibition by thymidine phosphate (Table 7-2; see also Table 1-2).

Independent Segregation of Biological Markers

Ideally, the biological profile of a virus should consist of independent markers. The most direct way of demonstrating this, short of mapping the responsible genes on the viral DNA, is to show whether or not pairs of characters segregate independently in genetic crosses between different viruses. Bedson and Dumbell (1964a,b) examined the segregation of various biological markers of variola virus in genetic crosses with rabbitpox virus and with cowpox virus. Evidence for the segregation of two additional markers was obtained from the work of Bedson (1982) and Rondle and Dumbell (1982). The results of the recombination experiments with variola major and cowpox virus are set out in Table 6-3. All markers studied segregated independently, both in this series of crosses (Table 7-3), and in crosses between alastrim virus and rabbitpox virus.

Structure of the Variola Virus Genome

Linear maps showing the location of endonuclease cleavage sites on the genome were constructed for two strains of variola virus for the

TABLE 7-2
*Inhibition of Thymidine Kinase Activity of
Various Orthopoxviruses
by Thymidine Triphosphate[a]*

Virus	Residual activity (%)
Vaccinia (4 strains)	31–84
Cowpox (3 strains)	65–73
Monkeypox (5 strains)	37–85
Camelpox (1 strain)	85
Ectromelia (1 strain)	61
Variola (11 strains)	3–12

[a] Data from Bedson (1982).

enzymes *Hind*III, *Xho*I, and *Sma*I (Mackett and Archard, 1979). Esposito *et al.* (1985) published *Hind*III maps for six strains of variola virus, and for a single strain, maps for *Bgl*I, *Kpn*I, *Sac*I, *Sal*I, *Sma*I, and *Xho*I. The *Hind*III maps of all strains of variola virus are similar to each other and are very different from those of vaccinia and monkeypox viruses, although the map of alastrim virus is rather different from all the others (see below).

A composite map of the genome of Harvey, the standard reference strain of variola major virus, has been constructed by cloning both the *Hind*III and the *Sac*I fragments of Harvey DNA into recombinant bacterial plasmids (Hamilton *et al.*, 1985) and then mapping on each recombinant the cleavage sites of eight endonucleases. The overlapping of the DNA inserts in the two sets of recombinants gave adequate confirmation for the combination of the results into a composite map. This map (Fig. 7-1) shows the location of 155 cleavage sites, for the endonucleases *Hind*III, *Sac*I, *Xho*I, *Sma*I, *Bam*HI, *Sal*I, *Pst*I, and *Eco*RI, but it does not include details within approximately 5 kb at one end and 9 kb at the other end of the genome, which represent the uncloned terminal fragments of variola virus DNA.

Variola Virus DNA Termini.   Vaccinia virus DNA has long inverted terminal repeats (see Chapter 2) and thus the right and left terminal

TABLE 7-3

Independent Segregation of 10 Marker Characters of Variola Virus in Recombinants with Cowpox Virus[a,b]

Variola virus character	Inverse of variola virus character									
	1	2	3	4	5	6	7	8	9	10
1. Pocks not ulcerated	−	+	+	+	+	+	+	+	. . .	+
2. A-type inclusions absent		−	+	+	+	+	+	+	+	+
3. Ceiling temperature <39.5°C			−	+	+	+	+	+	+	+
4. LS antigen diffusible				−	0	0	0	+	0	0
5. *d* antigen absent					−	0	+	+	+	0
6. Plaques in CEF slow (>2 days)						−	+	+	+	+
7. Plaques in CEF heavy rimmed							−	+	+	0
8. TK activity sensitive to TTP								−	+	+
9. Nonvirulent for chick embryo									−	+
10. Nonvirulent for rabbit										−

[a] Data from Bedson and Dumbell (1964b), Bedson (1982), and Rondle and Dumbell (1982).

[b] +, Combination of variola virus marker with inverse of other character found in eight recombinants tested; 0, Combination of indicated characters not found; . . . , appropriate test not done.

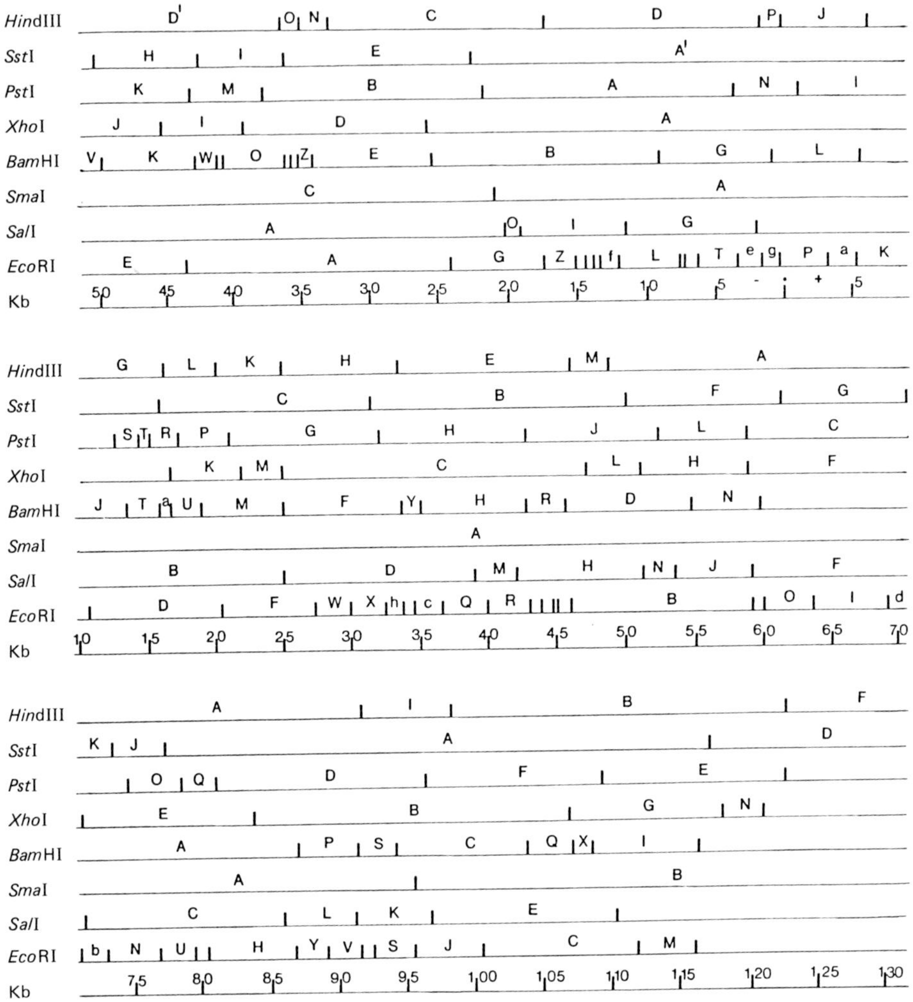

FIG. 7-1. *Fine-structure physical map of genome DNA of variola major virus Harvey (except for 5 kb at one terminus and 9 kb at the other). This is a composite map made by mapping cleavage sites of several endonucleases on a series of recombinant plasmids containing HindIII fragments of the Harvey virus DNA, and then aligning the resulting plasmid maps in the linear order and orientation of the Harvey virus genome. The map was confirmed by mapping a second series of overlapping fragments of Harvey virus DNA generated by cleavage with SacI and maintained as recombinants. The reference point for measurement is the junction between the HindIII "P" and "J" fragments. (From Hamilton et al., 1985.)*

fragments cross-hybridize strongly; the same is true of cowpox, monkeypox, and ectromelia viruses (Mackett and Archard, 1979). Cross-hybridization between opposite terminal fragments of variola virus was absent in strain Harvey (Dumbell and Archard, 1980), very weak in strain Butler (Mackett and Archard, 1979), and very weak in strain BSH-75 (Esposito and Knight, 1985). By using nine additional endonucleases, Esposito and Knight (1985) were able to show that the terminal repeat in strain BSH-75 was limited to no more that 0.5 kb.

Sequences conserved in the much longer terminal repeats of other orthopoxviruses are present at the right-hand end of the variola virus genome, for the right terminal *Hind*III fragment of variola virus DNA hybridized strongly with the terminal fragments of vaccinia or monkeypox virus DNA (Mackett and Archard, 1979). The left terminal *Hind*III fragment of variola virus genomes hybridized to a *Hind*III fragment of monkeypox virus DNA extending from 5.8 to 8.5 kb from the left terminus (Dumbell and Archard, 1980; Esposito and Knight, 1985).

Cloning of Variola Virus DNA. When the global eradication of smallpox had been confirmed, it became inappropriate to work with viable variola virus, even in maximum containment laboratories. This had been foreseen and DNA had been prepared from a selection of the available strains of variola virus so that studies could continue without the necessity for culture of the virus. The cloning of some of this material into recombinant plasmids had three advantages: it made work on variola virus DNA completely safe, it made available an indefinitely extendable amount of material, and it produced less cumbersome lengths of DNA for more detailed analysis. DNA from five strains of variola virus has been cloned: two isolates of variola major virus, two of alastrim virus, and one isolate of African variola minor virus. Cloning of the DNA from the cross-linked termini requires a different and more complex approach; this has been achieved for vaccinia and cowpox viruses, but not, as yet, for variola virus.

COMPARISON OF VARIOLA VIRUS WITH OTHER ORTHOPOXVIRUSES

Table 1-2 sets out a comparison of the biological properties of variola virus and the other nine known species of *Orthopoxvirus*, and Fig.1-1 compares the restriction maps of variola virus DNA and those of six other species of orthopoxvirus. It is useful to draw attention in this chapter to properties of variola virus that have been particularly useful in differentiating it from other orthopoxviruses.

Biological Characteristics

The most useful biological characteristics for species diagnosis of variola virus are the small dense white pocks (0.3–0.6 mm diameter) on the chorioallantoic membrane, with a low ceiling temperature (37.5°C for alastrim virus and 38.5°C for all other strains), the low virulence for mice and chick embryos, the failure to grow in rabbit skin, and the capacity to produce a cytopathic effect in PEK cells and hyperplastic foci in HeLa cells. When these characters were found in material obtained from a case of suspected smallpox they constituted positive confirmation of the diagnosis; indeed the recovery of typical variola virus pocks on the chorioallantoic membrane was usually accepted as diagnostic.

It is worth noting, however, that combinations of characters rather like those just described are found with both camelpox virus and tatera poxvirus (see Table 1-2 and Chapter 10). The source of the material usually removed any uncertainty about the diagnosis; neither camelpox virus nor tatera poxvirus has ever been found to produce disease in man and variola virus has never been recovered from animals under conditions where no suspicions arose of laboratory contamination. The three viruses can also be differentiated by other laboratory tests:

1. Only variola virus produces dense white pocks at all temperatures of incubation.

2. Variola virus produces hyperplastic foci and camelpox virus produces giant cells in several human and primate cell lines (Baxby, 1974).

3. Tatera poxvirus is serially transmissible in rabbit skin (Gispen, 1972); variola virus is not.

4. Tatera poxvirus is cytocidal for RK13 cells, in which variola virus produces hyperplastic foci (Huq, 1972).

5. Only variola virus produces a generalized disease in primates.

Differences in Antigens and Other Proteins

With viruses of other families, serological procedures are useful for differentiating species and sometimes strains. Cross-reactivity between the different species of orthopoxviruses is so extensive, however, that such tests are useful only when antisera are absorbed with homologous and heterologous viral antigens before testing (see Chapter 2). In practice, the only use to which such methods have been put has been in epidemiological investigations of human monkeypox and ecological studies of monkeypox virus (see Chapter 8).

Differences between the structural and intracellular polypeptides of variola virus and other orthopoxviruses are described in Chapter 2.

Differences in Genomic DNA

Identification of a virus by characterization of the genomic DNA is to be preferred to identification based on serological or biological properties, because it involves the whole genetic constitution of the virus and is not restricted to a small selection of gene functions. Digestion of genomic DNA with restriction endonucleases has been shown to be an excellent method for differentiating species of orthopoxvirus (see Fig. 1-1 and Chapter 2). The most important use of this technique in relation to variola virus has been to differentiate between monkeypox virus and its mutants, variola virus, and the "whitepox" viruses (see below).

Differences between the DNAs of Variola and Monkeypox Viruses. Because each causes a severe generalized poxvirus infection of man, special attention has been devoted to comparisons between the DNAs of variola and monkeypox viruses. Using cloned fragments of the DNA of these two viruses, Kinchington *et al.* (1984) compared the threshhold denaturation by formamide of homoduplexes and heteroduplexes. The method revealed a region of significant heterogeneity occurring in 4000–6000 bp out of 43,000–45,000 bp of the conserved region. The DNA sequences of these regions of heterogeneity are currently being examined.

Thymidine Kinase Gene of Various Orthopoxviruses. Esposito and Knight (1984) sequenced the thymidine kinase (TK) genes of variola and monkeypox viruses and compared their results with equivalent sequences determined for vaccinia virus (Hruby *et al.*, 1983; Weir and Moss, 1983). Differences were found at 41 nucleotide positions. Eight amino acid substitutions were theoretically involved; two of these might well have affected the kinetic properties of the enzyme. Comparisons between the TK genes of three orthopoxviruses, fowlpox virus, herpes simplex, a mammal, and a bird are discussed by Boyle *et al.* (1987) and illustrated in Fig. 2-8.

DIFFERENCES BETWEEN STRAINS OF VARIOLA VIRUS

Variola Major and Alastrim Viruses

Epidemiological evidence had clearly established the existence in southern Africa (de Korté, 1904) and the United States (Chapin, 1913, 1926) of a variety of smallpox much milder than the classic (Asian) smallpox. This led to the use of the terms of variola minor and variola major for the mild and the severe forms of smallpox, and in the Americas variola minor was often called "alastrim." The documented spread of alastrim around the world and the hypothetical spread of variola minor in Africa are illustrated in Fig. 11-5.

The virus isolated from either variety of smallpox produced pocks on the chorioallantoic membrane which were indistinguishable, nor was there any serological difference between them. Much effort was put into looking for a laboratory test to distinguish these two varieties of variola virus, since this would have practical importance at the outset of an outbreak following an importation and might also yield clues about the biological mechanisms that underlie virulence. Initially all experiments were done with alastrim virus and variola major virus. The first positive result was obtained by Dinger (1956), who showed that variola major virus was less readily eliminated by chick embryos that survived infection for 6 days. Then Helbert (1957) showed that strains of variola major virus had a lower LD_{50} for chick embryos than did alastrim virus and that variola major virus grew to higher titer in the livers of the chick embryos.

Demonstration of the greater virulence of variola major strains for the chick embryo was simplified by Nizamuddin and Dumbell (1961), who found that the ceiling temperature for growth of alastrim virus in the chick embryo was 37.5°C, compared to 38.5°C for variola major. Dumbell et al. (1961) further showed that elevated temperatures enabled the chick embryo to eliminate infection after an inoculum that would have proved fatal at a lower temperature. This finding suggested that growth of alastrim virus in internal organs might be limited by the temperature of the febrile patient. However, Dumbell and Wells (1982) found that the results obtained in chick embryos were not directly transferrable to a human cell system. In diploid cell cultures derived from human embryonic skin and muscle, the standard strains of variola major virus and alastrim virus had the same ceiling temperature for growth, although at elevated temperature variola major virus formed more hemagglutinin and liberated more virus into the medium than did the alastrim virus.

Further Analysis of Variola Minor Virus

All but two of the strains of variola minor virus used in the studies just described were derived from two outbreaks of variola minor, one in England in 1951–1952 and the other in Holland in 1953–1954. As discussed in Fenner et al. (1988), all the European outbreaks of variola minor except one in Italy just after the Second World War were ultimately derived from the Americas. Twenty-seven isolates of variola virus from São Paolo in Brazil were found to behave in the same way as the strains of variola minor virus just described (Downie *et al.*, 1963). This conformed to the clinical and epidemiological features of the disease in Brazil at the time (1960) that the specimens had been collected. However, 23 isolates from mild cases of smallpox in Tanzania (then Tanganyika) between 1961 and 1963 did not have the ceiling temperature of the Brazilian and European strains of variola minor. Careful comparative titrations on chorioallantoic membranes incubated at 35 and 38.3° C (Bedson *et al.*, 1963) showed that although the Tanzanian isolates suffered a bigger drop in titer at the elevated temperature than did the variola major virus isolates, the two distributions overlapped.

Tanzanian and other African isolates also killed chick embryos more slowly than typical variola major virus strains and differed in hemagglutinin production (Dumbell and Huq, 1975). It was suggested that characters shown by variola virus isolates in laboratory tests might be unrelated to human virulence and reflect the random differences that might arise in geographically separate and long continued cycles of transmission, and Dumbell and Huq (1975) tentatively proposed that variola virus could be grouped into Asian, African, and South American varieties. This grouping was further substantiated by the results of mortality tests in baby mice (F. Huq and K.R. Dumbell, unpublished observations).

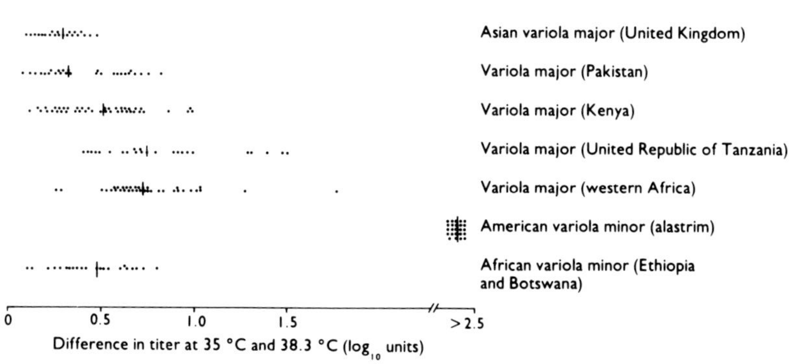

FIG. 7-2. *Differences in log titer when incubating each of 196 variola virus isolates from different geographic regions on the chorioallantoic membrane at 35°C and 38.3°C. The larger the difference in titer, the greater the sensitivity to a raised temperature (i.e., the lower the ceiling temperature). Each dot is the result of assays of a single isolate; the vertical line shows the median measurement for each group of isolates. Alastrim virus = strains of variola minor virus from South America and Europe, originally derived from the Americas. (Based on Dumbell and Huq, 1986.)*

Viruses from Cases of Variola Minor in Africa. The case–fatality rate of variola minor is generally taken as approximately 1%. Careful epidemiological studies in Ethiopia, Somalia, and Botswana during the smallpox eradication campaigns in these countries determined a case–fatality rate of less than 1% in each of these countries (Fenner *et al.*, 1988). Dumbell and Huq (1986) used several markers to compare a group of isolates of variola minor virus from these three countries with variola minor virus isolates (alastrim virus) from South America and Europe, with isolates of variola virus from elsewhere in Africa, from India, Bangladesh, and Pakistan and with well-attested variola major virus isolates from outbreaks following importations into Britain. Isolates of alastrim virus differed from all other variola viruses tested in having a ceiling temperature of less than 38°C (Fig. 7-2) and a characteristic profile in *Hin*dIII digests of their DNA. By contrast, although the group of African variola minor virus isolates differed in minor ways from typical Asian variola major virus, the degree of overlap was such that any individual isolate from Africa could not be confidently identified as being different from the variola major virus group (Table 7-4).

Variola Viruses from Different Endemic Areas

Variola viruses that were maintained for prolonged periods in separate chains of transmission could be expected to acquire slightly different characteristics. It is therefore not surprising that variola viruses endemic in different continents should show broad differences, as noted by Dumbell and Huq (1975). On a somewhat smaller scale, where differences between the variola viruses from adjoining areas were significantly greater than the differences found within each area, this implied that the areas were separate epidemiological units and that interaction between them was rare enough to allow the observed divergence to have arisen. Besides the biological properties just described, two products of variola virus infection of cultured cells show variations which have interesting geographic correlations; the hemagglutinin and a polypeptide of M_r 25K–27K.

Geographic Variations in Variola Virus Hemagglutinin Production. Orthopoxvirus hemagglutinin can be readily demonstrated by hemadsorption of sensitive fowl erythrocytes (Driessen and Greenham, 1959). Variola virus strains differ in their ability to produce hemadsorption on human embryonic fibroblast cultures at 39.5°C (Dumbell and Wells, 1982). This test had been used to characterize a number of variola isolates (Dumbell and Huq, 1975). Their results, with additional data (Table 7-5), showed that variola viruses from a wide zone of Africa

TABLE 7-4
Laboratory Characteristics of Isolates from Cases of Variola Major and Variola Minor[a]

Virus	D5[b]		$\log(X_{35°C}/X_{38.3°C})$[c]		HA/HEp2[d]		Hads 40[e]
	Median	Range	Median	Range	Median	Range	(mode)
Asian variola major strains	3.9	3.2–5.9	0.35	0.1–0.8	16	2–128	A (74%) B (26%)
African variola major strains	5.2	3.4–7.0	0.6	0.2–1.7	2	2–64	B (73%) A (27%)
Alastrim strains	6.4	5.6–7.0	2.5	2.5	2	2–16	C (85%) B (15%)
African variola minor strains	4.8	4.2–6.9	0.5	0.2–0.8	2	2–8	B (73%) A (27%)

[a] Data from Dumbell and Huq (1986).

[b] D5: The $\log_{10}$ dose of virus inoculum which will result in a harmonic mean survival time of 5 days for chick embryos inoculated on the chorioallantoic membrane at 12 days of age and incubated at 35°C.

[c] $\log(X_{35°C}/X_{38.3°C})$: The difference in titer (as $\log_{10}$ units) when virus is titrated on chick chorioallantoic membrane at 35°C and 38.3°C.

[d] HA/HEp2: The reciprocal end-point dilution of a hemagglutinin titration of the yield in 1.0 ml of 10^6 HEp2 cells inoculated with virus and incubated 3 days at 37°C.

[e] Hads 40: Production of confluent (A), scattered (B), or minimal (C) hemadsorption in cultures of human embryo fibroblasts which had been inoculated at a multiplicity of 1.0 pock forming unit per cell and incubated 48 hours at 40°C.

TABLE 7-5

Hemadsorption in Human Fibroblast Cell Cultures, Infected 48 Hours Previously with Various Isolates of Variola Virus at High Multiplicity and Incubated at 40°C[a]

		Number of isolates with hemadsorption		
Source of isolates	Dates	A (confluent)	B (focal)	C (absent)
Asian variola major virus				
UK major outbreaks	1946–1959	14	0	0
UK S. Wales	1962	1	0	0
UK Midlands	1962	0	4	0
India and Bangladesh	1961–1971	8	3	0
Pakistan—Karachi and Lahore	1961–1969	15	1	0
Pakistan—rural	1969	3	7	0
Indonesia	1968–1971	21	13	0
African variola major virus				
Kenya	1964–1967	25	18	0
Tanzania	1961–1965	2	20	0
Zaire	1963–1970	2	12	0
West Africa	1963–1970	0	27	0
South Africa	1961–1972	9	3	0
Variola minor virus				
Botswana and Ethiopia	1972–1973	6	16	1
Alastrim				
Europe	1953–1954	0	0	7
Brazil and Venezuela	1960–1962	0	5	20

[a] Includes data from Dumbell and Huq (1975, 1986).

differed from variola virus strains from urban areas in the Indian subcontinent. Kenya was an exception; less than half the the Kenyan isolates behaved like the other African strains of variola virus, perhaps because of the frequent importations of smallpox into Kenya from Asia (Seymour-Price *et al.*, 1960).

Indonesian isolates, which had been cultured from specimens taken during the most active phase of the smallpox eradication campaign and were largely from rural areas, also showed a mixture of traits. Isolates from cases in rural areas of Pakistan differed from those obtained from cases in major cities.

Further discrimination between strains was obtained by titrating the amounts of hemagglutinin produced when HEp2 cell cultures were inoculated with variola viruses from stocks prepared on the chorioallantoic membrane (Dumbell and Huq, 1986). Although the results were generaly similar to those of the hemadsorption test, greater consistency

TABLE 7-6

Differences in Polypeptides of Isolates of Variola Virus[a]

Isolates with 27K peptide	Isolates with 25K peptide
Variola major virus	Variola major virus
United Kingdom, 6 isolates	UK ex Pakistan, 2 isolates
India and Bangladesh, 4 isolates	Pakistan, 3 isolates
Indonesia, 4 isolates	Kuwait ex Pakistan, 3 isolates
Central Africa, 6 isolates	Iran ex Afghanistan, 4 isolates
Variola minor virus	India (Vellore), 2 isolates
Africa, 5 isolates	
Alastrim virus, 4 isolates	

[a] Data from L. Harper (personal communication, 1983).

was shown among the Kenyan isolates, and isolates from Togo and Upper Volta differed from isolates taken during the same years from other countries in western Africa.

When inoculated into a continuous line of rabbit kidney cells (RK13), most strains of variola virus produce small, transient, nontransmissible hyperplastic foci (Gispen, 1970) and no hemagglutinin (F. Huq and K. R. Dumbell, unpublished observations, 1971). Two strains of "whitepox" virus, later identified as variola virus from India (Dumbell and Kapsenberg, 1982) produced hemagglutinin in RK13 cells (Dumbell, 1974) and hemorrhagic lesions in the skin of a proportion of inoculated rabbits (Gispen, 1970). Subsequently K. R. Dumbell (unpublished observations, 1975) found that isolates of variola virus from Botswana produced hemagglutinin and cytolytic degeneration on first passage in RK13 cells. As described earlier, variola virus was not transmissible in RK13 cultures grown in medium containing calf serum but readily adapted to continued grown in RK13 cells grown and maintained in rabbit serum.

Geographic Variation in Variola Virus Polypeptides. Following the comparison of the intracellular polypeptides of various species of orthopoxvirus (see Chapter 2), L. Harper and H. S. Bedson (unpublished observations, 1978) examined 48 strains of variola virus by the same technique. Most of the peptide bands were conserved throughout all the strains, but two bands of apparent M_r 27K and 25K were mutually exclusive, with each isolate exhibiting one or the other. Four of the strains were alastrim virus and had the 27K polypeptide. Thirty-six of the remaining 44 strains could be assigned to specific endemic areas, as shown in Table 7-6. All of the strains which originated in Pakistan or

Iran (derived from Afghanistan: Fenner *et al.*, 1988) had a 25K polypeptide; all but two of the others had the 27K polypeptide. These two were the parallel isolates made in Holland in 1964 from monkey kidney cell cultures and shown subsequently to be due to laboratory contamination with a strain of variola virus isolated from smallpox scabs originating in Vellore, in Southern India (Dumbell and Kapsenberg, 1982).

Strains of Variola Major Virus from Cases of Differing Severity. Sarkar and Mitra (1967) attempted to show that variola viruses isolated from severe cases of variola major in Calcutta were more virulent than virus isolated from mild cases. They used virulence tests in baby mice and chick embryos as their markers and found that these were more often positive in isolates from the severe cases, with confluent rashes, than from cases from moderate severity, with a discrete rash. The level of severity was based on clinical presentation and the vaccination status was not recorded. In a subsequent study (Sarkar and Mitra, 1968) the vaccination status of the subjects was recorded, and it is apparent that many of the milder smallpox cases they saw had evidence of previous vaccination. Furthermore, epidemiological evidence suggests that most cases of smallpox in unvaccinated persons acquired from cases of hemorrhagic type smallpox (the most severe form of the disease) were not particularly severe (Rao, 1972). It is impossible to interpret the results of Sarkar and Mitra; one has to conclude that no satisfactory laboratory correlate of the virulence of variola virus for humans has yet been discovered.

UNUSUAL STRAINS OF VARIOLA VIRUS

Two unusual types of variola virus strains have been described: a unique example that failed to produce LS antigen in infected cells, and some strains that produced syncytia in cell cultures.

Strains Lacking a Major Antigen

Marennikova *et al.* (1976b) studied an isolate (K-5-67) from a smallpox outbreak in Kuwait that was described by Arita *et al.* (1970), and had originated from an importation from Pakistan. This isolate was typical of variola major virus in pock morphology, ceiling temperature, reaction in rabbit skin, and effects on various cell lines. However, the original specimen had not reacted in the usual way with anti-vaccinia serum in gel precipitation tests. Antigen preparations from K-5-67 virus cultures gave some bands with anti-vaccinia serum, but lacked the same strong line that differentiates cowpox virus from vaccinia virus in this test (Rondle and Dumbell, 1962), and is now known to be due to the LS antigen (see Chapter 2). The antigen is present in cowpox preparations in a nondiffusible form and can be demonstrated as a diffusing precipitin after trypsin treatment. Trypsin treatment did not reveal this line in preparations from cells infected with K-5-67 and absorption of antivaccinia serum with K-5-67 did not remove its ability to precipitate with material from cells infected with vaccinia virus or other strains of variola virus. Marennikova *et al.* (1976b) concluded that this antigen was absent from K-5-67.

Other specimens from the Kuwait outbreak had been sent to Britain for laboratory diagnosis, and two isolates (1628, 1629) were obtained from Dr. M. S. Pereira for further studies. K. R. Dumbell (unpublished observations) showed that only one of these two isolates (1628) lacked the antigen, and this strain was identical to K-5-67 in all tests. The other isolate (1629), like all other strains of variola virus that have been examined, produced the antigen. Restriction endonuclease digests of DNA from strains 1628 and 1629 with *Hind*III or with *Xho*I showed indistinguishable electropherograms. Tests on intracellular polypeptides produced in cultured cells (L. Harper and H. S. Bedson, unpublished observations) showed that all three isolates (1628, 1629, and K-5-67), had the 25K polypeptide described above, but strains K-5-67 and 1628 lacked a 96K polypeptide which was present in 1629 and in all other variola virus strains examined. This is the polypeptide which cross-reacts serologically with the protein of the A-type inclusion body of cowpox virus, and has been identified as the LS antigen of vaccinia virus (see Chapter 2).

Interpretation of this finding is complicated by the fact that some of the specimen sent to Britain arrived in vaccine vials and mixed cultures of small and large pocks were obtained on the chorioallantoic membrane (A. Macdonald, personal communication). Small pocks were individually picked by Dumbell for the isolates 1628 and 1629. The specimen from which K-5-67 was isolated was also contaminated with vaccinia virus (Dr. S. S. Marennikova, personal communication). Some of the specimens examined may therefore have contained recombinants between variola virus and vaccinia virus.

Strains Producing Syncytia

Tsuchiya and Tagaya (1972) reported that after 35 passages on the chorioallantoic membrane, the variola virus strain "Yamamoto" produced two types of foci in JINET cell cultures. Separate stocks were

isolated by limit dilution passage: strain "P," which showed cytopathology typical of variola virus, and strain "G," which produced multinucleated giant cells and progressed to a complete syncytial effect. Both P
and G strains had the pock morphology and other biological markers
typical of variola virus. No hemagglutinin was produced by strain G in
JINET or several other cell lines. G strain attained slightly higher titers of
virus than P when grown on either JINET cells or on the chorioallantoic
membrane, so presumably further continued serial passages of the
original material would eventually have resulted in a pure culture of
G. Tsuchiya and Tagaya found only the typical variola virus foci in cells
inoculated with strain Harvey at the third or fifth passage, and
K. R. Dumbell (unpublished observations) found no change in the
cytopathology produced by strain Harvey after 40 passages on the
chorioallantoic membrane. However, he found that after 70 passages on
the chorioallantoic membrane, another strain of variola virus, Hinden,
produced a syncytial effect in human embryonic fibroblast cultures and
also lost the ability to produce hemagglutinin.

USE OF INDIVIDUAL MARKERS AS
EPIDEMIOLOGICAL TRACERS

The range of biological characteristics described sometimes made it
possible to identify particular strains of variola virus that had been
involved in particular incidents, like laboratory-associated outbreaks.

Laboratory-Associated Outbreak in London, 1973

The index case in this outbreak was presumed to have been infected
on a day on which three strains of orthopoxvirus were handled in the
laboratory concerned. One of these was the standard variola virus strain
Harvey, the other two were "whitepox" viruses, which at that time had
not been shown to be strains of variola virus. Characterization of the
three viruses, together with the isolate from the index case, revealed
that the latter differed in two respects from the two "whitepox" viruses,
but did not differ from strain Harvey in the tests that could be made
(Dumbell, 1974), and it was concluded that the index case had been
infected with variola virus strain Harvey (Report, 1974).

Laboratory-Associated Outbreak in Birmingham, England,
1978

The virus isolated from the index case in this episode was unusual, in
that for the first time in a long experience of smallpox diagnosis,

virologists at the Central Public Health Laboratory, Colindale, London, reported recovery of a strain of variola virus which produced a syncytial effect in HeLa cells on primary isolation (M. S. Pereira, personal communication, 1978). This syncytial response was also observed in cells inoculated with virus after a few passages on the chorioallantoic membrane (Dumbell, in Report, 1980). The same unusual feature was shown by one of the strains of variola virus (strain Abid) which had been handled in the Birmingham laboratory on the days relevant to the infection. These two isolates had other characters in common, including the 27K rather than the 25K polypeptide (L. Harper and H. S. Bedson, unpublished observations), which together distinguished them from the other strains of variola virus under consideration (Report, 1980).

VARIOLA-LIKE VIRUSES FROM ANIMALS—THE "WHITEPOX" VIRUSES

The success of the smallpox eradication program depended on the absence of an animal reservoir of variola virus. In the course of studies on the nature and distribution of monkeypox virus, many specimens of animal tissues were examined in the WHO Collaborating Centers in Atlanta, (United States) and Moscow (USSR). From such material, four isolations were made in the Moscow laboratory of viruses which could not be distinguished from variola virus, either in their biological characters (Marennikova *et al.*, 1971, 1972b, 1976a; Shelukhina *et al.*, 1975) or in the restriction maps of their DNA (Esposito *et al.*, 1978; 1985; Dumbell and Archard, 1980). Two other strains of an orthopoxvirus resembling variola virus had been isolated in 1964 from normal cynomolgus kidney cells in the Public Health Laboratory in Bilthoven, the Netherlands. Until their status was clarified these six strains of orthopoxvirus were given the noncommital name of "whitepox" viruses.

After DNA mapping had been carried out it was clear that all the "whitepox" virus strains belonged to the variola virus species; the important question was whether they were derived from the tissues of the animals from which they were said they were isolated: cynomolgus monkeys for the Netherlands isolates and a chimpanzee, a monkey, a squirrel, and a multimammate rat for the Moscow isolates.

Seeking an explanation for the recovery of the "whitepox" viruses, Dr. Marennikova postulated that they may have arisen as "white clones" of monkeypox virus. Marennikova and Shelukhina (1978) recovered such "white clones" from organs of hamsters that had been inoculated some weeks earlier with the Moscow stock of monkeypox virus, and Marennikova *et al.* (1979) made similar isolations on the chorioallantoic membrane. All these isolates were identical, and were indistinguishable by biological tests from the Zaire and Utrecht "whitepox" virus isolates. Subsequently these reports were discounted, because they could not be repeated (Dumbell and Archard, 1980; Esposito *et al.* 1985), and because there was evidence of contamination in some of the parental monkeypox virus stocks. Further, the extensive differences between the genomes of variola and monkeypox viruses, described above, would preclude the rapid conversion of the one virus into the other.

The Netherlands "Whitepox" Viruses

These two viruses (64/7255 and 64/7275) were isolated in a diagnostic laboratory from cynomolgus monkey kidney cell cultures that had been inoculated with specimens from an outbreak of diarrhea in Holland. It was considered at the time that the two poxvirus isolates must have originated from the monkey kidney cells themselves. The viruses proved to be variola-like in their laboratory characters and two positive isolations from smallpox specimens from Vellore, India, had been made at the time. In 1982 detailed comparisons were made of these four viruses; in biological characters and in DNA structure the two "whitepox" viruses could be distinguished from one of the variola virus strains but not from the other (Dumbell and Kapsenberg, 1982). The DNA pattern of all four was somewhat unusual (Plate 7-2) and suggested a close relation between the four viruses. It was concluded that infection had been transferred while changing medium in a set of monkey kidney tube cultures and that these particular "whitepox" viruses were strains of variola virus derived from a case of smallpox in India.

The Moscow "Whitepox" Virus Isolates

The other four "whitepox" virus isolates had all been obtained from the organs of animals killed in Zaire. One (chimp 9) was isolated from a chimpanzee (Marennikova *et al.*, 1972b), one (MK-7-73) came from a monkey (Shelukhina *et al.*, 1975), and two (RZ-10-74 and RZ-38-75) were isolated from rodents (Marennikova *et al.*, 1976a). All four of these viruses were shown by DNA analysis to be typical strains of variola virus (Dumbell and Archard, 1980; Esposito and Knight, 1985; Esposito *et al.*, 1985). The biological profiles of strains of variola virus reported by Dumbell and Huq (1986) included those of two of the "whitepox" virus isolates, chimp 9 and MK-7-73. These two viruses were the only isolates from western Africa which had profiles indistinguishable from those characteristic of Asian isolates of variola virus. Dumbell and Huq considered that they could reasonably be excluded from the viruses

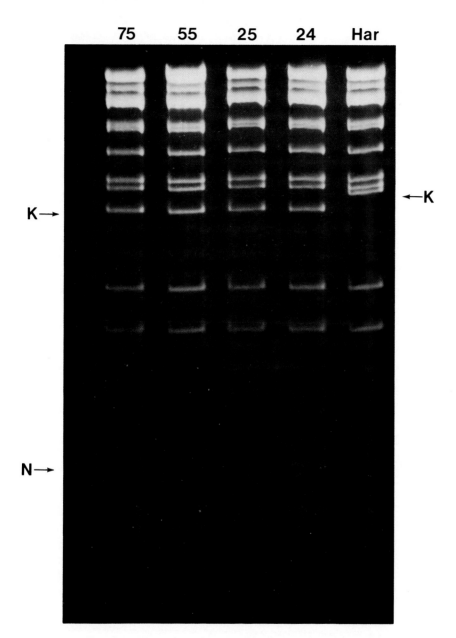

PLATE 7-2. *Use of restriction endonuclease digestion of genomic DNA as a method for species and strain diagnosis of orthopoxviruses. Comparison of digestion products of DNA of two "whitepox" viruses and three strains of variola virus, after digestion with* SalI. *75, 55, Netherlands "whitepox" virus strains; 25, 24, strains of variola virus from Vellore, India; Har, Harvey strain of variola virus. The identity of the pattern of one of the Vellore strains (24, note fragments "K" and "N") and the "whitepox" viruses confirmed the suspicion that the latter were contaminants of normal cynomolgus kidney cells with that strain of variola virus. (From Dumbell and Kapsenberg, 1982.)*

maintained by human chains of transmission in Zaire; their true origin remains unknown, but like the Netherlands isolates, they are most likely to have been laboratory contaminants.

LABORATORY AIDS TO THE DIAGNOSIS OF SMALLPOX

In endemic areas, or in the midst of an outbreak after an importation, a diagnosis of smallpox could usually be made on clinical grounds alone. When the diagnosis was in doubt and laboratory aid was sought, the most frequent problem was to decide between smallpox and chicken pox, and most of the early tests (Paul, 1915; Buddingh, 1938) were designed to achieve this. Before the widespread use of electron microscopy the only way of demonstrating virions was by intensive staining methods or by silver impregnation (Gispen, 1952). Virus particles, or "elementary bodies," could then be made out with the light microscope and a presumptive diagnosis might rapidly be obtained (see Plate 1-1). Considerable experience was required and even experts might be uncertain when the lesions had become pustular. More reliable methods were the use of complement fixation or precipitation tests to demonstrate orthopoxvirus antigen in fluid from the lesions or extracts of scabs, but precipitation tests were not reliable on specimens which took a long time to reach the laboratory (Nakano, 1973).

With developing knowledge of virology, the simplest method of differentiation was to demonstrate poxvirus particles in preparations made from vesicle fluid, scabs, or scrapings from early papular lesions. The examination of negatively stained preparations of such material in the electron microscope was quick, sensitive, and applicable at all stages of the lesions, from scrapings of early papules to extracts of scabs. The morphology of the virion was well preserved in specimens that had to be sent long distances and the test could also reveal herpesvirus particles, giving positive evidence of the most likely alternative diagnosis, chickenpox. It was the sheet-anchor of the very extensive laboratory diagnosis carried out in the Intensified Smallpox Eradication Program, especially in the certification of eradication (see Fig. 11-10).

With the discovery of human monkeypox in Africa in 1970, another dimension was added to the diagnostic problem (see Chapter 8). The best biological method to differentiate between these two species of orthopoxvirus was to isolate virus on the chorioallantoic membrane or in tissue culture, species diagnosis then being based on the type of pock on the membrane and the reaction of the rabbit to intradermal inoculation of the virus.

By the time that smallpox had been eradicated, the preferred method for differentiating between different species of orthopoxvirus was to characterize the DNA by digestion with restriction enzymes (see Plate 7-2). Initially this was carried out with DNA extracted from purified suspensions of virions, but Esposito *et al.* (1981) showed that it could be done rapidly and safely by extracting DNA from infected cell cultures by treatment with detergents.

CHAPTER 8

Monkeypox Virus

Monkeypox virus was discovered in 1958, when it was isolated from lesions of a generalized vesiculopustular disease among captive monkeys at the State Serum Institute, Copenhagen (von Magnus *et al.*, 1959). Between 1958 and 1968 a similar disease was recognized in several other colonies of captive primates in Europe and the United States, but it has not been reported since 1968. However, in 1970 it was found that a smallpox-like disease of humans living in tropical rain forest areas in several countries in western and central Africa was caused by monkeypox virus (Lourie *et al.*, 1972; Marennikova *et al.*, 1972a). Coming at a time when smallpox had just been eliminated from these countries (see Chapter 11), this discovery led to an intensive investigation of the virus and the human disease, coordinated by the World Health Organization and designed to assess its public health importance and to determine whether it represented a threat to the global smallpox eradication campaign (Jezek and Fenner, 1988).

TABLE 8-1

Outbreaks of Monkeypox in Captive Primates[a]

Country	Episode and reference	Virus isolation	Time	Species affected	Origin	Interval after arrival
Denmark	1a. von Magnus et al. (1959)	+	June 30, 1958	Cynomolgus	From Singapore by air	62 days
	1b. K. L. Fennestad (personal communication, 1980)	+	November 7, 1958		From Singapore by air	51 days
Netherlands	2. Peters (1966)	+	December 21, 1964	Index case: giant anteater; later orang-utan, gorilla, chimpanzee, gibbon, squirrel monkey, cercopithecus, marmoset	To zoo from dealer, then infections by contact in Blijdorp Zoo, Rotterdam	?
France	3. Milhaud et al. (1969)	+	November 29, 1968	Chimpanzee	Sierra Leone	11 days
United States	4. Prier et al. (1960); J. G. Prier (personal communication, 1970)	+	February, 1959	Cynomolgus; later rhesus also	Malaysia	"Newly arrived"

5. McConnell et al. (1962)	+	1962	Cynomolgus; serological positives in rhesus and African green monkey	?	9 months
6. C. Espana (personal communication, 1967)	+	December 1966 to March 1967	Indian and Malayan langurs, rhesus, cynomolgus and pigtailed macaques	India, Malaysia	2 years
7. A.H. Bruschner (personal communication, 1967)	. . .[b]	November, 1965	Cynomolgus	Malaysia and Philippines	?
8. M. Z. Brierly (personal communication, 1967)	. . .	1966	Rhesus	India	"Recently arrived"
9. J. H. Vickers (personal communication, 1967)	. . .	Before 1966	Rhesus	India	?

[a] Based on Arita et al. (1972).
[b] . . . , Not done.

MONKEYPOX IN CAPTIVE PRIMATES

Occurrence of Outbreaks

From the outset of the Intensified Smallpox Eradication Program in 1967, the close resemblance of the clinical manifestations of smallpox in humans and monkeys and monkeypox in captive primates focused the attention of the Smallpox Eradication Unit of the World Health Organization on monkeypox virus as a potential threat to the smallpox eradication campaign (Arita and Henderson, 1968). The World Health Organization therefore inquired of laboratories in Europe and North America which used monkeys (27 laboratories in 1968, 51 in 1970; Arita *et al.*, 1972) concerning the occurrence of monkeypox, asking specifically whether any human infections had occurred among laboratory workers or animal handlers. The ensuing investigations revealed four other reported outbreaks and four hitherto unreported outbreaks in captive primates (Table 8-1), but there were no reports of infection in humans. Monkeypox virus was recovered in six of these episodes. All except episode 3 occurred in Asian monkeys, although in some outbreaks African primates (and in episode 2, New World monkeys) were also infected.

The circumstances of these outbreaks have been summarized by Arita *et al.* (1972). One episode described in their paper, but omitted from Table 8-1, calls for special comment, namely the observation made by Gispen and Kapsenberg (1966) of the National Institute of Public Health in Bilthoven, the Netherlands, that monkeypox virus had been recovered from normal cynomolgus kidney cell cultures. Subsequent examination of the laboratory records led Dr. J. G. Kapsenberg (personal communications, 1980, 1983) to decide that this isolation was probably due to inadvertent laboratory contamination of the culture with monkeypox virus, which had been isolated in the same laboratory at about this time from animals infected in the Blijdorp Zoo outbreak (episode 2).

Seven of the nine outbreaks of monkeypox in captive monkey colonies between 1958 and 1968 occurred in monkeys shipped from Asia, leading to the suspicion that the reservoir of monkeypox virus was probably located in that continent. However, no orthopoxvirus antibodies were found in over 1000 monkey sera collected in India, Indonesia, Japan, and Malaysia, in collaborative serological surveys organized by the World Health Organization (Arita *et al.*, 1972). After the discovery of human monkeypox in Africa in 1970 (see below), sera were collected from monkeys and other animals in Zaire and several countries in western Africa. Monkeypox virus-specific antibodies were demonstrated in sera from seven species of monkeys and two species of squirrel, and monkeypox virus was recovered from the organs of a squirrel (see below).

Although primates from Asia, Africa, and South America (and an anteater from South America; Peters, 1966) have been infected with monkeypox virus in captivity, there is no evidence that the virus occurs naturally anywhere except in Africa. During the period 1958–1968, large numbers of primates were being imported into Europe and North America from Asia, and smaller numbers from western Africa, mainly for the manufacture and safety testing of poliovaccine. At that time the conditions under which monkeys were moved from their place of capture in Asia or Africa to the recipient laboratory in Europe or North America presented many opportunities for them to be infected with agents carried by other wild animals or by man while in transit (Kalter and Heberling, 1971). The cessation of outbreaks after 1968 can be ascribed to improvements in the conditions of shipment of primates at about that time, and the much more extensive use by laboratories of monkeys bred in captivity in Europe and North America.

Clinical Features

The clinical features of naturally occurring cases in cynomolgus monkeys have been described by von Magnus *et al.* (1959) and Sauer *et al.* (1960). No signs are detected until the rash appears, usually as a single crop of discrete papules over the trunk and tail and on the face and limbs, being particularly abundant on the palms of the hands and the soles of the feet (Plate 8-1). The papules become vesicular and then pustular and are often umbilicated. Scabs develop and fall off 7–10 days after the onset of the rash, leaving small scars. Circular, discrete ulcers about 2 mm in diameter often occur in the oropharynx.

The severity of signs varied among the several different primate species infected in the outbreak in the Blijdorp Zoo in Rotterdam (episode 2, Table 8-1). All species suffered from a generalized disease characterized by pocks on the skin, lips, and mucous membranes. Orangutans were particularly susceptible, several dying in the acute viremic stage, before the skin lesions were fully developed.

THE PROPERTIES OF MONKEYPOX VIRUS

In its morphology as seen with the electron microscope, the virions of monkeypox virus are indistinguishable from those of other orthopox-viruses (Nakano, 1985). In serological tests, monkeypox virus cross-

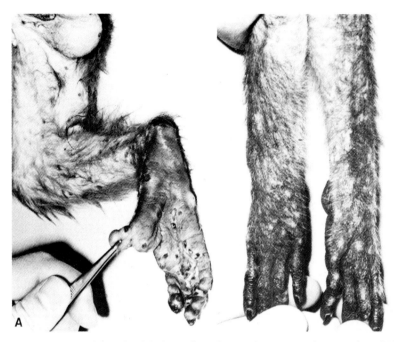

PLATE 8-1.   *Generalized lesions of monkeypox in a cynomolgus monkey. (A) Acute stage; pustules on the leg and sole of the foot. (B) Convalescent stage; healing pustules and scars. (From von Magnus et al. (1959), courtesy of Dr. K. L. Fennestad.)*

reacts with all other orthopoxviruses, but can be distinguished from vaccinia and variola viruses (the other orthopoxviruses that used commonly to infect humans) by tests with absorbed sera (Gispen and Brand-Saathof, 1974; Esposito *et al.*, 1977a,b).

The biological characteristics of monkeypox virus, compared with those of other orthopoxviruses, are summarized in Table 1-2, and the restriction endonuclease map of monkeypox virus DNA is compared with corresponding maps of DNAs of other species of *Orthopoxvirus* in Fig. 1-1.

Pathogenicity for Laboratory Animals

Monkeypox virus has a broad host range and produces lesions in most of the common laboratory animals and in many species of primate. In their initial description of monkeypox virus, von Magnus *et al.* (1959) were impressed with the similarity between the pocks it produced on the chorioallantoic membrane and those produced by variola virus. They were therefore concerned to determine whether the disease that they had observed in laboratory primates was caused by variola virus or not. The behavior of monkeypox virus in laboratory animals revealed two important differences from variola virus; it produced a large indurated lesion with a hemorrhagic center after intradermal inoculation in rabbits, and it could be passaged indefinitely in rabbit skin and in adult mice by the intracerebral route. Variola virus produces essentially negative results by both of these tests (see Chapter 7).

Effect on Chick Embryos.   After a 3-day incubation at 35°C, monkeypox virus produces grayish pocks with a hemorrhagic center on the chorioallantoic membrane, clearly distinguishable from the larger hemorrhagic pocks of cowpox virus and the opaque white pocks of variola virus (see Plate 4-2). At higher temperatures of incubation the distinction from variola virus pocks is not so clear. Strains of vaccinia virus vary greatly in their pock morphology; dermal strains lack the hemorrhagic center seen with monkeypox virus and the pocks of most neurovaccinia strains are larger.

The ceiling temperature for production of pocks on the chorioallantoic membrane by monkeypox virus (39°C) lies between that of cowpox and vaccinia viruses (40 and 41°C, respectively), and variola virus (37.5–38.5°C) (Bedson and Dumbell, 1961). Its lethality for the chick embryo is similar to that of cowpox and ectromelia viruses and intermediate between those of vaccinia and variola viruses (Bedson and Dumbell, 1961).

Rabbits and Laboratory Rodents.   The behavior of monkeypox virus in rabbits and laboratory rodents was studied by Marennikova and Shelukhina (1976b). After intravenous inoculation adult rabbits developed systemic disease with a generalized pustular rash, conjunctivitis, and rhinitis. Rabbits inoculated by scarification developed a localized papulopustular eruption, whereas intradermal inoculation induced a large indurated lesion with a hemorrhagic center. Generalized lesions also occurred in some of these animals. Young rabbits were much more susceptible than adult rabbits and developed generalized disease which was usually fatal, even after infection *per os* or by the intranasal route. Infection was transmitted by contact to uninfected baby rabbits within the same litter, or even in other cages in the same room (K. McCarthy, personal communication, 1971).

Mice were highly susceptible to infection by various routes, including footpad inoculation; 8-day-old mice usually died but some older mice survived. Adult rats appeared to be resistant but newborn rats died after intranasal inoculation. Hamsters appeared to be resistant, but after intracardiac inoculation lesions of the internal organs were produced. Virus could be reisolated from the kidneys for up to 3 weeks. Guinea pigs did not develop lesions or signs of infection after injection by intravenous, intraperitoneal, or subcutaneous routes (Marennikova and Shelukhina, 1976b), but injection in the footpad resulted in a local swelling and the development of a granulomatous lesion (Prier *et al.*, 1960).

T. Kitamura (personal communications, 1978, 1982) showed that the common African commensal rodent, *Mastomys natalensis*, was highly susceptible to intraperitoneal and intranasal inoculation of monkeypox virus.

Primates.   In outbreaks in captive primates (see Table 8-1), Asian monkeys (cynomolgus or rhesus) usually suffered from a generalized rash, whereas *Cercopithecus aethiops*, if present, apparently suffered only subclinical infections. Baboons (*Papio cynocephalus*), an African primate unlikely to be naturally infected, developed a generalized rash and some animals died (Heberling *et al.*, 1976). None of the several species of forest-dwelling primates that have yielded monkeypox-virus-specific sera (see below) has been inoculated experimentally.

Several species of primates not used as laboratory animals were exposed to infection with monkeypox virus in the Blijdorp Zoo outbreak in Rotterdam in 1964 (Peters, 1966). All were housed within a single large enclosure, with many separate apartments, into which two ant-eaters (*Myrmecophaga tridactyla*), which had not been quarantined, were introduced. Two weeks later both anteaters were ill, with a generalized rash, and infection with what proved to be monkeypox virus spread to many other animals in the monkey house. The responses of different species varied (Peters, 1966; Gispen *et al.*, 1967), as follows:

Orangutan (*Pongo pygmaeus*): all very ill and 6 out of 10 died, 5 of them before the rash had developed
Gorilla (*Gorilla gorilla*): two cases with generalized rash, one severe
Chimpanzee (*Pan troglodytes*): developed eruption on face and lips but no signs of general illness
Gibbon (*Hylobates lar*): one animal died after severe illness
Marmoset (*Callithrix sp.*): swelling of eyes and nose, one died with viremia
Squirrel monkey (*Saimiri sciureus*): one died with rash
Owl-faced monkey (*Cercopithecus hamlyni*): ill with eruption on lips, but recovered

No signs of illness were seen in primates of several other species: *Oedipoidas oedipus, Ateles paniscus, Ateles geoffreyi, Ateles fusiceps, Symphalangus syndactylus, Cebus capucinus, Cercopithecus mona,* and *Cercopithecus l'hoesti.*

Behavior in Tissue Culture

Monkeypox virus can replicate in a wide range of cultured cells, from simian, human, rabbit, bovine, guinea pig, and murine sources (McConnell *et al.*, 1962; Marennikova *et al.*, 1971). It produces a cytopathic effect consisting of rounding up, granulation, and condensation of cells and final detachment from the glass or plastic substrate. In suitable dilutions it usually produces discrete plaques (Rouhandeh *et al.*, 1967). When it was first discovered considerable attention was devoted to methods of differentiating monkeypox virus from variola virus. One characteristic useful for this purpose was that variola virus but not monkeypox virus produced plaques in pig embryo kidney cells (Marennikova *et al.*, 1971, 1972b) and grew to high titer in pig kidney cell lines (Y. Ueda and K. R. Dumbell, unpublished observations). Further, variola virus characteristically produces hyperplastic foci rather than plaques (see Chapter 7).

Comparison of DNA Maps of Strains of Monkeypox Virus

Esposito and Knight (1985) analyzed the DNA of 12 strains of monkeypox virus, 4 recovered from outbreaks in laboratory primates in Europe and North America and 8 from human cases from 4 different countries in western and central Africa. The *Hind*III restriction maps for the DNAs of these strains of monkeypox virus, and the DNA of two strains each of variola and vaccinia viruses, are compared in Fig. 8-1.

The DNAs of all strains of monkeypox virus are clearly different from those of both variola and vaccinia viruses. There is some intraspecies variation, which is associated with geographic origin, rather than the animal of origin or year of isolation. The differences are evident from the sizes of the fragments in the *Hind*III maps (Fig. 8-1), but are more clearly shown in *Pst*1 digests, in which three different patterns are evident: Zaire, Nigeria (and Benin), and strains originating in western Africa (M. Richardson and K. R. Dumbell, unpublished observations; Plate 8-2). The strains from all captive primates examined resemble those of the human cases in Liberia and Sierra Leone. Outbreaks in captive monkeys are more likely to have originated from West Africa, since exports of monkeys from Africa during the late 1950s and the 1960s (mainly *Cercopithecus aethiops*) were from countries in that part of the continent.

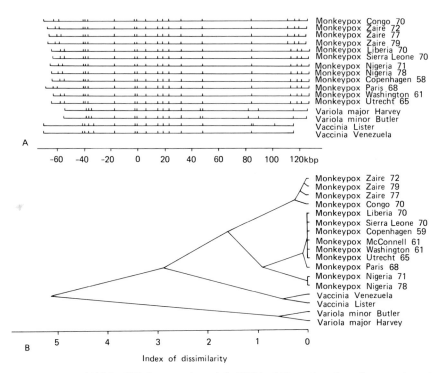

FIG. 8-1. (A) HindIII cleavage sites of the DNA of 12 strains of monkeypox virus, 2 strains of variola virus, and 2 strains of vaccinia virus. The upper eight strains of monkeypox virus were obtained from cases of human monkeypox; the lower four from primates, probably originating from countries of western Africa. (Data from Esposito and Knight, 1985.) (B) Dendrogram showing the similarities and differences between these orthopoxvirus strains, using centroid sorting strategy (Gibbs and Fenner, 1984). Three geographic groups can be distinguished: one from Zaire (from both human and animal sources), another from Nigeria, and a third from countries in western Africa (see also Plate 8-2).

White Pock Mutants

Like other orthopoxviruses that produce hemorrhagic pocks on the chorioallantoic membrane (see Chapter 4), monkeypox virus produces white pock mutants. These were first observed by Bedson (1964) and first reported by Gispen and Brand-Saathof (1972). They were shown closely to resemble the parental monkeypox virus in tests for species-specific antigen (Gispen *et al.*, 1976) and intracellular polypeptide patterns (Harper *et al.*, 1979). Unlike the white pock mutants of cowpox and rabbitpox viruses, at least some white pock mutants of monkeypox

L   ZS   ZH   C   SL   E   L   N   B   ZS

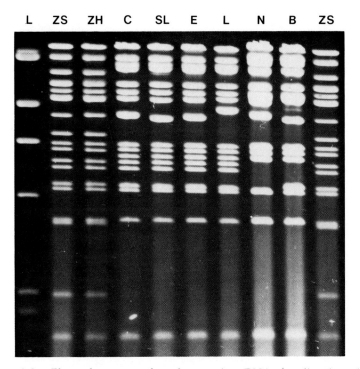

PLATE 8-2.   *Electropherograms of monkeypox virus DNA after digestion with the restriction endonuclease PstI. L, Lambda (size marker); ZS, Zaire squirrel; ZH, Zaire human; C, Copenhagen; SL, Sierra Leone; E, Espana; L, Liberia; N, Nigeria; B, Benin. Three patterns can be differentiated among strains from different geographic regions: Zaire (from human and squirrel); Nigeria (the Benin case was infected in Nigeria); and strains originating from other western African countries (Sierra Leone, Copenhagen, and Espana, the last two from outbreaks in captive monkeys, but probably derived from western Africa). (M. Richardson and K. R. Dumbell, unpublished observations, 1987.)*

virus produce hemorrhagic lesions on inoculation into the skin of rabbits (Gispen and Brand-Saathof, 1972). The white pock mutants of monkeypox virus were subjected to more detailed study after Marennikova *et al.* (1979) had claimed that monkeypox virus was the source of "whitepox" viruses, which were indistinguishable from variola virus (see Chapter 7).

In an attempt to reproduce Marennikova's results, but using a monkeypox virus stock from their own laboratory, Dumbell and Archard (1980) examined 11 white pock mutants of monkeypox virus.

Unlike Marennikova's "white clones," which uniformly resembled variola virus in their biological properties (and in their DNA maps; see Chapter 7), but like white pock mutants of rabbitpox virus (Gemmell and Fenner, 1960), they varied from one another. Some shared several biological characteristics with "whitepox" viruses, but the DNA maps of all of them resembled that of monkeypox virus DNA (Fig. 8-2). Esposito *et al.* (1985) confirmed this finding, and concluded that "spontaneous mutation of the DNA of monkeypox virus to the variola-like DNA of 'whitepox' virus was genetically impossible." The most likely interpreta-

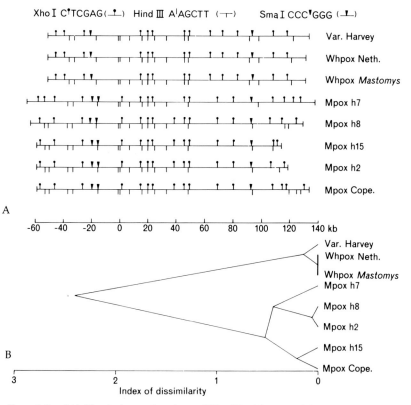

Fig. 8-2. *(A) Physical map locations of HindIII, XhoI, and SmaI cleavage sites in DNA from three strains of variola virus (Harvey; Netherlands "whitepox" 64-7255; Moscow RZ-10-74, a "whitepox" virus from Mastomys natalensis), wild-type monkeypox virus Copenhagen and its white pock mutants h2, h7, h8, and h15. (Data from Dumbell and Archard, 1980.) (B) Dendrogram showing the similarities and differences between these orthopoxvirus DNAs, using centroid sorting strategy (Gibbs and Fenner, 1984).*

tion of the "white clones" of monkeypox virus and of the "whitepox" viruses generally is that they resulted from laboratory contamination with variola virus (see Chapter 7).

Species Diagnosis

The biological characteristics used to identify monkeypox virus and, in material derived from human cases, to distinguish it from variola virus, are the hemorrhagic pock and high ceiling temperature on the chorioallantoic membrane, the production of a large hemorrhagic lesion after intradermal inoculation in rabbits, its wide host range, and its failure to grow in pig embryo kidney cells when first inoculated into these cells. The HindIII restriction map is characteristic of the species; PstI mapping can be used to distinguish West African, Nigerian, and Zairean strains.

Serological Diagnosis of Past Monkeypox Infection

An understanding of the ecology of monkeypox virus depends either on the isolation of virus from animals captured in the field or on serological surveys for monkeypox virus-specific antibodies. The isolation of virus from animals captured in the field is likely to be a rare event in orthopoxvirus infections, in which persistent infection does not occur, and in fact only one such isolation of monkeypox virus has been made, in spite of intensive efforts over several years (see below).

During the 1970s methods had been developed that enabled species-specific diagnoses of recent infection with monkeypox, vaccinia, and variola viruses to be made with hyperimmune or other highly potent sera, by absorption with appropriate viral suspensions and tests of residual antibody by gel precipitation (Gispen and Brand-Saathof, 1974), immunofluorescence (Gispen et al., 1976), radioimmunoassay (Hutchinson et al., 1977), and ELISA (Marennikova et al., 1981). For these tests, rather large quantities of high titer serum were required, and antibodies to the γ-globulin of the relevant species were thought to be necessary for radioimmunoassay adsorption test. These were available for monkeys but not for other species of wild animals.

However, as a result of experience with sera from persons known to have had human monkeypox, some of whom had been vaccinated years earlier, Dr. J. H. Nakano (personal communication, 1984) developed criteria that allowed a positive or presumptive diagnosis of monkeypox to be made in most suspected cases involving human sera. In mid-1985, Dr. Nakano (Fenner and Nakano, 1988) developed a method of carrying out radioimmunoassay-adsorption tests with sera from squirrels and some other species of wild animals, using *Staphylococcus* A protein instead of a species-specific anti-γ-globulin. This made it possible to test many animal sera from the field and has helped to elucidate the ecology of monkeypox virus.

HUMAN MONKEYPOX

Discovery of the Disease

The first case of human monkeypox was found in the Basankusu Hospital, Equateur Province, Zaire (Ladnyi *et al.*, 1972). The Basankusu Zone covers an area of about 20,000 km^2, and in 1970 had an estimated population of 62,000, mostly primitive farmers and hunter–gatherers living in small villages in dense tropical rain forest. The last known outbreak of smallpox in Basankusu Zone occurred in 1968 and comprised 70 cases with 18 deaths. Several suspected cases of smallpox were treated at the hospital in 1969, but none was confirmed. Two suspected cases were reported in 1970; one of these turned out to be chickenpox, and the other was the first case of human monkeypox. The patient, a 9-month-old boy, became ill with fever on August 22, 1970 and a rash developed 2 days later. He was admitted to hospital on September 1, the ninth day of the rash, which had the characteristic centrifugal distribution of smallpox. Crusts were collected for laboratory examination and sent through WHO headquarters in Geneva to the WHO Collaborating Center in Moscow, USSR. The patient recovered and was about to be discharged but on October 23 he developed measles (acquired while in hospital), and died 6 days later.

During 1970 the WHO Collaborating Center in Moscow had received a number of specimens from various provinces of Zaire (but not from Equateur Province) from which variola virus had been recovered. The virus from Basankusu Hospital produced pocks on the chorioallantoic membrane that were different from those of variola virus. More detailed studies of this isolate, including inoculation in rabbit skin, showed that it was monkeypox virus (Marennikova *et al.*, 1972a). Investigations of the epidemiological circumstances of the patient revealed that the child was the only unvaccinated member of his family, and that there had been no other cases of fever with rash recently in the village concerned nor in neighboring villages. Such an isolated case was most unlikely to be smallpox.

The discovery of human monkeypox in central Africa in September 1970 was followed by the realization that four cases of suspected smallpox in Liberia and one case in Sierra Leone in 1970, and one each in

Nigeria and Côte d'Ivoire in 1971 (Foster *et al.*, 1972), were cases of human monkeypox (Lourie *et al.*, 1972). A series of coordinated laboratory and field studies was organized by the World Health Organization to determine the incidence of the disease, to study its clinical features and epidemiology and to search for the animal reservoir or reservoirs of the virus (Fenner *et al.*, 1988).

Incidence and Distribution

All known cases of human monkeypox have occurred in tropical rain forest areas of West and Central Africa (Fig. 8-3). Of 404 cases reported between 1970 and 1986, 89% were in small villages (<1000 inhabitants), and 10% in larger villages (1000–5000 inhabitants); only 4 cases were reported from towns of over 5000 inhabitants. Even the last-named population groups had ample opportunities for direct contact with animals killed in the rain forests.

Between 1972 and 1981, many more cases had been reported from Zaire than from any other country (Table 8-2), probably because the number of people living in villages in tropical rain forests is much

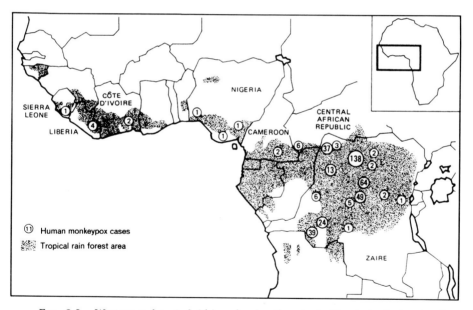

FIG. 8-3. *Western and central Africa, showing the extent of tropical rain forest and the locations where cases of human monkeypox have occurred, with numbers of cases reported between 1970 and 1986. (From Jezek and Fenner, 1988.)*

TABLE 8-2

Human Monkeypox: Areas of Tropical Rain Forest and Annual Numbers of Cases Reported in Countries in Western and Central Africa: 1970–1986[a]

	Cameroon	Central African Republic	Côte d'Ivoire	Liberia	Nigeria	Sierra Leone	Zaire	Totals
Population in thousands (1980):	8,554	2,290	8,247	1,871	80,555	3,296	28,532	124,791
Rain forest (1980)[b]	17,920	3,590	4,458	2,000	5,950	740	105,650	140,308
Percentage of rain forest[c]	9.5	1.9	2.4	1.1	3.2	0.4	56.2	74.7
Cases of monkeypox								
1970	—	—	—	4	—	1	1	6
1971	—	—	1	—	2	—	—	3
1972	—	—	—	—	—	—	5	5
1973	—	—	—	—	—	—	3	3
1974	—	—	—	—	—	—	1	1
1975	—	—	—	—	—	—	3	3
1976	—	—	—	—	—	—	5	5
1977	—	—	—	—	—	—	6	6
1978	—	—	—	—	1	—	12	13
1979	2	—	—	—	—	—	8	10
1980	—	—	—	—	—	—	4	4
1981	—	—	1	—	—	—	7	8
1982	—	—	—	—	—	—	40	40
1983	—	—	—	—	—	—	84	84
1984	—	6	—	—	—	—	86	92
1985	—	—	—	—	—	—	62	62
1986	—	—	—	—	—	—	59	59
Total cases:	2	6	2	4	3	1	386	404

[a] From Jezek and Fenner (1988).
[b] Thousands of hectares. Source: Food and Agriculture Organization (1981).
[c] In western and central Africa; 25.3% of total occurs in six countries in which human monkeypox has not been reported.

greater there. From 1982 onward many more cases were reported from Zaire than in previous years. This was partly due to the intensive health institution-based surveillance system that had been developed in enzootic foci in that country by the World Health Organization in collaboration with the government of Zaire (Jezek and Fenner, 1988), but there appears to have been a real increase in incidence in 1983 and 1984. Since then the incidence in Zaire has fallen, although there are now many more unvaccinated children than formerly. The explanation is not clear; it may be due to fluctuations in the extent of infection in the animals from which human infections were acquired.

Clinical Features

A description of the clinical features of human monkeypox based on 47 cases diagnosed up to the end of 1979 (Breman *et al.*, 1980) has been refined in the light of experience since then (Jezek *et al.*, 1987c; Jezek and Fenner, 1988). Clinically, human monkeypox closely resembles discrete ordinary-type or occasionally modified-type smallpox. No case has yet been seen, among the cases diagnosed in the years 1970–1986, that is comparable to flat-type or hemorrhagic-type smallpox. The only clinical feature that differentiates human monkeypox from smallpox is the pronounced lymph node enlargement seen in most cases of monkeypox (Plate 8-3), sometimes only in the neck or inguinal region, but more often generalized. Lymph node enlargement occurs early, and has often been observed at the time of onset of fever, usually 1–3 days before the rash appears. Lymph node enlargement was observed in 90% of 98 cases in which its presence or absence was recorded and was a presenting sign, preceding the rash, in 65% of these cases.

The eruption begins after a prodromal illness lasting 1–3 days with fever, prostration, and usually lymph node enlargement. As in smallpox, the lesions develop more or less simultaneously and evolve together at the same rate through papules, vesicles, and pustules before umbilicating, drying, and desquamating. This process usually takes about 2 to 3 weeks, depending on the severity of the disease. The distribution of the rash is mainly peripheral. Severe eruptions can cover the entire body, including the palms and the soles. Most pustules are about 0.5 cm in diameter, and lesions have been noted on the oral mucous membranes and the tongue. Subclinical cases also occur, in unvaccinated as well as vaccinated subjects (see below).

Laboratory Confirmation. To investigate this newly discovered disease, it was important to obtain laboratory confirmation of the clinico-epidemiological diagnosis, initially because of the possible occurrence of

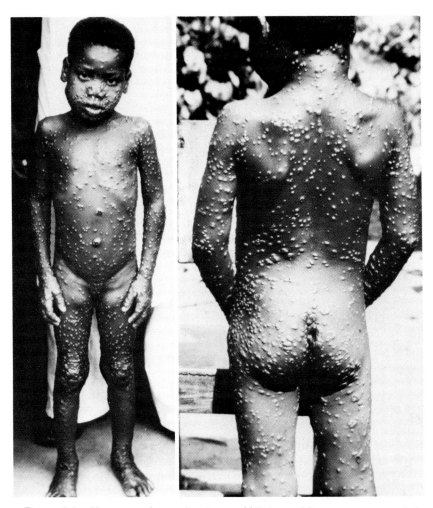

PLATE 8-3. *Human monkeypox in a 7-year-old Zairean girl, acute stage, seventh day of rash. Note bilateral inguinal lymphadenopathy and enlarged submaxillary lymph nodes on right side. Pustular lesions on lips also occur inside the mouth as ulcerated lesions: the enanthem. (From Breman et al., 1980; courtesy World Health Organization.)*

smallpox and later because of the suspicion that "whitepox" virus (see Chapter 7) might cause infection of humans. Working through the Smallpox Eradication Unit of the World Health Organization, all laboratory investigations were carried out in the WHO Collaborating Centers at the Centers for Disease Control, Atlanta, and the Research Institute for Viral Preparations, Moscow. The methods of laboratory diagnosis were the same as those used in smallpox, supplemented by serology in cases in which viral isolation was not possible. In spite of unavoidable delays in the collection and transmission of specimens, virus was recovered from the majority of samples taken from cases eventually diagnosed as human monkeypox. Virtually all cases found positive by either electron microscopy or culture were positive by the other method, but 22% of the cases that were seen too late to obtain lesion material could only be confirmed serologically.

Severity and Case–Fatality Rates. On the basis of the number of skin lesions and the severity of systemic symptoms, cases in Zaire were classified as mild, moderate, or severe (Table 8-3). The majority of cases (51.8%), and the majority of severe cases among the unvaccinated (58.7%), occurred among unvaccinated children in the age group 0–4 years. Few cases were diagnosed in vaccinated subjects, only 43 of the 338 patients (12.7%) having a visible vaccination scar. The youngest of these was a 5-year-old boy who had been vaccinated shortly after birth and developed monkeypox late in 1983, i.e., about 5 years after vaccination.

All 33 deaths occurred in unvaccinated children between 7 months and 8 years of age (Table 8-4), the overall case–fatality rate among unvaccinated persons being 11.2%. The case–fatality rate for the age group 0–4 years (15.0%) was almost three times that in unvaccinated children aged 5–14 years (5.7%).

Complications and Sequelae. The principal complications were due to secondary bacterial infection (Table 8-5); bronchopneumonia occurred late in the course of the illness in 34 patients, of whom 19 died. Encephalitis, which was fatal, occurred in one 3-year-old girl.

Desquamation of crusts leaves areas of hypopigmentation. Hyperpigmentation follows after a few months and usually diminishes with time. As in smallpox, pitting scars may develop, most frequently on the face; they tend to diminish in prominence with time. Secondary infection of the lesions is common and this may play a role in scarring. About half of the scars from lesions seen initially on the face and body were detectable 1 to 4 years after the acute illness. In a few cases corneal lesions have caused unilateral blindness.

Subclinical Infections. Some cases of monkeypox in vaccinated subjects were extremely mild, with very few skin lesions. By analogy with smallpox (Heiner *et al.*, 1971), it was to be expected that many infections in vaccinated subjects would be subclinical. A more important question

TABLE 8-3

Human Monkeypox in Zaire, 1981–1986: Severity of Illness by Vaccination Status and Age[a,b]

Vaccination scar	Age group (years)	Clinical severity							
		Mild		Moderate		Severe		Total	
		Number	Percentage	Number	Percentage	Number	Percentage	Number	Percentage
Absent	0–4	10	5.8	35	20.2	128	74.0	173	58.6
	5–9	11	10.4	15	14.1	80	75.5	106	35.9
	10–14	1	6.2	5	31.2	10	62.5	16	5.4
	Total:	22	7.5	55	18.6	218	73.9	295	100
Present	0–4	0	—	1	—	1	—	2	4.6
	5–9	1	10.0	3	30.0	6	60.0	10	23.2
	10–14	2	25.0	2	25.0	4	50.0	8	18.6
	>15	13	56.5	4	17.4	6	26.1	23	53.5
	Total:	16	37.2	10	23.3	17	39.5	43	100

[a] Data from Jezek and Fenner (1988).
[b] Mild: less than 25 skin lesions; no incapacity and no need for special care. Moderate: 25 to 99 skin lesions; incapable of most physical activity but not requiring nursing care. Severe: 100 or more skin lesions; fully incapacitated and requiring medical care.

TABLE 8-4

Age- and Sex-Specific Case–Fatality Rates (CFR) among Unvaccinated Monkeypox Patients; Zaire, 1981–1986[a,b]

Age group	Males			Females			Total		
	Cases	Deaths	CFR (%)	Cases	Deaths	CFR (%)	Cases	Deaths	CFR (%)
0–2	41	7	17.1	50	10	20.0	91	17	18.7
3–4	42	5	11.9	40	4	10.0	82	9	11.0
5–6	37	4	10.8	24	1	4.2	61	5	8.2
7–9	31	2	6.2	14	0	—	45	2	4.4
>10	11	0	—	5	0	—	16	0	—
Total:	162	18	11.1	133	15	11.4	295	33	11.2

[a] Data from Jezek and Fenner (1988).
[b] All deaths occurred in unvaccinated patients less than 10 years old.

TABLE 8-5

Complications in Cases of Human Monkeypox; Zaire, 1981–1986[a]

| | Vaccination scar | | | |
| | Absent | | Present | |
Complications	Number	Percentage	Number	Percentage
None	178	60.3	38	88.4
Secondary bacterial infection of skin				
(boils, abscesses, septic dermatitis)	46	15.6	2	4.6
Bronchopneumonia, pulmonary distress	34	11.5	2	4.6
Vomiting, diarrhea, dehydration, marasmus	22	7.5	0	0
Keratitis, corneal ulceration	11	3.8	1	2.3
Necrosis of skin	2	0.7	0	0
Encephalitis	1	0.3	0	0
Septicemia	1	0.3	0	0
Total observed:	295	100	43	100

[a] From Jezek and Fenner (1988).

was the extent to which inapparent infections occurred in unvaccinated persons. Data bearing on this problem emerged from the intensive surveillance activities in Zaire in 1981–1986 (Jezek and Fenner, 1988). During a 3-year period, 3711 contacts of 338 confirmed cases of human monkeypox were examined and questioned, often on several occasions. Sera were taken from 74% of the unvaccinated contacts and 6% of the vaccinated contacts, and tested at the WHO Collaborating Centers in Atlanta and Moscow (Table 8-6). The laboratory tests showed that 159 (18%) of the contacts examined had been infected with monkeypox virus. Of the 136 seropositive unvaccinated contacts, 109 had a history or lesions compatible with human monkeypox and 69 of these appeared to be secondary cases, resulting from transmission of infection from another human case. Twenty-seven of these unvaccinated subjects (19.8%) gave no history and had no lesions suggestive of human monkeypox, and were therefore classed as cases of subclinical infection. The majority of such cases occurred in children aged between 2 and 10 years who had been household contacts of a severe case of human monkeypox. Only two subclinical cases were recognized in vaccinated subjects, but many fewer laboratory tests were carried out in them and no special effort was made to detect subclinical infections among vaccinated persons in a way comparable to the studies of Heiner et al. (1971) with variola major in Pakistan.

TABLE 8-6

Evidence of Infection with Monkeypox Virus among Close Contacts of Cases of Human Monkeypox in Zaire, 1981–1986[a]

Vaccination scar	Type	Contacts		Total Number	Laboratory evidence of monkeypox		
		Number examined	Laboratory tests		Clinical disease		Subclinical infection
					Coprimary case	Secondary case	
Absent	Household	559	426	98	29	49	20
	Other	472	348	38	11	20	7
	Total:	1031	774	136	40	69	27
Present	Household	1551	78	20	1	18	1
	Other	1129	40	3	0	2	1
	Total:	2680	118	23	1	20	2

[a] Based on Jezek and Fenner (1988).

Large-scale serological surveys of unvaccinated persons in Zaire (Jezek *et al.*, 1987a; see below) also revealed a few cases of subclinical infection.

Epidemiology

Although the clinical features of human monkeypox are similar to those of discrete ordinary-type smallpox, the epidemiology is quite different. Human monkeypox occurs mainly as single or occasionally multiple sporadic cases, in small villages in dense tropical rain forest, in a limited part of Africa, among villagers who are engaged for at least part of their time as hunters and gatherers. It is thus a zoonosis, infection usually being contracted from a wild animal. However, person-to-person infection occurs in a minority of cases.

Two observations made in the early 1980s have a bearing on the epidemiology of the disease. Mutombo *et al.* (1983) a reported a bizarre case in which a 6-month-old infant in a small village in the tropical rain forest in Zaire was abducted by a chimpanzee, but rescued after sustaining a superficial wond on the lower leg and a fractured femur. The infant developed typical monkeypox, fever beginning 6 days after the incident and a rash 7 days later. Monkeypox virus was isolated from crust material. Lymphadenopathy began in the left inguinal region and eventually became generalized, but the time of its appearance in relation to other symptoms was not determined. Although not proved, it is a reasonable hypothesis that the infant acquired monkeypox from the chimpanzee.

The other observation concerns monkeypox among Pygmies who live in the tropical rain forests in the southern part of the Central African Republic, adjoining Zaire, among whom Khodakevich *et al.* (1985) discovered a cluster of five cases of monkeypox, confirmed by virus isolation. The Pygmies living in the rain forests readily recognized the disease when shown the monkeypox recognition card, whereas the Bantus, and Pygmies living in agricultural settlements, had never seen a disease like that. Interrogation through interpreters revealed that the forest Pygmies had a special name for the disease and believed that it was acquired from animals and not from humans.

Age and Sex Distribution. The ages of patients in Zaire varied between 3 months and 69 years. The majority were infants and children, 86% being below 10 years of age and 52% below 5 years. When analyzing the epidemiology of human monkeypox it is useful to distinguish between infections acquired from an animal source (primary cases) and those due to person-to-person infection (secondary cases)

TABLE 8-7

Human Monkeypox: The Age and Sex Incidence of Primary and Secondary Cases in Zaire,
1981–1986[a]

Age group (years)	Primary cases[b]				Secondary cases[c]			
	Males	Females	Total	(%)	Males	Females	Total	(%)
0–1	8	8	16	6.5	3	8	11	11.8
1–2	21	26	47	18.3	9	8	17	18.3
3–4	33	34	67	27.3	10	7	17	18.3
5–6	32	17	49	20.1	6	9	15	16.1
7–9	30	6	36	14.7	6	10	16	17.2
10–14	14	7	21	8.6	2	1	3	3.2
>15	4	5	9	3.7	4	10	14	15.1
Total:	142	103	245	100.0	40	53	93	100.0
Percentage:	58.0	42.0			43.0	57.0		

[a] From Jezek and Fenner (1988).
[b] Presumed to have been infected from an animal source.
[c] Presumed human-to-human infection.

(Table 8-7). Among primary cases, the highest rate of infection occurred in children aged 3–4 years (27%), followed by those aged 5–6 years (20%), and then those 1–2 years old (19%). Only 4% were in adults.

Although the sex ratio in the general population was 96.7 males per 100 females, there was a preponderance of males (58%) over females (42%) among the primary cases, especially in the age group 5–14 years. The age distribution of secondary cases was quite different, adult patients being relatively more common. There was also a significant preponderance of females among secondary cases (57%), with 10 cases in adult females compared with only 4 among adult males.

Young unvaccinated boys and adult females seem to be at special risk, the former perhaps because they trap small arboreal rodents and play much more with carcasses of killed monkeys brought home by hunters, the latter because of infection of a mother by her sick child or an older sister by nursing her younger sibling.

Seasonal Distribution. Breman *et al.* (1980) reported a preponderance of cases in Zaire in the dry season, but with the institution of more intensive surveillance since 1981 the incidence of cases has been found to be much the same throughout the year. The monthly incidence of primary cases varied a good deal from year to year (Table 8-8); there was no clearly evident seasonal pattern. Secondary cases showed the same absence of a seasonal effect.

TABLE 8-8

Human Monkeypox: Monthly Incidence of Primary and Co-primary Cases in Zaire, 1981–1986, Calculated from Date of Onset[a]

Year							Month						
	January	February	March	April	May	June	July	August	September	October	November	December	Total
1981	—	1	1	1	—	1	1	—	—	—	1	—	6
1982	5	—	—	—	1	2	4	1	3	2	2	4	24
1983	—	1	3	2	7	5	8	6	9	2	6	9	58
1984	6	8	11	4	9	9	3	4	5	2	1	—	62
1985	—	2	1	1	2	8	6	9	6	6	4	2	47
1986	4	4	8	3	6	6	5	6	1	1	2	2	48
Total:	15	16	24	11	25	31	27	26	24	13	16	17	245

[a] Data from Jezek and Fenner (1988).

Sources of Infection of Sporadic Cases. Epidemiological investigations in Zaire indicated that wild animals were the probable source of infection for some 70% of patients and person-to-person infection was suspected to have occurred in the remaining 30% (see Table 8-7). Since monkeypox virus has a wide host range and evidence of infection in African wild animals has been obtained from chimpanzees, several species of monkey, and two species of squirrel (see below), the disease is probably transmitted to humans by more than one species of wild animal. It was virtually impossible to determine by case–control studies which animals might be the source of infection, because the whole of the population in affected localities had had multiple daily contacts with the same varieties of wild animals, in the settlements, agricultural areas, or nearby forests (see Fig. 8-4). Species with which patients had had multiple close contacts (within 3 weeks before onset of rash), through hunting, skinning, playing with the animals, or eating the carcasses, included various types of monkeys (65%), squirrels (12%), antelopes and gazelles (12%), terrestrial rodents (9%), and other animals (3%). Seventy-one percent of suspected monkeys associated with patients belonged to the genus *Cercopithecus*, 12% to *Colobus*, and 8% to *Cercocebus*. Two-thirds of suspected rodents were squirrels and the rest, *Cricetidae*. The majority of animals suspected of being the source of infection were apparently healthy.

The small villages in tropical rain forest in which cases of human monkeypox occur are not closely surrounded by high forest on all sides (Fig. 8-4). Usually they consist of groups of houses along roads through

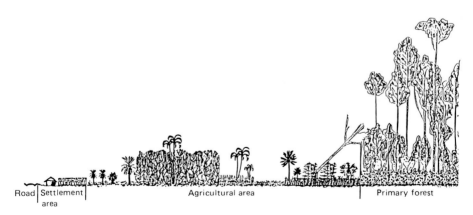

Road | Settlement area | Agricultural area | Primary forest

FIG. 8-4. *Diagram illustrating the different ecological zones around villages in the tropical rain forest in Bumba Zone, Zaire. (From Khodakevich et al., 1987a.)*

TABLE 8-9

Human Monkeypox: Occurrence of Primary and Secondary Cases in Zaire,
1981–1986[a]

Year	Primary cases[b]		Person-to-person infection (human generation)				Total
	Sporadic primary case	Presumed coprimary case	1	2	3	4	
1981	6	—	1	—	—	—	7
1982	22	2	13	3	—	—	40
1983	47	11	19	3	3	1	84
1984	52	10	18	6	—	—	86
1985	40	7	11	—	—	—	62
1986	36	12	7	3	1	—	59
Total:	203	42	69	19	4	1	338

[a] From Jezek and Fenner (1988).
[b] Presumably infected from an animal source.

the forests, with extensive agricultural areas around the settlement itself, consisting of gardens and secondary forest, often with many oil palms, which provide food much favored by certain squirrels. Beyond this, perhaps 3–5 km away, is the primary rain forest. Each of the three zones—settlement, agricultural area, and forest—has a characteristic fauna (Khodakevich *et al.*, 1987a). Domestic animals and commensal rodents frequent the immediate environs of the houses, terrestrial and arboreal rodents and bats are found in the agricultural areas, and larger animals, including monkeys, inhabit the rain forest.

Different age groups in the population differ in the degree to which they move through these areas. Children below the age of 2 years are rarely let out of their mother's sight; between the ages of 3 and 5 years they go with their mothers to the agricultural area, and after the age of 5 years they go on their own to this area and hunt for small animals. Only the men, and boys over 15 years of age, hunt in the forest for large animals including monkeys, antelopes, and porcupines. Persons of all age groups would be exposed to infection from wild animals brought to the household for food. Hunters and children old enough to capture small animals such as squirrels and rats in the agricultural area might conceivably be exposed to additional risk. Very few primary cases have occurred in hunters (the great majority of whom had been vaccinated), whereas a high proportion of primary cases, but a somewhat lower proportion of secondary infections, have occurred in children (especially

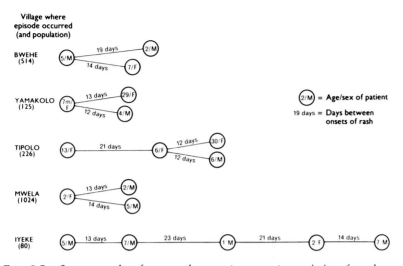

FIG. 8-5. *Some examples of presumed person-to-person transmission of monkeypox. All occurred among close family contacts, mostly children, who lived in small villages in the tropical rain forest. Since patients could presumably remain infectious for about a week after the onset of the rash, the intervals between cases could be longer than the usual incubation period (presumed to be about 12 days; range 7–21 days). (From Fenner* et al., *1988.)*

boys) aged between 5 and 9 years (see Table 8-7). This may have been related to the relatively high incidence of monkeypox virus infection among squirrels captured in the agricultural areas (see below).

Person-to-Person Spread. Most cases of human monkeypox (48%) have occurred as single sporadic infections. However, sometimes cases occurred in clusters, suggesting either multiple infections from a common source—a coprimary case (if the dates of onset lay within the presumed minimum incubation period of 7 days)— or person-to-person transmission. The distribution of sporadic cases, presumed coprimary cases, and presumed secondary or subsequent person-to-person infections in Zaire in 1981–1986 are shown in Table 8-9. Intervals of 7 and 21 days between the dates of appearance of the rashes in persons in close family contact were taken as the limits for presumed person-to-person spread. In the 6 years 1981–1986 during which intensive surveillance was operating in Zaire, 93 out of 338 cases (27.5%) appeared to have been due to transmission from person to person. Examples of the type of pattern observed are shown in Fig. 8-5. An extreme example involving four probable successive person-to-person infections has been described by Jezek *et al.* (1986a).

Secondary Attack Rate. The most widely used measure of the risk of transmission of infectious diseases is the secondary attack rate, i.e., the proportion of individuals exposed to an index case who become ill within the accepted incubation period, in relation to the total number of exposed contacts. In Zaire, during the 6 year period 1981–1986, 2278 persons were identified as close contacts of 245 patients (203 primary and 42 coprimary cases) suspected to have been infected from animal sources (Jezek and Fenner, 1988). Vaccination scars were seen on 1555 contacts (68%); the other 723 (32%) were regarded as unvaccinated. One thousand four hundred and twenty (62%) of these contacts lived in the same households as primary or coprimary cases and the remaining 858 (38%), living outside the affected house, had casual face-to-face contact with an infectious monkeypox patient. Since there were 69 first generation secondary cases, the observed crude secondary attack rate among contacts of the primary and coprimary cases was 0.030 (69 per 2278), that is an overall 3% probability of becoming infected from a human source.

Secondary attack rates were strongly related to two variables: vaccination status, which influenced susceptibility, and place of residence, which influenced the degree of contact with the monkeypox patient. The attack rate among unvaccinated contacts (7.47%) was almost eight times higher than that among vaccinated contacts (0.96%), and the overall risk of attack for household contacts (3.73%) was twice as high as the risk for those living outside the affected house but in casual contact with the monkeypox case (1.86%). Combining these factors, the highest secondary attack rate (9.3%) was found among unvaccinated persons living in the same household as a monkeypox patient; their attack rate was seven times higher than the corresponding rate for vaccinated household members (1.3%). Among household contacts, those who had direct physical contact with the infected person, by playing with the patient or sharing the same bed, or those who provided nursing care, had an increased risk of subsequent attack, emphasizing the importance of intimate contact in the spread of the disease. These figures are much lower than those for smallpox, in which the overall first generation secondary attack rates in household contacts were 58.4% for unvaccinated persons and 3.8% among vaccinated contacts (see Chapter 11).

Although the number of monkeypox patients reported increased substantially during 1981–1986 (48 cases in 1970–1980 and 338 cases in 1981–1986) the secondary attack rates for both periods were quite similar, indicating that there had been no increase in the transmissibility of the virus over that period.

A Stochastic Model of Person-to-Person Transmission in Monkeypox. Using the data obtained from observations on person-to-person spread among contacts of 147 primary or coprimary cases of human monkeypox in Zaire in 1982–1984 (Jezek *et al.*, 1986b), Jezek *et al.* (1987b) developed a stochastic model for person-to-person infections with monkeypox virus assuming overall vaccination rates of 70% (the rate in the observed populations), 50%, and zero. Although the expected numbers of generations and of cases infected by contact increased with falling vaccination rate, the model suggested that the person-to-person infectivity of monkeypox was such that the disease always died out, after a maximum number, in the simulation, of 11 generations. This result supports an argument based on historical data—namely that monkeypox virus has been enzootic in animals of the tropical rain forests in Zaire for centuries, without ever establishing continuous person-to-person infection in a population that had been almost completely unvaccinated until about 1967.

The Prevalence of Monkeypox Virus Infection in Humans. In an attempt to discover the prevalence of monkeypox virus infection of humans in tropical rain forest areas in various parts of central and western Africa, serological surveys of persons without vaccination scars were carried out in 1981 in Congo, Côte d'Ivoire, Sierra Leone, and Zaire. Cases of monkeypox had been reported from all these countries except Congo, which borders Zaire and has a large area (over 21 million ha) of tropical rain forest. Specimens of serum collected from allegedly unvaccinated persons were tested for orthopoxvirus antibodies by hemagglutination-inhibition (HI) or immunofluorescence tests (Table 8-10). Of 10,300 sera tested, 15.4% gave positive results. Supplementary examination of many of these sera by neutralization and ELISA tests were in good agreement. The intention was to subject sera containing orthopoxvirus antibodies demonstrable by the screening test to further assay by either a radioimmunoassay adsorption test or an ELISA adsorption test. However, because the titers were too low, only 420 of the 1583 positive sera could be tested; of these 73 gave results indicating that the subjects had been infected with monkeypox virus. None of the sera from the Congo gave a positive result; the proportions of all sera designated as monkeypox virus positive varied from 0.70% for Côte d'Ivoire to 1.01% for Sierra Leone.

Follow-up visits to Côte d'Ivoire and Sierra Leone in June–July 1982 to examine those who had monkeypox virus antibody in their sera showed that some specimens had inadvertently been taken from vaccinated subjects. However, none of the 13 subjects investigated had unequivocal evidence of past vesiculopustular disease (by history or residual pockmarks). If any of them had been infected with monkeypox virus, as the

TABLE 8-10

Human Monkeypox: Results of Serological Survey among Allegedly Unvaccinated Persons Inhabiting Villages in Tropical Rain Forest Areas of Four Countries of Western and Central Africa, 1981

Country	Number of tested	Positive by HI test		Positive for monkeypox virus antibodies by RIAA test		
		Number	Percentage	Number tested	Number positive	Percentage of total sera
Congo[a]	1,433[b]	231	16.1	78[c]	0	0.0
Côte d'Ivoire[d]	2,840	369	13.0	93	20	0.70
Sierra Leone[d]	2,567	320	12.5	71	26	1.01
Zaire[d]	3,460	663	19.2	178	27	0.78
Total:	10,300	1,583	15.4	420	73	0.71

[a] Maltseva et al. (1984); Marennikova et al. (1984).
[b] Tested by ELISA test with monkeypox antigen.
[c] Tested by ELISA–adsorption.
[d] Dr. J. H. Nakano, unpublished observations.

serological results indicated, the infections was subclinical, or so relatively mild as to have been forgotten.

Because surveillance was much better in Zaire, it was possible to
obtain more information about the possible frequency of subclinical
infection from the survey in Kole zone, in which some 400 localities were
visited, involving about 10,000 households and about 50,000 persons,
who were examined for vaccination scars and facial pockmarks (Jezek *et
al.*, 1987a). Only 15% of those investigated had no vaccination scar, and
1.3% of these subjects had facial skin changes suggesting a past attack of
a vesiculopustular disease. Of a total of 3460 serum samples collected
from persons without vaccination scars, 27 sera showed evidence of
monkeypox-virus-specific antibodies by the radioimmunoassay adsorption test. The subsequent field investigation of 19 of these subjects who
were less than 15 years old revealed that 12 of them had experienced
vesiculopustular disease or fever with lymphadenopathy in the past, 1
had a possible vaccination scar, and there were no signs or history of a
disease like human monkeypox in the other 6 children. The prevalence
rate of monkeypox-virus-specific antibodies showed significant differences in different age groups; it was 4 times higher in children aged 5–9
years (13.1 per 1000) than in those aged 0–4 years (3.3 per 1000).

The lack of a serological test that was sufficiently sensitive and specific
to permit the diagnosis of a previous monkeypox virus infection without
resorting to serum absorption made it impossible to determine the
prevalence of human infections with monkeypox virus from these
surveys. The significance of the overall orthopoxvirus-positive antibody
rate of 15.4% remains obscure. However, follow-up studies in three
countries supported the view that emerged from intensive surveillance
in Zaire (see Table 8-6), namely that some infections of unvaccinated
humans with monkeypox virus are subclinical.

THE ECOLOGY OF MONKEYPOX VIRUS

Understanding of the epidemiology of primary cases of human
monkeypox (i.e., those derived from an animal source) depends on a
knowledge of the ecology of the virus, including which animals act as
reservoir and incidental hosts and how the virus is transmitted from one
animal to another. Initially, studies of this problem were focused on
monkeys. Serological surveys of Asian monkeys were negative, but
monkeypox virus-specific antibodies have been found in eight species of
monkeys that occur in central and western Africa.

Because members of each species of monkey usually move in small self-contained troops, and because monekypox virus does not cause persistent infections and is not transmitted by flying arthropods, it seems unlikely that nonhuman primates are the reservoir hosts. From 1973 onward, attention was directed to a wider range of wild animals, especially terrestrial and arboreal rodents, some of which occur in populations that are sufficiently large to support enzootic monkeypox virus infection.

Serological Surveys of Captive African Primates

Altogether 1447 sera of African primates held in various laboratories in Africa, Europe, and the United States were tested for orthopoxvirus antibodies by either HI or neutralization tests; all were negative (Arita *et al.*, 1972). With the possible exception of sera from 25 gorillas and 167 chimpanzees, all these sera were obtained from animals captured in countries which have not reported cases of monkeypox, and the monkeys tested belonged to species occurring in the savanna rather than in tropical rain forests.

Serological Surveys of Primates from West Africa

Breman *et al.* (1977) examined primate sera that had been collected in western Africa for a yellow fever survey. HI and neutralization tests were done on 206 sera obtained from 27 different sampling zones in Côte d'Ivoire, Mali, and Upper Volta, which were situated in forest and heavily wooded preforest and in the savanna. Out of 195 sera, 15 (8%) were orthopoxviruspositive by HI and 44 (21%) by neutralization tests. The testing of three HI-positive sera from forest-dwelling monkeys (one *Colobus badius* and two *Cercopithecus petaurista*) by immunofluorescence after absorption showed that they contained monkeypox virus-specific antibodies (Gispen *et al.*, 1976).

Another survey included 692 sera obtained from a variety of animals from Chad, Côte d'Ivoire, Liberia, Nigeria, Senegal, Sierra Leone, and Upper Volta between 1970 and 1972 (Dr. J. H. Nakano, unpublished observations, 1973). One hundred and fifty eight (23%) gave positive results by the HI test and 50 out of 186 (27%) were positive by neutralization. Among sera from nonhuman primates included in the 692 samples, 92 out of 334 (28%) gave positive HI results; 35 out of 147 sera tested by neutralization (24%) gave positive results. Positive HI titers were observed with occasional serum samples obtained from a variety of other animals, including squirrels, rodents, ungulates, and warthogs. Subsequently, 273 of the monkey sera were tested by radioimmunoassay adsorption. Seven sera contained monkeypox-virus-specific antibodies: two *Cercopithecus petaurista*, two *Cercopitheus aethiops*, two *Cercopithecus nictitans*, and one *Colobus badius* (Dr. J. H. Nakano, personal communication, 1986). The most interesting result was that obtained with *C. aethiops* from Côte d'Ivoire. Not only is this monkey typically an inhabitant of the savanna rather than the tropical rain forest, but it is the species that was exported from West Africa to North American and European countries on a large scale during the period when monkeypox was occurring in captive monkeys in these countries, and animals of this species may have been the source of infection of Asian monkeys during transit.

Tissues from 648 animals of 73 species obtained in Liberia and Nigeria in 1971 were tested for orthopoxviruses by 2 serial passages in primary monkey kidney cells, with negative results (Dr. J. H. Nakano, personal communication, 1983).

Studies on Material from Zaire, 1971–1979

Since most cases of human monkeypox had occurred in Zaire, attempts to determine the reservoir host or hosts of the virus were subsequently concentrated in that country, mostly in places where human monkeypox cases had occurred.

Between 1971 and 1975 serological and virological investigations concerning a wild animal reservoir of monkeypox virus were carried out at the WHO Collaborating Center in Moscow. Some 200 sera from areas distant from what is now recognized as the monkeypox enzootic area (see Fig. 8-3) were virtually all negative, whereas monkey sera from Zaire collected in 1971 and 1973 showed 14 out of 81 positive by the HI test and 11 out of 65 by the neutralization test (Marennikova *et al.*, 1975a). Subsequently another collection of sera from Zaire yielded 24 HI-positive monkey sera out of 117 tested, and 26 HI-positive rodent sera out of 245 tested.

Attempts were made to isolate virus on the chorioallantoic membrane from the kidneys of primates, rats, and squirrels collected in Zaire. None yielded monkeypox virus, but "whitepox" virus was said to have been obtained from four specimens and vaccinia virus from one specimen (see Chapter 7).

In July 1979 a large-scale ecological survey was organized in Zaire by the WHO Smallpox Eradication unit. Sera and organs were obtained from a wide variety of wild animals. The animal species were identified by expert zoologists and the sera and organs were tested at the WHO Collaborating Center in Atlanta. In all, 1331 sera from 45 species of wild animals were tested by HI as a screening test for orthopoxvirus

TABLE 8-11

Results of Hemagglutination-Inhibition, Radioimmunoassay, and Radioimmunoassay Adsorption Tests on Monkey and Squirrel Sera Collected in Zaire in July 1979[a]

Species	Hemagglutination-inhibition test[b]		Radioimmunoassay[b]		Monkeypox virus-specific antibodies[c]	
	Number tested	Number positive	Number tested	Number positive	Number tested	Number positive[d]
Monkeys						
Allenopithecus nigroviridis	10	7	10	8	8	7
Cercocebus albigena	3	0	3	0	0	—
Cercocebus galeritus	11	5	11	2	2	2
Cercopithecus ascanius	94	30	93	20	20	13
Cercopithecus mona	37	11	37	4	4	2
Cercopithecus neglectus	10	1	10	0	0	—
Cercopithecus nictitans	47	10	47	1	1	1
Cercopithecus pogonias	14	7	14	0	0	—
Colobus pennanti	10	3	7	0	0	—
Perodicticus potto	5	1	5	0	0	—
Squirrels						
Funisciurus anerythrus (and *F. isabella*)	48	10	44	6	6	6
Heliosciurus rufobrachium	58	25	51	0	0	—

[a] Based on unpublished observations by J. H. Nakano.
[b] Using vaccinia virus antigens.
[c] By radioimmunoassay adsorption.
[d] Discrepancies between number tested and number positive due to nonspecific reacting material.

antibodies; 227 sera (13.2%), from a wide range of animals, gave positive results (Dr. J. H. Nakano, personal communications, 1983, 1986). All 50 sera from *Rattus* spp. were negative.

The subsequent testing of certain sera by radioimmunoassay adsorption cast doubt on the significance of the positive results obtained by the HI test, since none of 25 HI-positive sera of the squirrel *Heliosciurus rufobrachium* gave positive results by radioimmunoassay (Table 8-11). On the other hand, additional radioimmunoassay adsorption tests on monkey and squirrel sera from this collection revealed positive results in five species of monkeys and in squirrels of the genus *Funisciurus* (Dr. J. H. Nakano, personal communication, 1986).

Kidneys and spleens from 930 of the animals from the 1979 Zaire study, including all of the monkeys, were passaged in Vero cells, and the monkey material was also tested on the chorioallantoic membrane, with negative results (Dr. J. H. Nakano, personal communication, 1983).

Studies in Zaire, 1985–1986

Ecological investigations in Zaire were renewed in 1985. Attention was concentrated on animals found around the houses and in the adjacent agricultural area (see Fig. 8-4) near villages in which cases of human monkeypox had recently occurred. An early and exciting result was the recovery of monkeypox virus from a diseased squirrel *(Funisciurus anerythrus)* (Khodakevich *et al.*, 1986). This species of squirrel is quite common in the agricultural area adjoining the villages, where it feeds on oil palm seeds.

Subsequent studies on sera from terrestrial rodents and goats found near houses and squirrels found in the agricultural area revealed many monkeypox-virus-specific sera in two species of squirrel (*Funisciurus anerythrus* and *Heliosciurus rufobrachium*), but none in the animals found around the houses (Table 8-12). In other parts of Zaire where cases of human monkeypox had occurred, monkeypox virus-specific antibodies were found in 13% of *Funisciurus anerythrus,* 20% of *Funisciurus* spp., 13% of *Heliosciurus rufobrachium,* 10% of *Cercopithecus ascanius,* and 4% of *Cercopithecus pogonias* (Jezek and Fenner, 1988), where as in Bas-Zaire, an area from which no cases of human monkeypox have been reported, monkeypox virus-specific antibodies were found in 48% of *Funisciurus lemniscatus* and 28% of *Heliosciurus gambianus;* however, the local customs forbade the use of squirrels and other small rodents for food.

The investigations just described show that monkeys of 10 species, belonging to 4 genera, have been shown to be infected with monkeypox virus under natural conditions:

TABLE 8-12

Results of Hemagglutination-Inhibition, Radioimmunoassay, and Radioimmunoassay Adsorption Tests on Sera from Animals Living in the Settlements and Agricultural Areas Adjacent to Selected Villages in Zaire, 1985–1986[a]

Species	Hemagglutination-inhibition test[b]		Radioimmunoassay[b]		Monkeypoxvirus-specific antibodies[c,d]	
	Number tested	Number positive	Number tested	Number positive	Number tested	Number positive
Terrestrial rodents[e]	579	180[f]	415	0	0	—
Goats	121	0	121	0	0	—
Cats	65	11	65	4	4	—
Squirrels						
Heliosciurus rufobrachium	39	8	39	7	7	7
Funisciurus anerythrus	352	41	337	92	83	80

[a] Based on unpublished observations by J. H. Nakano.
[b] Using vaccinia virus antigens.
[c] By radioimmunoassay adsorption.
[d] Discrepancies between number tested and number positive due to nonspecific reacting material.
[e] Various species found around houses.
[f] Nonspecific.

Cercopithecus, six species: *C. ascanius, C. aethiops, C. mona,*
   C. nictitans, C. petaurista, C. pogonias, Allenopithecus
   nigroviridis
Cercocebus galeritus
Colobus badius
Pongidae: Pan troglodytes

Two genera and four species of tree squirrels have been shown to be
infected with monkeypox virus under natural conditions:

Funisciurus (rope squirrels): *F. anerythrus* and *F. lemniscatus*
Heliosciurus (sun squirrels): *H. rufobrachium* and *H. gambianus*

Conditions Relevant to Human Infection with
Monkeypox Virus

From the point of view of the chances that animals of any of the
species known to be naturally infected with monkeypox virus may be
involved as sources of human infection, it is important to know their
habits and habitats. Table 8-13 illustrates the vertical stratification in the
tropical rain forest and its environs of wildlife species known to be
infected with monkeypox virus.

Based on experience gained from the ecological surveys carried out in
Zaire, the following conclusions can be drawn concerning the natural
history of monkeypox virus (Jezek and Fenner, 1988):

1. Monkeypox virus circulates among mammals inhabiting arboreal
levels of the forest; terrestrial rodents and domestic animals do not
participate in this cycle

2. Squirrels, especially those of the genus *Funisciurus,* are important
hosts and circumstantial evidence indicates that they are the reservoir
host in the secondary forests surrounding human settlements (Khodakevich *et al.,* 1987b).

Depending on the vegetation, various species of *Funisciurus* may serve
as a reservoir host: *F. anerythrus* in northern Zaire (Bumba zone) and
central Zaire (Ikela zone) and *F. lemniscatus* in Bas-Zaire

3. Squirrels of the genus *Heliosciurus* usually occupy a higher stratum
of the forest than *Funisciurus,* but share habitats at some levels. Two
species, *H. gambianus* and *H. rufobrachium,* were found to be infected,
but at a lower incidence than *Funisciurus*

4. Monkeypox virus-specific antibodies were found in primates
which dwell on forest strata also occupied by squirrels. However,

TABLE 8-13

Habits and Habitat of Animals Known to Be Infected with Monkeypox Virus[a]

Genus and species	Habitat and habits
Cercopithecus	Arboreal and diurnal; all strata colonized except floor; groups of 40–50 composed of families of 4–5; remain in one area; sedentary; eat vegetables, sometimes insects (ants), birds' eggs, fruits, and oil palm nuts; gestation 130 days (single birth); longevity about 20 years
Allenopithecus nigroviridis	Forest, swamp forest: eat fruits, seeds, and insects
C. ascanius	Forest galleries
C. aethiops	Savanna and woodland, open country; forest gallery to sleep; bands of 6–20, mix with *C. petaurista*
C. mona	Forest galleries, lower and middle strata
C. nictitans	Forest, woody savanna; highest galleries; mix with *C. mona*
C. petaurista	Forest, fringe savanna; lower and middle forest canopies
C. pogonias	Upper strata of the forest
Cercocebus	High forest and galleries; live in small troops of 4–12 in low story of the forest and also in clearings; feed on fruits and seeds (oil palm nuts) and kernels
C. galeritus	Live in lower story of forest, largely terrestrial, often descend trees to ground
Colobus	Arboreal; live exclusively in forested areas; troops of up to 25; eat leaves, fruits, and nuts; coat highly prized by hunters
C. badius	Large monkey, living in upper strata of forest galleries
Pongidae	
Pan troglodytes	Diurnal; both terrestrial and arboreal; habitats vary from rain and swamp forests to woodland and savanna; highest density in forests with open canopy; no stable groups; feed on fruits, nuts, shoots, bark, eggs, and insects
Funisciurus	Diurnal, medium-size tree squirrels living in primary and secondary forest formations, invade oil palm plantations; live singly or in pairs; diet fruits, nuts, oil palm nuts, birds' eggs, insects
F. anerythrus	Densely spread in all types of secondary forest formations, where it makes nests at 2–10 m above the ground, coming to ground in search of food
F. lemniscatus	
Heliosciurus	Usually occupy a higher stratum of the forest than *Funisciurus* spp., but share habitat at some level with them

[a] From Jezek and Fenner (1988), modified from Dorst and Dandelot (1969).

comparison of their lifespan and antibody prevalence rates with those of squirrels suggested monkeys were infected with monkeypox virus at least by an order of magnitude less frequently than *Funisciurus*

5. There is no evidence that arthropods participate in the circulation of monkeypox virus

6. The absence of cases of human monkeypox does not necessarily mean that the virus does not circulate in wildlife in that area, since customs or nutritional habits may restrict contact between infected animals and humans

7. Conditions that facilitate the occurrence of human infections include the following:

  a. enzootic circulation of monkeypox virus in wild animals in the agricultural areas and forest surrounding human settlements

  b. use of meat of wild animals as an important source of animal protein in human diet

  c. close contact with wild animals, including trapping, killing, skinning, playing with carcasses, and consumption of raw or poorly cooked meat.

SUMMARY

Laboratory studies show that monkeypox virus is a distinct species of *Orthopoxvirus*. First reported as the cause of epizootics among captive monkeys in laboratory colonies in Europe and the United States and in an epizootic in a zoological garden in the Netherlands, it was found in 1970 to be the causative agent of a generalized human infection that clinically resembled smallpox.

Unlike smallpox, however, human monkeypox occurs only in persons living in small villages in tropical rain forests in countries of western and central Africa, where hunting is an important method of obtaining food. The vast majority of reported cases have been found in Zaire, during an intensive surveillance campaign based on health institutions that has been in operation there since late in 1981. The majority of cases can be attributed to infection from an animal source, but person-to-person infection sometimes occurs, mainly between unvaccinated children. The longest chain of transmission observed so far is an incident in which there appeared to be four serial person-to-person infections.

The results of serological tests used for screening sera from wild animals (hemagglutination inhibition, neutralization, and immunofluorescence) suggest that orthopoxvirus infections are relatively common in a wide variety of animal species in West and Central Africa. Besides monkeypox virus, orthopoxviruses known or suspected to occur in this region included variola and vaccinia viruses (in humans), and taterapox virus. Another orthopoxvirus causes Uasin Gishu disease in horses in Kenya and Zambia (see Chapter 10); however, no antibodies to orthopoxviruses were found in sera from 300 monkeys (species not recorded) trapped in Kenya (K.R. Dumbell and R. Reith, unpublished observations, 1978). Other, unknown orthopoxviruses may also occur in African wild animals in the region. The results of tests for monkeypox virus-specific antibodies indicate that at least eight species of primates and four species of squirrel can be naturally infected with monkeypox virus. The squirrels may act as reservoir hosts; any wild animal acutely infected with monkeypox virus could act as a source of human infection.

Even in the parts of Zaire in which it appears to be most common and is best reported, monkeypox is a rare disease (338 known cases in a population of about 5 million during the 6 years 1981–1986). However, serological studies suggest that occasionally subclinical infections occur among unvaccinated as well as vaccinated persons. There is no reason to believe that it is a new disease or that its frequency is increasing. Indeed, it appears to be disappearing from countries in West Africa, probably because of ecological changes associated with development, and it is becoming less common in Zaire in spite of the waning immunity in those previously vaccinated and the increasing numbers of young, unvaccinated people.

CHAPTER 9

Ectromelia Virus

The disease that is now called mousepox was first described and its viral etiology established by Marchal (1930), following investigation of an unusually high mortality in mice received from commercial breeders by the National Institute of Medical Research at Hampstead, England. She called it "infectious ectromelia" because many mice that recovered from the disease had an amputated foot (Plate 9-1). Subsequently the disease was recognized in laboratory mouse colonies in many countries in Europe, and in Japan and China, but did not appear to occur as an enzootic disease among laboratory mice in the Americas or Australia. However, from time to time it was imported into the United States from Europe with mouse stocks or mouse cell lines or tumors, or as an unknown virus sent for identification, and sometimes caused disastrous outbreaks. This led to a prohibition of work on the virus in the United States from the early 1950s until about 1980, when research was instituted in high-containment laboratories in the National Institutes of Health and in Yale University, with the aim of defining more accurately its behavior in various pure lines of mice. In consequence, most research on mousepox has until recently been carried out in laboratories in Europe, Australia, and Japan, rather than in the United States. In Australia, in particular, mousepox was used as a model for studies on

PLATE 9-1. *Mouse that had recovered from mousepox, showing scars on face and amputation of hind foot, a lesion which led Marchal (1930) to propose the name "infectious ectromelia" for the disease.*

the pathogenesis of generalized infections, the cellular immune response to viral infections, and experimental epidemiology.

PROPERTIES OF ECTROMELIA VIRUS

Strains of Virus

Isolates from Manchester, England (McGaughey and Whitehead, 1933), Paris (Schoen, 1938), Germany (Kikuth and Gönnert, 1940), Japan (Ichihashi and Matsumoto, 1966), the USSR (Andrewes and Elford, 1947), and the United States (Allen *et al.*, 1981) were found to be serologically indistinguishable from the original Hampstead strain of Marchal (1930). Two strains, Hampstead and Moscow, have been extensively used for experimental studies. Other strains that have been used are designated Ishibashi and NIH-79. Their histories are given below.

Hampstead Strain. The original isolation was made by Marchal (1930). This virus was used in Greenwood's epidemiological experiments (Greenwood *et al.*, 1936) and in most of the early European work with ectromelia virus. A mouse-passaged Hampstead strain retained its high virulence, but egg passage led to a substantial reduction in its virulence for mice (Fenner, 1949b), and it was then more readily adapted to growth in the rabbit cornea (Paschen, 1936).

TABLE 9-1

Comparison of the Virulence of Several
Strains of Ectromelia Virus by Footpad
Inoculation of BALB/cByJ Mice[a]

Strain	LD_{50} (pfu/ml)[b]
Washington University	1.0×10^1
St. Louis 79	2.8×10^1
Moscow	3.9×10^1
NIH 79	9.3×10^1
Ishibashi I-III	4.3×10^4
Hampstead egg	—[c]

[a] From Wallace and Buller (1986).
[b] Titrated in BS-C-1 cells.
[c] No deaths with largest dose.

Moscow Strain.   This strain was isolated in the laboratory of Professor V.D. Soloviev in Moscow and is highly virulent and highly infectious (Andrewes and Elford, 1947). It was used extensively by a succession of Australian workers (Fenner, Mims, Roberts, Blanden).

Ishibashi Strain.   Ichihashi and Matsumoto (1966) showed that this strain, which was recovered from an outbreak in Ishibashi in about 1950 (Y. Ichihashi, personal communicaiton, 1986), differed from the Hampstead strain in that it produced larger plaques in chick embryo fibroblasts, and the A-type inclusion bodies failed to occlude virus particles, i.e., they are V^- (see Chapter 4). The Ishibashi I-III strain, which has been passaged at least 30 times in cultured cells, was less virulent than all other strains tested except for Hampstead egg (Table 9-1).

Munich 1 Strain.   Recovered from a sick mouse in an outbreak in Munich in 1976, this strain became highly attenuated after over 300 passages in chick embryo fibroblasts (Mahnel, 1983).

NIH-79 Strain.   This strain was recovered from an outbreak in the United States National Institutes of Health in 1979 (Allen *et al.*, 1981), and has been used in investigations at the National Institutes of Health and Yale University. Its virulence is the same or slightly higher than that of the Moscow strain (Table 9-1), and the A-type inclusion bodies occlude virus particles (V^+).

It is not known whether these strains, and others not listed here, represent different isolates from a hypothetical wildlife source of the virus (perhaps *Mus musculus*, perhaps some other rodent), or whether they have differentiated since mice were domesticated. The distribution throughout Europe and to Japan and China could have resulted from the worldwide exchange of mice that was begun by mouse fanciers early in the nineteenth century and greatly expanded when mice were used for biological research, beginning early in the twentieth century (Morse, 1981).

Morphology and Chemical Composition of the Virion

The virions of ectromelia virus are indistinguishable from those of other orthopoxviruses (see Plate 9-4). Like other orthopoxviruses, the infectivity of ectromelia virus is relatively resistant; dried spots of infected blood retained infectivity for 11 days at room temperature (Bhatt and Jacoby, 1987c). However, the virus is rapidly inactivated by procedures commonly used in diagnostic and research work. Small differences have been detected between the genome DNA and certain antigens of ectromelia virus and those of other orthopoxviruses.

Restriction Mapping of Ectromelia Virus DNA. Ectromelia virus was included in groups of orthopoxviruses whose DNAs were analyzed by restriction endonuclease mapping (Mackett and Archard, 1979; Esposito and Knight, 1985; see Fig. 1-1). The two strains available (Moscow and Hampstead) had identical maps when tested with *Hind*III, *Xho*I, and *Sma*I, both being distinctly different from those of all other species of *Orthopoxvirus*. The genome of ectromelia virus (about 200 kbp) is somewhat larger than that of vaccinia virus (about 180 kbp). Buller (1986) reported that a mutant of the NIH-79 strain obtained from a persistently infected cell line was less virulent than wild-type virus and had a 10-kbp deletion from the left-hand end of the genome.

Antigenic Properties. Although the size and shape of the virions had suggested that ectromelia virus might be a poxvirus (Barnard and Elford, 1931; Ruska and Kausche, 1943), its definitive classification as an *Orthopoxvirus* was made by Burnet (1945), who showed that extracts of ectromelia virus-infected cells agglutinated the same limited group of chicken red blood cells as vaccinia virus hemagglutinin, and that this reaction could be specifically inhibited with vaccinia-immune serum. Burnet and Boake (1946) then showed that mice could be immunized against ectromelia by vaccination with vaccinia virus.

Subsequently, serological comparisons were made between a number of orthopoxviruses (McCarthy and Downie, 1948; McNeill, 1968; see Chapter 4). Whatever tests were used, there was extensive cross-reactivity between all members of the genus, but titers between homologous viruses and anitsera to them were higher than with heterologous combinations. Using suspensions of virions of ectromelia and vaccinia viruses as antigens for an ELISA, Buller *et al.* (1983) found that ectromelia-immune serum gave higher titers with the homologous antigen, whereas vaccinia-immune sera gave identical titers with both antigens.

Biological Characteristics

Behavior in Laboratory Animals.   In contrast to vaccinia, cowpox, and monkeypox viruses, ectromelia virus has a narrow host range, as far as clinical signs are concerned, but inoculation of large doses of virus can cause subclinical infections in several species of laboratory animal.

Laboratory Mouse. The laboratory mouse can be infected by all routes of inoculation. The pathogenesis of the disease produced by footpad inoculation is described in Chapter 4, and the clinical signs, pathology, and genetic differences in the response of mice to infection with ectromelia virus elsewhere in this chapter. Feral mice (*Mus musculus*) are also susceptible (McGaughey and Whitehead, 1933). Roberts (1986) found that feral mice obtained from three widely separated regions in Australia were equally susceptible to infection, but that mice from one area were much more likely to die than mice from the other two areas.

Other Species of Mouse. The production in the United States in recent years of laboratory colonies of several species of mouse other than *Mus musculus domesticus* gave Buller *et al.* (1986) the opportunity to study their response to infection with ectromelia virus (Table 9-2). Three types of response were observed: complete resistance (*Pyromys platythrix* and *Coelomys pahari*), high susceptibility, with no footpad lesion but an overwhelming systemic infection (*Mus caroli*, *Mus cookii*, and *Mus cervicolor popaeus*), and susceptibility to infection, with swelling of the inoculated foot and generalization of variable severity. In order to determine whether *Pyromys platythrix* and *Coelomys pahari* were completely resistant to infection, several animals of these species and several C57BL/6J (the most resistant strain of laboratory mouse) were inoculated with much larger doses of virus by both the intraperitoneal and footpad routes. the response of C57BL/6J mice followed the usual pattern (see below)—death after intraperitoneal inoculation, and local replication to high titer, but limited replication in the liver and spleen, after inoculation in the footpad. In the other two species there was very limited replication in the foot, no signs of disease after intraperitoneal inoculation, and a low but significant development of neutralizing antibodies.

Rat. Burnet and Lush (1936a) showed that when large doses of virus were inoculated intranasally into the rat, inapparent infection and

TABLE 9-2

Mortality in Various Genera and Species of Mouse following Footpad Inoculation with a Small Dose of Ectromelia Virus[a]

Genus and species	Origin	Footpad lesion	Mortality (%)	Lesions of liver and spleen
Pyromys platythrix	India	−	0	−
Coelomys pahari	Thailand	−	0	−
Mus caroli	Thailand	−	100	+ + +
Mus cookii	Thailand	−	100	+ + +
Mus cervicolor popaeus	Thailand	−	100	+ + +
Nannomys minutoides	Kenya	+	24	+ to + + +
Mus musculus musculus	Czechoslovakia	+	43	+ to + + +
Mus musculus domesticus	Maryland, United States	+	22	+ to + + +
Mus spretus	Spain	+	64	+ to + + +
Mus spretus	Morocco	+	35	+ to + + +

[a] Data from Buller *et al.* (1986).

replication of the virus occurred in the cells of the olfactory mucosa, with the development of neutralizing antibody.

Intradermal inoculation of large doses of ectromelia virus in the skin of the rat produced no lesions, or at the most a tiny papule, but the hemagglutination-inhibition (HI) titer of the sera of such animals had risen to a high level 14 days later (F. Fenner, unpublished observations). Intraperitoneal inoculation was also without effect except for the production of antibody. Mooser (1943) found that when peritoneal fluid from moribund mice which were infected with both ectromelia virus and *Rickettsia prowazeki* was inoculated intraperitoneally into rats, there was a severe rickettsial peritonitis, but characteristic ectromelia virus inclusion bodies could also be found in cells in smears of the peritoneal fluid.

Iftimovici *et al.* (1976) suggested that when housed with infected mice, young rats can be naturally infected with ectromelia virus; however, the infection among the rats died out spontaneously.

Rabbit. Most early workers found that the rabbit was resistant to infection, although Paschen (1936) reported that virus which had been passed several times on the chorioallantoic membrane produced infection of the rabbit's cornea, with the production of inclusion bodies, and could then be passed in the rabbit skin. Burnet and Boake (1946) found that two rabbits inoculated intravenously with a large dose of virus died 6 and 10 days later, hemagglutinin being demonstrable in suspensions of the liver and spleen. They found that intradermal inoculation of the virus resulted in the appearance of indurated papules. Subsequent experiments (F. Fenner, unpublished observations) showed that papules developed about 4 days after the intradermal inoculation of a large dose of virus and sometimes ulcerated a few days later, to be followed by the appearance of HI antibodies in the serum. Christensen *et al.* (1967) suggested that ectromelia virus could be used to immunize rabbits against rabbitpox, although because of the danger to mouse stocks,it was preferable to use other strains of vaccinia virus for this purpose.

Guinea Pig. Paschen (1936) found that the plantar surface of the guinea pig foot and the cornea were susceptible to infection with egg-passaged ectromelia virus but not with mouse liver virus. Intradermal inoculation of egg membrane preparations of the virus regularly produced local indurated lesions in the footpad (F. Fenner, unpublished observations), and HI antibodies could be found in the sera 14 days later. Intraperitoneal inoculation of large doses caused no symptoms but was followed by antibody production.

Hamster. According to Flynn and Briody (1962), Syrian hamsters are not susceptible to infection with ectromelia virus.

Man. The only conscious attempt at infection of humans with ectromelia virus (F. Fenner, unpublished observations) was scarification by the multiple-pressure technique of the arms of two men who had been vaccinated with vaccinia virus several times previously. In both a small papule appeared on the second day, became slightly vesicular in one of them on the fourth day, and disappeared by the eighth day. No change occurred in the ectromelia HI antibody titer of the serum, a not unexpected result in view of Nagler's (1944) observation that after revaccination with vaccinia virus an increase in vaccinia HI antibody occurred only if there had been a good "take," with definite vesiculation.

Packalén (1947) has shown that the Laigret–Durand strain of "mouse-pathogenic murine typhus rickettsia," like the epidemic typhus strain studied by Mooser (1943), owed its mouse pathogenicity to its inadvertent contamination with ectromelia virus. He considered that the strain received by his laboratory in 1940 was probably pure ectromelia virus at that time, so that subsequent experiments on "mouse pathogenic murine typhus rickettsiae" (Ipsen, 1945; Kling and Packalén, 1947) were really experiments with ectromelia virus. If this was so, active ectromelia virus, either alone or mixed with rickettsiae, has been inoculated subcutaneously into hundreds of thousands of humans in doses of up to 1000 "mouse units" (Laigret and Durand, 1939, 1941; Packalén, 1945). No local or general reaction of any significance was reported (Laigret and Durand, 1941).

Chick Embryo. Infection of the chick embryo was first described by

Paschen (1936) and Burnet and Lush (1936b). Both workers grew the virus on the chorioallantoic membrane, and Burnet and Lush showed that if dilute suspensions of virus were inoculated, separate foci developed, which could be counted (see Plate 4-2). This pock-counting method of titration of the virus was exploited by Fenner (1948a) for quantitative studies of mousepox infection. Chorioallantoic inoculation with large doses of virus was usually followed by death of the embryo 4–5 days later, with scattered areas of necrosis in the liver and spleen.

Serial passage of ectromelia virus on the chorioallantoic membrane modified its biological characteristics. Paschen (1936) found that egg-passaged virus was more suitable for infection of the rabbit and guinea pig cornea and skin than was mouse liver virus. Serial chorioallantoic passage of the Hampstead strain of ectromelia virus (50–60 passages intermittently over a period of 10 years) resulted in greatly reduced virulence for mice (Fenner, 1949b). On the other hand, no change occurred in the high virulence of the Moscow strain of virus after 20 passages on the chorioallantoic membrane.

Growth in Cultured Cells. Ectromelia virus replicates in HeLa cells and human amnion cells, L cells, mouse fibroblasts, chick embryo fibroblasts, and BS-C-1 cells (a continuous African Green monkey kidney cell line). Ichihashi and Matsumoto (1966) found that the Ishibashi strain of ectromelia virus produced much larger plaques in chick embryo fibroblasts than the Hampstead strain. Plaque production in L cells (in laboratories in Australia) and in BS-C-1 cells (in the United States) are currently the preferred methods of assay.

Using a cell culture system in which mouse hepatocytes maintained effective gluconeogenesis, Lees and Stephen (1985) showed that ectromelia virus infection of such cells produced changes similar to those found *in vivo*. Gluconeogenesis was inhibited, and B-type but not A-type inclusion bodies developed, but cell rounding of the kind usually found in orthopoxvirus-infected cells (see Chapter 4) did not occur. This system may prove useful for the study of biochemical changes of virus-induced damage in differentiated cells.

Since there is now an extensive interlaboratory traffic of hybridoma cells and mouse ascitic fluid, contamination of such material with ectromelia virus (and other viruses) is a matter of some concern. and at least one outbreak of mousepox has been attributed to such a source (Bhatt *et al.*, 1981). Buller *et al.* (1987c) found that each of six lymphoma and T-cell hybridoma cell lines supported ectromelia virus replication, but only two out of seven B-cell hybridomas were susceptible.

Intracytoplasmic Inclusion Bodies. Like all orthopoxviruses, ectromelia virus produces B-type inclusion bodies in infected cells (see

Chapter 3). These are the only type of inclusion body seen in infected liver cells, both *in vivo* and *in vitro* (Lees and Stephen, 1985); up to eight inclusion bodies (usually one to three) may occur within a single cell. Most other types of cell infected with ectromelia virus also contain A-type inclusion bodies (see Plate 9-4).

Electron microscope examination of cells infected with different strains of ectromelia virus showed that with the Hampstead strain, all the mature viral particles were finally occluded within the A-type inclusion body (Marchal body)—a situation designated V^+ (see Chapter 4). With another strain (Ishibashi) these inclusion bodies were devoid of virions, i.e., V^- (Matsumoto, 1958; Ichihashi and Matsumoto, 1966).

Possible Wildlife Sources of Ectromelia Virus

It is not clear whether ectromelia virus is uniquely a disease of the mouse (*Mus musculus domesticus*), which accompanied it during its domestication, or whether there are wildlife sources of the virus, in *Mus musculus* or other animal species. The principal evidence bearing on this problem is an unconfirmed report by Gröppel (1962) that certain wild mice in Germany, living far from possible exposure to laboratory mice, were infected with ectromelia virus. In a study of viral infections of wild rodents in Britain, Kaplan *et al.* (1980) found ectromelia virus HI antibodies in field mice (*Apodemus* spp.) and voles (*Microtus agrestis*), but this result is just as likely to be due to infection with cowpox virus (see Chapter 6). The problem warrants further study.

MOUSEPOX IN LABORATORY MICE

Geographic Distribution

Since 1930 mousepox has been recognized as a relatively common infection of laboratory mouse colonies in Europe, Japan, and China. It has been so common in some European countries that few of the many outbreaks were reported; for example Iftimovici *et al.* (1976) suggest that in Romania in the 1970s "most of the laboratory mouse colonies are either latently or manifestly infected with ectromelia virus." Recovery of the virus from laboratory mice has been described in England (Marchal, 1930; McGaughey and Whitehead, 1933; Fairbrother and Hoyle, 1937; Carthew *et al.*, 1977), France (Schoen, 1938; Guillon, 1975), Germany (Kikuth and Gönnert, 1940; Schell, 1964), the USSR (cited in Andrewes and Elford, 1947), Switzerland (Mooser, 1943), and Sweden, Czechoslovakia, and Israel (Briody, 1959). In response to recent inquiries, authorities in three countries in Europe (France: Dr. J. C. Guillon, personal communication, 1986; Germany: Professor H. Mahnel, personal communicaton, 1986; and the USSR: Dr. S. S. Marennikova, personal communication, 1986) reported that mousepox was still enzootic in some colonies of laboratory mice, but that its incidence had decreased as management practices had improved.

Mousepox is still common in laboratory colonies of mice in the People's Republic of China (Dr. Jiang Yutu; personal communication, 1979) but is now much less prevalent than formerly in Japan (Dr. M. Nakagawa; personal communication, 1979), because of better management and the extensive use of specific pathogen-free (SPF) and "clean" conventional colonies of mice. Vaccination was never used in Japan for the control of mousepox.

In contrast, mousepox does not seem ever to have been enzootic in mouse colonies in the United States. However, the virus has been unwittingly imported into the United States from Europe with mice or mouse tissues on several occasions, sometimes causing devastating outbreaks (Melnick and Gaylord, 1953; Trentin, 1953; Briody, 1955; Dalldorf and Gifford, 1955; Briody et al., 1956; Bhatt et al., 1981; Whitney et al., 1981), and sometimes being spread from one locality in the United States to another (Briody, 1955; Laboratory Animal Science, 1981; Wallace, 1981).

With the development of specific pathogen-free (SPF) mouse colonies and more effective health monitoring, ectromelia virus-free stocks of widely used mouse strains have been established. However, the simultaneous large-scale development of many strains of mice with special genetic features, which are usually not produced as SPF animals, and the extensive international exchange that occurs in these animals or in tumor material, hybridomas, and viruses derived from them have increased the risk of importation of the virus into previously uninfected mouse stocks. As a consequence, there have been periodic efforts to limit the spread of the virus between stocks of mice in different laboratories (Shope, 1954; Report, 1973; Anslow et al., 1975; Wallace, 1981).

Clinical and Pathological Features

Disease in Susceptible Mice.   Early workers (Marchal, 1930; McGaughey and Whitehead, 1933; Schoen, 1938) described two forms of the disease, a rapidly fatal form in which apparently healthy mice died within a few hours of the first signs of illness, and showed extensive necrosis of the liver and spleen at autopsy, and a chronic form characterized by ulcerating lesions of the feet, tail, and snout. Fenner

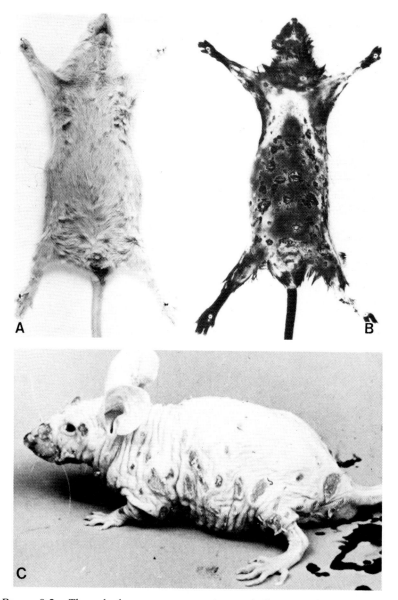

PLATE 9-2. *The rash of mousepox, as seen in genetically susceptible mice: (A and B) In outbred WEHI mice 14 days after infection by footpad inoculation, with and without depilation; (C) in a naturally infected hairless mouse (not athymic). (C, Courtesy of Zentralinstitut für Versuchstiere, Hanover, Federal Republic of Germany.)*

(for review, see 1949d) showed that in every case of mousepox in susceptible outbred WEHI mice there was a stage in which virus replicated to high titer in the liver and spleen (see Chapter 4). Some mice died at this stage, but if they survived they almost invariably developed a rash (Plate 9-2) which occurred over the whole body, not only on the hairless extremities.

It is now clear that these "classical" syndromes occurred only in nonimmune mice belonging to a susceptible strain. Maternal immunity can largely prevent the occurrence of clinical disease among genetically susceptible young mice (see below), and infection is usually inapparent among genetically resistant strains of mice. However, all strains of mice appear to be equally susceptible to infection (Wallace and Buller, 1985; Bhatt and Jacoby, 1987a).

Pathological Changes in Genetically Susceptible Mice. The pathological changes in "classical" mousepox, which are quite characteristic and are found in animals infected by contact or by subcutaneous or intradermal inoculation, have been described and illustrated by Allen *et al.* (1981) and Fenner (1982). Other kinds of lesions occur after oral infection and intraperitoneal or intranasal inoculation. These are described below, because oral infection may be of epidemiological importance and ectromelia virus is sometimes unwittingly passaged by intraperitoneal and intranasal routes.

Skin Lesions. Fenner (1947b) suggested that infection through skin abrasions was the usual mode of natural infection in mousepox, and that a local lesion (the "primary lesion") developed at the site of entry of the

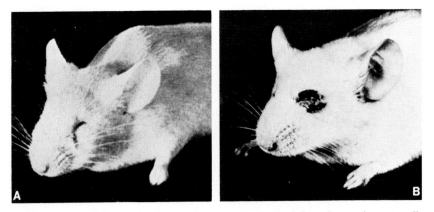

PLATE 9-3. *The primary lesion of mousepox on the left eyebrow of a naturally infected mouse (A) 8 days and (B) 14 days after infection.*

virus. This lesion appeared first as a localized swelling surmounted usually by a minute breach of the surface (Plate 9-3). It rapidly increased in size, with pronounced edema of the surrounding tissues. Later a hard adherent scab formed and fell off after a week or two, leaving the site of the primary lesion marked by a deep hairless scar which often persisted for life.

The earliest primary lesions that could be recognized macroscopically were the sites of advanced histological changes, for viral replication had by then been in progress for several days. Usually there was no obvious breach of the skin surface, but the dermis and subcutaneous tissue were edematous and there was widespread lymphocytic infiltration of the dermis. Similar lesions occurred in the foot after footpad inoculation, and A-type inclusion bodies could be seen in almost every epithelial cell (Plate 9-4). Necrosis of the epidermal cells was followed by ulceration of the surface and necrosis. The exudate dried and formed a scab beneath which healing occurred.

If the mouse did not die from acute liver damage (see below), a generalized rash developed (see Plate 9-2). Individual lesions of the rash appeared 2 or 3 days after the development of the primary lesion, and when first seen on the shaved skin they were slightly raised pale areas 2–3 mm in diameter. They increased in number and size, ulcerated, and

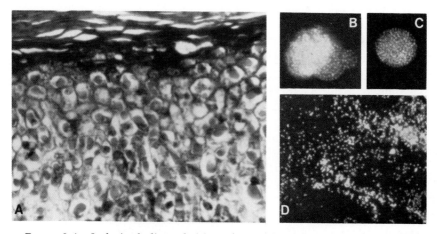

PLATE 9-4. *Inclusion bodies and virions of ectromelia virus. (A) Eosinophilic A-type inclusion bodies in skin of foot after footpad inoculation (Mann's stain). (B and C) Isolated inclusion bodies, photographed by ultraviolet light and showing virions occluded within protein matrix (see electron micrographs, Plate 4-1). (D) Isolated virions, photographed by ultraviolet light. (B, C, and D from Barnard and Elford, 1931.)*

in animals that survived healed by the end of the third week after infection. Conjunctivitis and blepharitis occurred frequently during the stage of the rash and in severe cases ulcers could be found on the tongue and buccal mucous membrane.

Histologically the first changes of the rash were seen on the same day as the primary lesion was detected. A few localized areas of the epidermis appeared hyperplastic with dark-staining nuclei surrounded by vacuoles; occasionally these cells contained intracytoplasmic inclusion bodies. The areas of proliferation and edema increased in size until they became macroscopically visible as pale, slightly raised macules. Numerous A-type inclusion bodies were then present in the epidermal cells. Fresh foci appeared in the intervening regions of previously normal skin, and by the next day necrosis of the superficial cells of the early lesions had commenced. Massive necrosis followed quickly, with accompanying widespread edema and lymphocytic infiltration of the dermis, and the papules were converted to ulcers with closely adherent scabs. The lesions progessed in size and number for a couple of days and then healing commenced and was complete in a few days.

Lesions of the Liver. The liver and spleen were invariably invaded during the incubation perioed and virus replicated to high titer there (see Chapter 4). The liver remained macroscopically normal, even in cases which would prove fatal, until within 24 hours of death, when it appeared enlarged and studded with many minute white foci (Plate 9-5A). In animals which died of acute mousepox, the necrotic process extended rapidly and at the time of death the liver was enlarged with many large semiconfluent necrotic foci (Plate 9-5B). The fat content of such livers was about 13% of the liver weight, compared with the normal figure of 3–4%. In mice which survived, the liver usually returned to its normal macroscopic appearance, but occasionally numerous white foci could be found along the anterior border of the median lobe.

Histologically, little change was apparent until macroscopic changes had appeared, that is within a day or so of death in rapidly fatal cases, although with fluorescent antibody staining it could be shown that infection always occurred first in the littoral cells of the hepatic ducts, from which it spread to contiguous parenchymal cells (Mims, 1959). Numerous scattered foci of necrosis then appeared throughout the liver parenchyma and in fatal cases these rapidly extended until they became semiconfluent (see Plate 9-5B). At the margins of the necrotic areas the hepatic polygonal cells usually showed active regeneration with many multinucleate cells, even in fatal cases. The portal tracts showed slight infiltration with lymphoid cells. A-type inclusion bodies were never

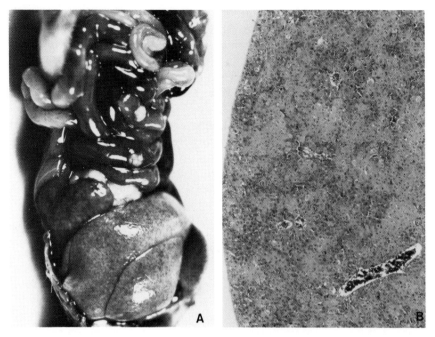

PLATE 9-5. *(A) Swelling and necrosis of the spleen and liver, enlarged Peyer's patches, and hemorrhagic small intestine in a susceptible BALB/c mouse killed when moribund. (B) Section of the liver, showing extensive irregular necrosis, with little inflammatory reaction. (A, from Allen et al., 1981.)*

found in infected hepatic parenchymal cells, although electron microscope studies showed that these cells contained very large numbers of virus particles.

The necrotic foci occasionally seen after recovery from infection consisted of hyaline necrotic tissue. In the liver of most animals which survived and showed no macroscopic lesions at autopsy there were small accumulations of lymphocytes, usually around branches of the portal tract but occasionally in the parenchyma of the liver.

The necrotic process in the liver, which was the dominant histological finding, was focal and random in distribution, and showed no regular relationship to the normal liver architecture. Liver regeneration commenced early and was active, especially in nonfatal cases, and fibrosis did not occur.

Lesions of the Spleen. The spleen showed macroscopic changes at least a day earlier than the liver and higher titers of virus were found in the spleen each day until death, when the titers of the spleen and liver were approximately the same. Fluorescent antibody studies (Mims, 1964) revealed that virus probably reached the spleen in infected lymphocytes, which initiated infection in cells of the follicles. While infected follicles were destroyed by the spreading infection, neighboring follicles showed the proliferative response characteristic of antibody production in the spleen.

The spleen was at first engorged; later rather pale, slightly depressed areas of necrosis appeared either in isolated patches or were semiconfluent. In surviving mice obvious lesions of the spleen were much more common than were lesions of the liver, and varied in severity from small raised plaques about a millimeter in diameter to strands of fibrous tissue which after severe attacks almost completely replaced the normal splenic tissue. These changes constitute the most reliable autopsy evidence that a mouse has recovered from an attack of mousepox.

Histologically the early changes consisted of lymphoblastic hyperplasia of the follicles and congestion of the sinuses of the red pulp, and focal necrosis with fragmentation of the lymph follicles. The necrosis rapidly extended, and the red and white pulp was characterized by transformation to endothelial-type cells. Very extensive necrosis of the spleen occurred in some mice which ultimately recovered. The plaques seen on the spleen in mice which had recovered from mild attacks of mousepox consisted of localized areas of hyperplasia of the serosal cells, and the scarred areas consisted of fibrous tissue.

Lesions of Other Organs. The regional lymph nodes draining the site of the primary lesion were enlarged from the time that the primary lesions could be detected, and they usually showed localized areas of necrosis, with pyknotic nuclear debris in a featureless background. Sometimes these necrotic areas were almost confluent, and often numerous A-type inclusion bodies were present. In the later stages the majority of lymph nodes showed lymphoblastic hyperplasia and small foci of necrosis.

In fatal cases of mousepox the gut was often engorged (see Plate 9-5A) and the lymphoid follicles enlarged. Greenwood et al. (1936) reported that a careful histological survey of the intestines showed that small necrotic foci with typical inclusion bodies occurred in about 65% of acutely fatal cases of mousepox. Briody (1959) commented upon the frequency with which a hyperemic or blood-filled small intestine was observed in genetically highly susceptible strains of mice during epizootics of mousepox in the United States.

No other organs were regularly affected in natural mousepox, but occasionally, especially in very young mice, there were hemorrhagic foci in the bladder and in the kidneys, where widespread focal necrosis occurred, especially in the region of the convoluted tubules. A-type inclusion bodies could be seen in the epithelial cells and sometimes they appeared to occur free in the lumina of the tubules. Later, in animals which survived, the necrotic areas were replaced by areas of fibrosis and cellular infiltration.

Differential Susceptibility of Different Strains of Mouse. *Natural Outbreaks.* When outbreaks of mousepox occurred in the United States in the 1950s, it was observed that the ability of ectromelia virus to produce acute death with visceral lesions and no rash, a systemic disease with a generalized rash, or a subclinical infection depended on the genotype of the mouse. For example, in an outbreak in Buffalo, New York in 1954–1955, case–fatality rates among different strain of mice were as follows: DBA/1, 84%; A, 84%; C3H; 71%; C57BL/6, <1% and AKR, <1% (Briody *et al.*, 1956). Thus mousepox need not necessarily cause a disease with obvious signs and a high mortality; it can occur among genetically resistant mice for months without being recognized, until circumstances arise that lead to infection of genetically susceptible mice.

Other factors also play a role in determining the severity of signs, notably passive immunity. For example, in European laboratory mouse colonies, Guillon (1975) noted that the first outbreaks seen after introduction of ectromelia virus into a colony of susceptible mice were characterized by acute deaths with predominantly visceral lesions. Subsequently, if enzootic disease was established in the colony, few acute deaths were observed (although they may have occurred in young mice that were then eaten), and the situation was dominated by subacute, chronic, and inapparent infections. Studies of enzootic mousepox in genetically susceptible mice in England (Fenner, 1982) suggested that the change was largely due to maternal immunity, which protects suckling mice from death but allows infection with concomitant active immunization (see below).

Experimental Studies. One of the major advances in studies of mousepox in the last decade has been the elucidation of the influence of genotype on susceptibility. The first experimental study on the genetics of the resistance of mice (Schell, 1960a,b) was based on the observation that in outbreaks in the United States in the 1950s, C57BL mice were highly resistant. He found no difference in the infectivity endpoint of a viral suspension titrated in susceptible WEHI and resistant C57BL mice, but the titer of virus in the footpad of C57BL mice ceased to rise 6 days after infection, and the highest titers in blood, liver, and spleen were 2 to 3 log units lower than in the susceptible mice. Cultured cells from resistant and susceptible strains were equally susceptible to infection

TABLE 9-3

Comparison of Infectivity and Lethality
of Ectromelia Virus in Inbred
Strains of Mice[a,b,c]

Mouse strain	Lethality (LD_{50})	Infectivity (ID_{50})[d]
C57BL/6J	1.0	7.0
AKR/J	1.0	. . .[e]
BALB/cByJ	6.9	7.1
DBA/2J	6.2	6.2
A.By/SNJ	6.3	7.0
C3H/HeJ	7.1	7.2

[a] Data from Wallace and Buller (1985).
[b] Expressed as negative log base 10.
[c] NIH-79 strain virus stocks with titer of $10^{6.3}$ plaque-forming units/ml in BS-C-1 cells.
[d] Infectivity defined by positive ELISA (titer $\geq$ 20).
[e] . . . , Not done.

with ectromelia virus, but neutralizing antibody, delayed type hypersensitivity, and active immunity were demonstrable 1 or 2 days earlier in C57BL than in WEHI mice. The strain differences were best demonstrated after footpad inoculation or in natural epizootics, for C57BL mice develop obvious signs and often die after intranasal, intracerebral, or intraperitoneal infection (see below).

Comparisons between mice of different genotypes were expanded by Wallace and Buller (1985) and Bhatt and Jacoby (1987a). Using footpad inoculation with the NIH-79 strain of ectromelia virus, both groups of investigators found that all mouse strains tested were equally susceptible to infection, but infection was often lethal in BALB/c, DBA, and C3H mice whereas few mice of strains C57BL/6 or AKR died (Table 9-3). Further experiments were carried out with BALB/c and C57BL/6 strains, as prototypes of genetically highly susceptible and highly resistant mice.

1. *Signs and Pathological Findings in BALB/c Mice.* The disease in BALB/c mice infected by footpad inoculation or in contact experiments closely resembled that described in outbred WEHI mice (see above), but rashes were rarely seen (Bhatt and Jacoby, 1987a), although conjunctivitis was common in mice that did not die of acute hepatitis. However, about 30% of these mice survived oral infection (gastric inoculation with a soft plastic tube), and developed chronic lesions consisting of severe scab-

bing of the tail and feet and conjunctivitis (Wallace and Buller, 1985). Immunohistochemical studies showed that in addition to the sites of viral replication identified by Fenner, focal areas of infection occurred in bone marrow, ovary, uterus, Peyer's patches and adjacent intestinal epithelium, nasal turbinates, and oral mucosa (Jacoby and Bhatt, 1987). Sites regarded as particularly important in relation to shedding of virus and transmission are the skin lesions, nasal turbinates, oral mucosa, and small intestine.

2. *Signs and Pathological Findings in C57BL/6 Mice.* The only sign of disease after footpad inoculation of C57BL/6 mice was swelling of the foot. Virus was not detected in the blood on days 3 through 9, but clearly transient viremia must have occurred, since serial histological and immuno-histochemical examination showed that there were a few necrotic foci in the liver, around which polymorphonuclear leukocytes and mononuclear cells accumulated (Jacoby and Bhatt, 1987). Small quantities of viral antigen and small areas of necrosis were seen in the liver, bone marrow, and nasal mucosa, but not in the spleen.

Genetics of Resistance. The number of resistance genes detected in breeding experiments depends on the route of inoculation of the virus and the genotype of the crossed parents. Schell (1960b) and Ermolaeva *et al.* (1974) demonstrated that resistance was dominant over susceptibility and was largely determined by a single gene which was not linked to sex or certain color genes. The overriding importance of this gene has been confirmed (O'Neill and Blanden, 1983; Wallace *et al.*, 1985), gene action being expressed via cell-mediated immune responses in which cytotoxic T cells and macrophages attack virus-infected cells and thus limit the spread of infection. Using matched C57BL/6J (resistant) and BALB/c (susceptible) mice, O'Neill and Blanden (1983) showed that viral titers in lymph nodes and spleen were consistently higher, and rose earlier, in the BALB/c mice. Using chimeric H-2-compatible strains of resistant and susceptible mice, they found that overall mortality and time to death were determined by the genotype of the irradiated host. Thus the effect of the C57BL background which confers resistance is not expressed through radiosensitive lymphomyeloid cells. Recent investigations (O'Neill and Brenan, 1987) showed that cytotoxic T cells could be found in the lymph nodes draining the site of primary infection 2 days earlier in C57BL mice than in susceptible BALB mice.

A minor role for *H-2* genes has been demonstrated by O'Neill *et al.* (1983), who showed that several C57BL/10 strains of mice, recombinant at the *H-2* locus, differed in their resistance to footpad inoculation with ectromelia virus. Chimera experiments showed that the *H-2*-linked

TABLE 9-4

Percentage Mortality after Footpad Inoculation
of Ectromelia Virus in Inbred, Selected
F_1 and Backcrossed Mice[a]

Mice	Female	Male	Total
C57BL/B6 (B6)	0	0	0
AKR/J (AK)	0	0	0
A/J (A)	90	100	97
BALB/c ByJ (C)	50	100	69
DBA/2J (D2)	80	82	81
(B6 × A) F_1	0	7	2
(B6 × C) F_1	0	0	0
(B6 × D2) F_1	0	0	0
(AK × D2) F_1	0	0	0
(B6 × A) F_1 × A	40	51	46
(B6 × C) F_1 × C	0	19	10
(B6 × D2) F_1 × D2	7	17	12
(AK × D2) F_1 × D2	9	21	14

[a] Data from Wallace *et al.* (1985).

genes influenced resistance through radiosensitive lymphomyeloid cells. Death in susceptible B10 strains was not dose dependent and was not associated with massive liver necrosis (the cause of death in other susceptible strains of mice).

Briody (1966) had identified AKR as another resistant strain of mouse. Wallace *et al.* (1985) investigated the genetics of virulence in crosses between resistant C57BL/6J and AKR/J and several susceptible strains, and backcrosses of the F_1 progeny with the susceptible parents (Table 9-4). They confirmed that resistance in the C57BL strain was controlled by a single autosomal dominant gene, which they provisionally termed *Rmp-1* (resistance to mousepox). The results in other backcrosses were more complex, and there were sex-related differences in the frequencies of death in each backcross population, male mice being more susceptible than females. Two or more independently assorting loci may be involved in these crosses. Since the ID_{50} doses of ectromelia virus for all parental strains of mice were the same (see Table 9-3), the *Rmp-1* alleles do not affect resistance to infection, but must play a role in controlling the spread of virus through the body (see Chapter 4).

Response to Oral Infection. Fenner (1947b) considered that the oral route was probably unimportant in natural transmission except when cannibalism occurred and mice that had died of acute mousepox were eaten. However, this mode of infection may be important for reasons that he did not suspect, namely in maintaining inapparent infection in either susceptible or resistant strains of mice. Gledhill (1962a,b) found that genetically susceptible mice could regularly be infected *per os* if large doses of virus were used, and about 20% were infected by relatively small doses. In contrast to infection by other routes, mice infected orally with small doses sustained inapparent infections, and ectromelia virus could be isolated from the tail skin and feces of a small proportion of such mice for up to 4 months after infection.

Reinvestigating this problem with genetically defined strains of mice, Wallace and Buller (1985) found that about 30% of BALB/c mice survived after oral infection whereas all died after footpad inoculation. Virus could be recovered from the feces of recovered mice for up to 5 weeks after infection. In other experiments they found that virus could occasionally be recovered from orally infected C57BL mice, in which infection was inapparent. Further, transmission sometimes occurred from these mice during the third and fourth weeks after infection. The sources of virus in the feces were probably the microscopic lesions in the small intestine and Peyer's patches reported by Jacoby and Bhatt (1987).

Lesions after Intraperitoneal Inoculation. In some laboratories intraperitoneal inoculation was the standard method of passaging ectromelia virus, and sometimes mousepox was first seen when mouse cells or tissue extracts, unknowingly contaminated with ectromelia virus, were inoculated intraperitoneally. The lesions differ considerably from those observed in the natural disease, and it is obvious that some descriptions of the pathological changes in mousepox are based upon the results of intraperitoneal inoculation. There is, of course, no primary skin lesion, but in acutely fatal cases the necrosis of the liver and spleen resembles that found after natural infection or footpad inoculation. In addition there is usually some increase in intraperitoneal fluid and a considerable amount of pleural fluid, and the pancreas is often edematous. In animals which survive the acute infection the signs of general peritonitis are much more pronounced. There is a great excess of peritoneal and pleural fluid, the peritoneal surfaces of the liver and spleen are covered with a white exudate, the walls of the gut are thickened and rigid, and there is often fat necrosis in the intraperitoneal fat. Later extensive adhesions between the abdominal viscera develop. Animals infected by the intraperitoneal route which survive long enough develop a characteristic rash. The survival times in fatal cases inoculated intraperitoneally are usually 2 or 3 days shorter than when the same dose of virus is inoculated into the foot.

Schell (1960b) found that the resistance of C57BL mice was overcome by intraperitoneal inoculation of the virus, even small doses killing all C57BL mice as quickly as WEHI mice. On the other hand, Bhatt and Jacoby (1987a) recorded only a 36% case–fatality rate in C57BL mice inoculated by the intraperitoneal route.

Lesions after Intranasal Inoculation. In the 1930s and 1940s, one common way in which ectromelia virus used to be encountered in laboratory practice was during the mouse lung passage of influenza virus. Some workers (e.g., Kikuth and Gönnert, 1940) suggested that ectromelia virus acquired pneumotropic properties after serial lung passage, because deaths from pneumonia then occurred with little macroscopic evidence of involvement of the liver and spleen. However, when small doses of virus are inoculated intranasally there is usually little change in the lungs except patchy congestion. The survival times of fatal cases are then approximately the same as in fatal cases inoculated in the footpad with the same dose of virus, and the changes in the liver and spleen are those characteristic of acute naturally acquired mousepox. With larger doses of virus, congestion of the lungs is more pronounced and consolidation may occur, and when very large doses are given death occurs with patchy or complete consolidation of the lungs and little change in the liver and spleen. It is this picture which led investigators to speak of the pneumotropism of the virus, and the detailed histological description given by Kikuth and Gönnert (1940) is of such lungs. There was early exudation into the alveoli and bronchi and small foci of necrosis occurred of the bronchial epithelium. Inclusion bodies were seen in the bronchial epithelial cells, in histiocytes, in cells of the pleural epithelium, and eventually in the alveoli. When the lungs, liver, and spleen of such fatal cases were examined for their virus content, it was found that the viral titers of the apparently normal liver and spleen were very high, just below the threshold at which demonstrable necrosis occurred (F. Fenner, unpublished observations; Ipsen, 1945). Indeed such necrosis was found in a few mice which survived for 6 instead of the usual 4 or 5 days after the intranasal inoculation of a large dose of virus.

Using fluorescent antibody staining, Roberts (1962a) showed that either macrophages or alveolar mucosal cells were initially infected but it was the macrophages that carried virus to the pulmonary lymph nodes and thus to the blood stream, from which it was taken up by the liver and spleen in which replication then proceeded. "Pneumotropism" is thus a laboratory artifact, due to the fact that the local reaction which occurs after the intranasal inoculation of very large doses of virus kills the animal before there is time for the characteristic changes in the liver and spleen to occur.

Schell (1960b) found that C57BL mice were even more susceptible than WEHI mice to the intranasal inoculation of large doses of virus, a result that he ascribed to the more vigorous immune response and correspondingly more severe pneumonia in C57BL mice.

Immune Responses

Active Immunity. Early investigations showed that mice that had recovered from mousepox were immune to reinfection (Marchal, 1930; Greenwood *et al.*, 1936). Fenner (1949c) showed that 2 weeks after infection mice were solidly immune to reinfection by footpad inoculation of the virus. This immunity declined slowly but even a year after recovery replication of the virus after footpad inoculation was confined to the local skin lesion and in only very occasional animals could virus be isolated from the spleen. When replication of the virus in the foot occurred, with consequent swelling, the HI titer usually rose significantly, but if no replication occurred there was no antibody rise. Long-continued epidemics showed the epidemiological importance of this durable immunity, for in only 3 mice out of 168 which had recovered from mousepox did any sign of reinfection or recurrence occur, and in none of these did it proceed beyond a local lesion in the foot (Fenner, 1948d).

Serological Changes. Sera from mice which have recovered from mousepox contain antibodies that can be recognized by a variety of serological tests: neutralization, hemagglutination inhibition (HI), precipitation, ELISA, immunofluorescence, etc. HI tests were long used for serological surveys of mouse sera in the United States (Briody, 1959), but during outbreaks in the United States in 1979–1980 it was found that the HI test occasionally gave both false positive and false negative results (see *Laboratory Animal Science*, 1981). Collins *et al.* (1981) and Buller *et al.* (1983) prefer the ELISA method for large-scale screening tests.

Passive Immunity. Tests on fetuses obtained just before birth, and exchange of newborn mice between nonimmune and immune mothers, showed that newborn mice received some antibody via the placenta but much more in the milk, for at least the first 7 days after birth (Fenner, 1948c). Maternal antibody declined to undetectable levels by the seventh week after birth, but before this it conferred protection against death but not against infection with moderate doses of ectromelia virus, which were invariably lethal in normal young mice. Maternal immunity may play an important role in the maintenance of enzootic mousepox in mouse colonies, especially among genetically susceptible mice.

Cell-Mediated Immunity.   Mice that have recovered from mousepox exhibit delayed-type hypersensitivity (Fenner, 1948a; Owen *et al.*, 1975). Understanding of the role of cell-mediated immunity in mousepox has been greatly expanded by the investigation of Blanden and his colleagues (for reviews, see Blanden, 1974; Cole and Blanden, 1982). The full story is complicated and involves different classes of T cells and the *H-2* antigens. Briefly, mechanisms controlling viral growth in the major visceral target organs (liver and spleen) are operating 4–6 days after footpad inoculation. Cell-mediated immune responses occur soon after infection, e.g., virus-specific cytotoxic T cells are detectable 4 days after infection and reach peak levels in the spleen 1–2 days later, while delayed hypersensitivity is detectable by the footpad test 5–6 days after infection. In contrast, significant neutralizing antibody is not detectable in the circulation until the eighth day.

Mice pretreated with anti-thymocyte serum die from otherwise sublethal doses of virus, due to uncontrolled viral growth in target organs. These mice have impaired cell-mediated responses, but normal neutralizing antibody responses, elevated interferon levels in the spleen, and unchanged innate resistance in target organs.

Very large doses of interferon or immune serum transferred to preinfected recipients are relatively ineffective against the established infection in target organs, though high antibody titers can be demonstrated in the sera of the recipients of immune serum. On the other hand, immune spleen cells harvested 6 days after donor immunization transfer specific and highly efficient antiviral mechanisms which rapidly eliminate infection from the target organs of the recipients, although neither antibody nor interferon is detectable in the recipients. The active cells in the immune population are theta positive and immunoglobulin negative (i.e., they are T cells). The kinetics of their generation and the requirement for sharing of *H-2K* or *H-2D* genes between donor and recipient identify them as cytotoxic T cells. Depending on the mouse strain, helper T cells, which recognize antigens dependent on the *I* region of the *H-2* complex, may be important in the generation of cytotoxic T cells, but are not important at effector level.

Monocytes of immune T cell recipients, labeled with tritiated thymidine before T cell transfer, appear in foci of infection in the liver after T cell transfer, and prior irradiation of immune T cell recipients in a regimen designed to reduce blood monocyte levels significantly reduces the antiviral efficiency of the transferred cells. Further, subpopulations of immune spleen cells labeled with tritiated thymidine show immunologically specific localization in liver lesions of cell recipients.

These findings support the idea that blood-borne cytotoxic T cells specific for virus-induced antigenic changes in infected cell surface membranes enter infectious foci and retard viral spread by lysing infected cells before the maturation and assembly of progeny virions. This T cell activity attracts blood monocytes which contribute to the elimination of infection by phagocytosis and intracellular destruction of virus, or by becoming unproductively infected. Macrophage activation and locally produced interferon may increase the efficiency of virus control and elimination, but are less important than T cells.

Modes of Spread

Ectromelia infects mice by all routes of inoculation, hence virus-contaminated serum, ascites fluid or mouse cells, tumors or tissues constitute a risk to laboratory colonies previously free of infection, if such material is inoculated into mice in the course of experimental work (Bhatt et al., 1981; Whitney, 1974).

The most usual source of natural infection is via minor abrasions of the skin (Fenner, 1947b), which may occur from direct contact between mice, from contaminated bedding, or during manipulations by animal handlers. Infection may also occur by the respiratory route (Werner, 1982), but probably only between mice in close proximity to each other (Wallace et al., 1981).

Besides the obvious sites of virus shedding, Jacoby and Bhatt (1987) demonstrated viral antigen, and hence presumably the presence of infectious virus, in several other sites that could contribute to transmission, viz., the intestinal tract, the urogenital tract, the upper respiratory tract, and the oral tissues. Shedding from such sources could be important in maintaining transmission among mice which showed no signs of disease, e.g., genetically resistant C57BL mice. Behavioral traits such as grooming, barbering, sniffing, tail pulling, and biting could contribute to transmission (Bhatt and Jacoby, 1987b).

Mousepox as an Epizootic Disease

Outbreaks of mousepox have occurred in laboratory mice when ectromelia virus has been inadvertently introduced into colonies of susceptible animals ("natural epizootics"), and deliberately planned epizootics have been used as a tool for the experimental study of epidemiology.

Natural Epizootics. Mousepox is notorious for the damage it can cause if unwittingly introduced into large colonies of laboratory mice. As noted above, mousepox has been a recurrent problem in European laboratories, where outbreaks were not regarded as unusual enough to justify detailed investigation and reporting. The position was quite different in the United States, because of the rarity of outbreaks and the large size of the laboratory mouse industry there, and several reports of outbreaks have been published (Melnick and Gaylord, 1953; Trentin, 1953; Briody, 1955, 1959; Briody et al., 1956; Whitney, 1974; *Laboratory Animal Science*, 1981). The most interesting feature of these epizootics was the striking differences in the clinical signs and mortality in different strains of mice, already described. This variability in clinical response emphasizes the need for reliable laboratory methods of diagnosis, and different procedures are now recommended for the confirmation of a suspicious case and the screening of mouse stocks (see below).

In spite of the impression gained from early outbreaks in the United States, there is no convincing evidence of aerosol transmission of ectromelia virus between cages or rooms. This was most clearly demonstrated in the 1979–1980 outbreaks in the National Institutes of Health (Wallace et al., 1981). Late in 1979, mousepox was diagnosed in three experimental mouse rooms. The prevalence of infection in all three rooms was low: 3% of 939, 3% of 541, and 1% of 789 mice, respectively. Although all rooms housed susceptible strains of mice, spread between cages was minimal and occurred slowly. Infected mice were found in only a few cages, which were located close to each other. Transmission between cages was almost certainly due to transfer of virus by workers handling the animals.

Experimental Epidemiology. In the early 1920s the distinguished British bacteriologist, W. W. C. Topley, embarked upon a long-term study of experimental epizootics in mice housed in specially designed cages (Topley, 1923). Initially he used a variety of bacteria, but with Marchal's discovery of ectromelia virus he undertook studies with this agent, which have been summarized in Greenwood et al. (1936). Their studies of long-continued epizootics (1.75 and 3.25 years) in herds of mice maintained by the regular addition of normal mice suffered from the fact that the only indication of infection was death. In similar experiments carried out after the demonstration of the nature of the virus and the pathogenesis of the disease, Fenner (1948d) constructed life tables for mice exposed to a virulent and an attenuated strain of virus. The characterisic virulence of each strain was maintained throughout the 190 and 290 days of the two experiments. Anderson and May (1979) have used the results of the long-term experiments by Greenwood and Fenner in an analysis of the population biology of infectious diseases.

Greenwood *et al.*, (1936) also used closed epidemics to test the effects of vaccination with small doses of active ectromelia virus. Fenner considerably extended and refined this approach by regularly examining mice for primary lesions and rash, as well as observing mortality, and demonstrated the protective effects of prior vaccination with vaccinia virus (Fenner, 1947a; Fenner and Fenner, 1949) and the differing virulence and infectivity of three strains of ectromelia virus (Fenner, 1949b).

Enzootic Mousepox

Although now much less common, mousepox appears to have been enzootic in many mouse-breeding establishments in Europe and Japan. A variety of mechanisms probably operated to maintain the virus, without so disrupting the mouse-breeding program as to make control mandatory. One important factor was probably the high level of genetic resistance and trivial symptomatology exhibited by many mouse genotypes, such as C57BL and AKR, already described.

Another factor, which could minimize the severity of enzootic mousepox among genetically susceptible mice, is maternal antibody. Fenner (1948c) showed that antibody was transmitted from mousepox-immune dams *in utero,* and also in the milk for at least the first week after birth. It was very effective in saving young mice of a highly susceptible stock from death due to virulent ectromelia virus infection. A combined serological and clinical study of a large mouse-breeding colony in England in 1950 (F. Fenner and E. M. B. Fenner, unpublished results) suggested that only a few breeding cages were actively infected at any time, and that infection in the colony was maintained in part by the handling of mice and bedding from infected and noninfected cages by the animal attendants. Infection of young mice which had sucked immune mothers resulted in nonfatal disease, but the lesions on such animals were infectious. When young mice from several litters were assembled in large cages after weaning, prior to their dispatch to the laboratory, these animals acted as a source of infection. Since the protective effect of maternal antibody is lost within 4 weeks (Fenner, 1948c), mice not infected as weanlings were highly susceptible to the disease and deaths were frequent. An epizootic could thus be set up in mice awaiting shipment to the laboratory, and the large number of severe cases which then occurred acted as a further source from which virus might be spread to "clean" cages.

A third possibility, clearly involving mouse genotype, is chronic, clinically inapparent infection. The most convincing evidence of this is the report by Gledhill (1962a,b) of intestinal infection with excretion of virus in the feces and lesions in tail skin; a finding confirmed by Wallace and Fuller (1985). Some authors believe that mice may suffer from latent infection that can be activated by various kinds of stress, including the inoculation of tissue homogenates, but reported instances of activation are as readily explained by the existence of unsuspected actively circulating ectromelia virus in the mouse colonies. Gledhill (1962b) was unable to demonstrate "activation" in mice which he knew harbored a persistent infection.

The occurrence of clinically inapparent infections is dependent upon host genotype, the route of infection, and the effects of maternal antibody. Such cases are important both as potential sources of virus for natural mouse-to-mouse spread and as a source of virus that might be unwittingly transferred by subinoculation.

LABORATORY DIAGNOSIS

To most scientists who work with laboratory mice, mousepox is an important but unwelcome disease. The clinical and pathological features are pathognomonic in nonimmune, genetically susceptible mice (acute death and massive necrosis of the liver and spleen, or sickness with blepharitis and skin lesions), but infection of genetically resistant mice may be inapparent. Since it can spread widely within a colony and can easily be distributed between experimenters in different laboratories and different countries (*Laboratory Animal Science*, 1981), methods of diagnosis are required that will allow rapid recognition of a suspected case, and reliable retrospective diagnosis by serological methods is essential if the disease is to be effectively controlled.

Diagnosis of a Suspicious Case

Mousepox can be diagnosed by the microscopic or electron microsope examination of the tissues of suspected cases, the diagnostic features being the distinctive eosinophilic cytoplasmic inclusion bodies (Marchal, 1930; see Plate 9-4), the presence of specific antigen (Jacoby and Bhatt, 1987), or the poxvirus particles that can be visualized in secions of the liver (Allen *et al.*, 1981). Passage of extracts of infected tissues on the chorioallantoic membrane should produce confluent lesions which when ground up exhibit characteristic hemagglutination which is inhibited by vaccinia virus-immune serum (Burnet and Boake, 1946); suitably diluted suspensions will produce small pocks. Virus can also be recovered in cell cultures of various kinds: mouse embryo cells (Osterhaus *et al.*, 1981), chick embryo fibroblasts (Ichihashi and Matsumoto, 1966), or BS-C-1 cells (Wallace and Buller, 1985). Inoculation of genetically susceptible nonimmune mice will produce the signs described earlier, whereas vaccinated mice of the same strains should prove resistant. Gel-diffusion tests with scab material have also been used for rapid diagnosis (Carthew *et al.*, 1977).

Following the pattern established for the diagnosis of smallpox in the Smallpox Eradication Program of the World Health Organization, rapid diagnosis could be most readily effected by searching for virus particles with the electron microscope in negatively stained preparations of homogenized tissues (or scabs) from suspected cases. No other poxvirus is likely to be found in laboratory mice, and the specific diagnosis could be readily confirmed by inoculation of mice or tissue culture.

Contamination of Material with Ectromelia Virus. Several examples have been reported, and others have undoubtedly occurred, in which ectromelia virus has been confused with other viruses potentially present in human or other material inoculated into mice (Fairbrother and Hoyle, 1937; Mooser, 1943; Dalldorf and Gifford, 1955; MacCallum *et al.*, 1957; Palmer *et al.*, 1968). The most bizarre such incident was the production of a "live vaccine against murine typhus" by serial intranasal passage of *Rickettsia mooseri* (Laigret and Durand, 1941) and its administration to hundreds of thousands of persons in north Africa. Packalén (1947) demonstrated that this "typhus vaccine" owed its mouse pathogenicity to its high content of ectromelia virus.

Screening Tests for Mouse Stocks

The late Dr. B. A. Briody, who developed a Mouse Pox Service Laboratory in 1956 under contract to the National Cancer Institute, National Institutes of Health, United States, demonstrated the value of the HI test, using vaccinia virus to provide the hemagglutinin (Briody, 1959). Positive results were obtained in several epizootics in the United States, and with sera from several enzootically infected European stocks. On the basis of the serological examination of over 100,000 serums from various colonies of mice throughout the United States, Briody (1966) concluded that enzootic mousepox did not occur in that country at that time. But international transfer of mice or mouse cells or tumor material poses a constant threat that it might be introduced, as was evident in 1974 in laboratories of the National Institutes of Health in Bethesda, following the importation of a tumor from London, to which it had earlier (1972) been sent from Prague, from a laboratory where mousepox was then enzootic (Whitney *et al.*, 1981).

The consequences of the reported discovery of mousepox in a mouse colony are so serious that it is unwise to rely on any one serological test for diagnosis (Collins *et al.*, 1981). Some mutants of vaccinia virus fail to produce hemagglutinin (Cassel, 1957; Fenner, 1958) and similar variants might occur with ectromelia virus; Guillon (1975) noted that several strains of ectromelia virus were only slightly hemagglutinogenic upon isolation. Although many years of experience in the United States demonstrated that practical value of the HI test for surveillance of mousepox, experience with false-positive (Collins *et al.*, 1981) and false-negative (Manning and Fisk, 1981) results during investigations carried out during the 1979–1980 outbreaks in the United States has not been reassuring. Collins *et al.* (1981) suggest that the ELISA method should be used as the basic screening test, and that any positive results should be confirmed by another serological test, such as immunofluorescence; Buller *et al.* (1983) have confirmed the value of the ELISA. Antibody to ectromelia virus can be differentiated from anti-vaccinia virus antibody by using both ectromelia and vaccinia virus antigens in the HI test (Fenner, 1947a) or ELISA (Buller *et al.*, 1983). If the HI-negative IHD-T strain of vaccinia virus is used for vaccination the HI test will remain negative.

PROPHYLAXIS AND CONTROL

The problem of the control of mousepox in laboratory colonies of mice has analogies with the problems of the control of smallpox until that disease was eradicated. The approach will be different in countries (or laboratories) where mousepox is still enzootic in breeding colonies and in countries (or laboratories) free from the disease.

Most well-managed laboratory colonies of mice are now free of mousepox, and the aim is to keep them free of the disease while still importing mice (for genetic purposes) or mouse organs or tissues (as sources of arboviruses, tumors, or hybridomas).

Exclusion of Ectromelia Virus from Laboratories

Quarantine and regulation of the importation and distribution of ectromelia virus or materials infected with it is mandatory in the United States in order to protect the large, uninfected mouse colonies of that country (Shope, 1954; Anslow *et al.*, 1975). However, there is some doubt as to how well such regulations apply (Briody, 1959), and in any case they offer no protection against unsuspected sources of infection. Briody (1959) suggested that all mice imported into the United States from other countries should be effectively immunized with vaccinia virus. Only those mice which develop the typical lesion in 6 to 7 days following intradermal scarification of the tail with vaccinia virus should be admitted to the colony, since failure to react might indicate prior infection with ectromelia virus. Upon arrival, the mice should be isolated from the main colony, kept under rigid quarantine, and carefully observed. This period of isolation and quarantine should persist until the first generation offspring have been effectively vaccinated. Healthy, genetically susceptible mice should be placed as sentinel animals in pens with the imported mice for a few weeks to reveal the possible presence of ectromelia virus. Where the imported mice are to be used as breeding stock, this test necessitates segregating the males and females during this period of contact, or alternatively, introducing only ovariectomized females as susceptible contact mice. Tumors or hybridomas derived from mice obtained from other countries should be especially suspect; after testing for the absence of ectromelia virus, such material should be passed for at least two transplant generations in vaccinated mice before being used.

Arbovirologists face a particularly difficult problem, in that they conduct extensive international exchanges of mouse brain suspensions and mouse sera and ascitic fluids. Bhatt *et al.* (1981) have described how mousepox was introduced into the experimental mice and the mouse colony of the Yale Arbovirus Research Unit in 1966 and the drastic measures that were required for its eradication. To prevent further episodes, all mouse tissues were tested for an agent that could be neutralized by potent vaccinia-immune rabbit serum before being passed through quarantine restrictions. No further entries of ectromelia virus occurred between 1968 and 1975, when the procedure was altered, for economic reasons. Since 1975 all samples were first tested for ectromelia virus by passage in infant mouse brain, and between 1975 and 1980 the virus was recognized in six samples coming from four continents.

Eradication of Mousepox from an Infected Colony

If mousepox is present in a colony, three procedures for control and eradication may be followed, each of which should be associated with surveillance and restriction of the movement of mice: slaughter and disinfection of the animal rooms, vaccination, or quarantine and serological surveillance. Within large institutions, where many groups of investigators use mice which are maintained in different rooms or buildings, a combination of methods may be required (Small and New, 1981). Also, special efforts may be required to preserve genetically valuable mouse stocks in which mousepox may be enzootic.

Slaughter and Disinfection. The NIH Committee of which Dr. Richard E. Shope was Chairman recommended that infected colonies should be destroyed and materials (viral stocks, tumors, etc.) derived from such animals incinerated (Shope, 1954). Animal rooms and materials, air-conditioning ducts, etc., must be appropriately disinfected. Colonies of mice suspected of infection should be isolated and quarantined.

During outbreaks at eight institutions in the United States during 1979–1980, slaughter and disinfection were employed at some stage in each institution, but the extent and timing of the destruction of mouse stocks varied, depending upon their value to the investigator, the effectiveness of quarantine measures and the physical location of the infected colonies (*Laboratory Animal Science,* 1981). The usual method of disinfection of animal quarters is with formaldehyde vapor generated from formalin or by heating paraformaldehyde.

Vaccination. *Vaccinia Virus.* Vaccination with vaccinia virus, by any route of inoculation, will protect mice against serious disease or death from mousepox, although the protection is not absolute, especially against virulent stains of ectromelia virus, and it wanes with time (Fenner, 1947a, 1949c). Closed epidemics (Fenner and Fenner, 1949) showed that vaccination provided substantial but not complete protection against the naturally spreading disease. In experiments with highly susceptible BALB/c mice, Buller and Wallace (1985) showed that each of three different strains of vaccinia virus inoculated by scarification provided protection against clinical disease, but not against infection. Further, cage transfer experiments showed the 4–8 weeks after vaccination, vaccinated mice that had been exposed to mousepox were capable of transmitting virus to nonvaccinated cagemates. In experiments that confirmed these results, Bhatt and Jacoby (1987d) showed that protection against disease persisted for at least 9 months.

The practical use of vaccination for the control of mousepox in infected colonies has been elaborated by Trentin and Ferrigno (1957) and Briody (1959), and by several European scientists (Salaman and Tomlinson, 1957; Zeller and Reckzeh, 1965a,b; Munz *et al.,* 1974, 1976). Although the differential titer of HI antibodies to ectromelia and vaccinia viruses can be used to determine whether mice have been infected with vaccinia virus, ectromelia virus, or both (Fenner, 1947a, 1949a), Briody's (1959) procedure of vaccinating with a strain of vaccinia virus (IHD-T; Cassel, 1957) that does not produce hemagglutinin simplified subsequent serological studies. This advantage is rendered obsolete by the development of the ELISA method, and other strains of vaccinia virus may be preferable for vaccination if this serological procedure is used (Buller and Wallace, 1985).

During the epizootics in the United States in 1957 and 1958 a number of different control procedures were used by breeders (Briody, 1959). In some laboratories the newly established mouse colony was immunized with vaccinia virus for one or two generations; in others vaccination was established on a continuing basis after cleanup of the epizootic situation. By routine vaccination of all mice in their colony over a period of about 2 years, Trentin and Ferrigno (1957) claimed to have eradicated ectromelia virus from it without the slaughter of infected mice. Within any given mouse colony, Briody suggested that universal vaccination represented the most effective weapon for the prevention of mousepox and should guarantee freedom from the epizootic disease. He suggested that universal vaccination should be practiced as soon as exposure to ectromelia virus was known or suspected.

Intradermal scarification of the dorsal surface of the base of the tail is rapid and effective (Salaman and Tomlinson, 1957; Briody, 1959; Flynn, 1963), but some European workers have suggested oral vaccination (Munz et al., 1974), vaccination by aerosol (Munz et al., 1976), or intraperitoneal inoculation (Zeller and Reckzeh, 1965a,b). Oral vaccination with vaccinia virus proved quite unreliable. Aerosol dispersion of vaccinia virus presents problems in achieving safety (for man) and effective dosage for the mice. Zeller and Reckzeh (1965b) conclude that their experiences over a 5-year period show that vaccination of all mice received by an institute will eliminate the occurrence of mousepox. They believe that the expenditure of time and money on vaccination and revaccination by intraperitoneal injection is low in comparison with other methods of vaccination and with the losses which may arise from uncontrolled intercurrent mousepox infection. Mahnel (1985a) reported that intraperitoneal inoculation of the MVA strain of vaccinia virus, which is highly attenuated for man (Stickl et al., 1974) and is therefore without risk to animal handlers, was very effective for prophylaxis and emergency vaccination of mice against mousepox.

The strain of vaccinia virus, the age and genotype of the mouse (Owen et al., 1975), the concentration of virus, and the route of inoculation all influence the response to vaccination, which may vary from lethal infection at one extreme to failure to produce an antibody response at the other. In general, strains used for human vaccination seem to be satisfactory for mice.

Attenuated Ectromelia Virus. Mahnel (1983) found that after 300 passages in chick embryo fibroblasts, ectromelia virus was so attenuated for mice that it no longer produced disease after the inoculation of large doses by intradermal or oral–nasal routes or in drinking water, although it was immunogenic. Subsequently, he showed that this virus could be administered in the drinking water. Two exposures for 30 hours to drinking water containing at least 10^6 TCID/ml was recommended for vaccination of young mice (Mahnel, 1985b).

Quarantine and Serological Surveillance. Present-day knowledge suggests a third method of control in a country such as the United States, if mousepox were to be unwittingly introduced into a situation in which irreplaceable breeding stocks were exposed to infection. Since infection spreads slowly and usually remains localized (Wallace and Buller, 1985; Wallace *et al.*, 1981), slaughter of valuable animals is not justified. Rather than run the risk of establishing inapparent enzootic infection with ectromelia virus in mice vaccinated with vaccinia virus, Buller and Wallace (1985) suggest that control could be achieved by serological surveillance, using the ELISA method and removing only positive reactors. This procedure could be followed up by breeding from mice that had recovered from mousepox (Bhatt *et al.*, 1985; Bhatt and Jacoby, 1987b).

CHAPTER 10

Other Orthopoxviruses

As outlined in Chapter 1, up to the present time 10 species of the genus *Orthopoxvirus* have been recognized. Some of these are among the best studied of all viruses; others have been almost totally neglected. This chapter is concerned with the latter group, which comprises camelpox virus, raccoon poxvirus, tatera poxvirus, the virus of Uasin Gishu disease, and vole poxvirus.

CAMELPOX VIRUS

From an economic point of view, camelpox is possibly the most important remaining orthopoxvirus disease. Its study has been neglected because it presents no risk to man or domestic animals in developed countries. Camelpox virus attracted some attention during the Intensified Smallpox Eradication Program of the World Health Organization, because of descriptions of the then newly discovered virus as "smallpox-like" (Baxby, 1972). However, further studies (Bedson, 1972; Baxby, 1974; Marennikova *et al.*, 1974; Baxby *et al.*, 1975; Esposito and Knight, 1985) clearly showed that camelpox virus was a distinct species of orthopoxvirus, and careful investigations showed that it rarely if ever caused disease in humans (Jezek *et al.*, 1983).

Camelpox in Camels

Although the causative virus was not isolated until 1970 (Sadykov, 1970), camelpox has long been recognized as a generalized pox disease of camels, found especially among young animals (Leese, 1909; Dalling *et al.*, 1966). However, as was observed with cowpox in cattle (see Chapter 6), what is called "camelpox" may be caused by other pox-viruses: occasionally vaccinia virus (Krupenko, 1972) or a parapoxvirus (Roslyakov, 1972; Jezek *et al.*, 1983). The most detailed descriptions of the clinical features and epidemiology of camelpox are provided by Baxby *et al.* (1975), Kriz (1982), and Jezek *et al.* (1983).

Geographic Distribution. Camelpox virus has been recovered from dromedary camels (*Camelus dromedarius*) suffering from camelpox in Iran (Ramyar and Hessami, 1972), Iraq (Al Falluji *et al.*, 1979), Kenya (Davies *et al.*, 1975), Somalia (Kriz, 1982), and the USSR (Sadykov, 1970; Marennikova *et al.*, 1974). The disease probably occurs throughout areas of Africa and Asia where dromedary camels occur (Dalling *et al.*, 1966), but it has not so far been recognized in bactrian camels (*Camelus bactrianus*). It does not occur among feral dromedary camels in Australia.

Clinical Features. Five days after the intradermal inoculation of camelpox virus into young camels, papules appeared at the inoculation site and progressed through vesicles and pustules to become crusted by 9–10 days (Baxby *et al.*, 1975). Generalized skin lesions appeared 9–11 days after the inoculation and progressed in the same way as the primary lesions; the animals had recovered within 4 weeks. Two uninfected camels kept in contact with the inoculated animals developed generalized skin lesions 13 days after exposure.

In epizootics of camelpox in Somalia, Jezek *et al.* (1983) noted that the onset was marked by low-grade fever and prostration lasting 1 to 3 days, followed by the development of eruptions of the skin and mucous membranes. The lips and mucosa of the nose became swollen and the skin lesions passed rapidly from papules into vesicles and pustules. With the appearance of the rash the animal became afebrile. The rash was usually concentrated around the mouth, nose, eyes, and on the mucous membranes of the oral cavity (Plate 10-1A and B), with fewer lesions on the head, neck, and forelegs. Occasionally, especially in very young animals, skin lesions occurred all over the body (Plate 10-1C). The illness lasted 2 to 5 weeks, with an average duration of 18 days. Occasionally animals suffering from the generalized disease suffered from secondary bacterial infection which progressed to a fatal septice-mia; more often localized abscesses developed. If the conjunctivae were involved, unilateral or bilateral blindness sometimes occurred. As Leese

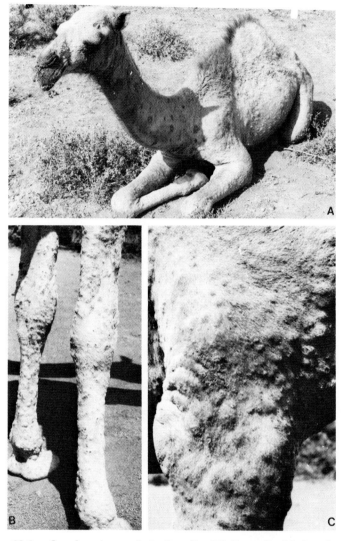

PLATE 10-1. *Camelpox in camels in Somalia. (A) Generalized lesions in a young camel. (B and C) Lesions on forelimbs. (Courtesy Dr. Z. Jezek.)*

(1909) had noted in the Punjab long before, these complications were more common and severe in the rainy season.

Epidemiology. Kriz (1982) collected epidemiological information among 30 camel herds in southwest Somalia, in which 295 cases of camelpox were seen among 1052 animals. The incidence of disease and

TABLE 10-1

Sex and Age Distribution of Camelpox in 30 Outbreaks in Somalia in 1978[a]

	Cases		Deaths		Case–fatality rate (%)
	Total number	Per 1000	Total number	Per 1000	
Sex					
Male	68	313	6	27.6	8.8
Female	227	272	10	11.9	4.4
Age[b]					
1 year	52	. . .[c]	7	. . .	13.5
2 years	87	. . .	5	. . .	5.7
3 years	46	. . .	2	. . .	4.3
4 years	71	. . .	2	. . .	2.8

[a] Data from Kriz (1982).
[b] There were also 39 cases of unknown age, with no deaths.
[c] . . . , Data not available.

the case–fatality rate were higher in male than female camels, and the incidence and severity were highest in young animals (Table 10-1). A more extensive survey in 1978 showed that camelpox occurred in camel herds throughout Somalia, multiple continuous chains of transmission being identified in the northeast and southwest regions (Jezek *et al.*, 1983).

Properties of Camelpox Virus

Camelpox virus is a typical orthopoxvirus in its morphology, hemagglutinin production, and serological cross-reactivity. Like variola virus, it has a narrow host range in experimental animals, and under certain conditions it produces pocks on the chorioallantoic membrane that are rather like those of variola virus. Much of the limited amount of experimental work carried out with the virus was directed to differentiating camelpox virus from variola virus.

Although investigators have found that strains of camelpox virus from Iran, Iraq, Kenya, Somalia, and the USSR were similar in their biological characteristics, Tantawi (1974) claimed that there was no cross-protection in camels or rabbits between strains of camelpox virus recovered from Egypt (Tantawi *et al.*, 1974) and the Iran strain of camelpox virus or vaccinia virus. This finding suggests that the Egyptian strains may have been parapoxviruses rather than orthopoxviruses, since viruses of both genera may cause lesions described as "camelpox" (Roslyakov, 1972; Jezek *et al.*, 1983).

Host Range in Animals. Camelpox virus has been recovered in cultures of lamb kidney cells (Ramyar and Hessami, 1972), but inoculation on the chorioallantoic membrane, on which it produces dense white pocks (see Plate 4-2), is probably the optimum method for isolation and initial characterization. Although the pocks were initially described as "smallpox-like," Marennikova *et al.* (1974) noted that they were smaller and flatter than those of variola virus and that if the eggs were incubated at 34.5 instead of 37°C, the pocks were even flatter and a hemorrhagic center developed. The ceiling temperature for pock production is 38.5°C (Baxby, 1972).

Camelpox virus produces only a slight erythematous reaction after intradermal inoculation into rabbits (Baxby, 1972) and no reaction in guinea pigs (Al Falluji *et al.*, 1979), but a localized pustule develops after intradermal inoculation in rhesus monkeys (Baxby, 1972; Al Falluji *et al.*, 1979). As might be expected, very small doses of camelpox virus are infectious for camels, producing a local lesion followed by generalization. In contrast, even large doses of variola virus produce only a trivial local lesion after intradermal inoculation in camels, but viral replication occurs since the animals develop hemagglutinin-inhibiting antibodies and are protected against challenge infection with camelpox virus (Baxby *et al.*, 1975).

Results obtained after the intracerebral inoculation of mice varied; Marennikova *et al.* (1974) found camelpox virus to be more pathogenic than variola virus but less pathogenic than monkeypox virus, whereas Al Falluji *et al.* (1979) reported that camelpox virus had no effect on mice. In view of the significance of infection of man with orthopoxviruses for the prosecution of the smallpox eradication campaign, Kriz (1982) and Jezek *et al.* (1983) made a careful study of the possible infection of humans with camelpox virus, with essentially negative results. Recently, however, J. H. Nakano (personal communication, 1987) found that serum from a Somali woman who had a vesiculopustular eruption (diagnosed as suspected smallpox) gave a radioimmunoassay adsorption profile that was suggestive of camelpox virus infection. The woman had been associated with camels, but no information was available about the possible occurrence of camelpox among them.

Thus, although the range of animals tested has been limited, camelpox virus falls into the group of orthopoxviruses with a narrow rather than a wide host range in animals. In nature it probably occurs only in camels.

Behavior in Cultured Cells. Although difficult to differentiate from variola virus in camel, rabbit, and chick cells (Mirchamsy and Ahourai, 1971; Baxby, 1972), camelpox virus produces multinucleate giant cells in

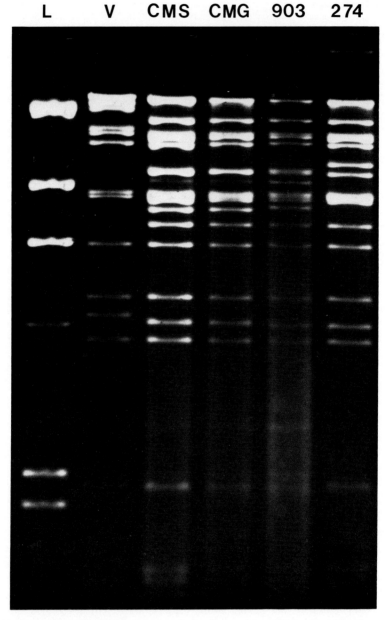

PLATE 10-2. *Electropherograms of camelpox virus DNA after digestion with* HindIII. *L, bacteriophage* λ *DNA digested with* HindIII *(size marker); V, vaccinia virus (Dairen strain); CMS, CMG, strains of camelpox from Iran, 1970 (Baxby, 1972); 903, 274, strains of camelpox virus from Somalia, 1978 (From Dr. J. H. Nakano.) (K.R. Dumbell, unpublished observations.)*

HeLa, BS-C-1 (transformed Green monkey kidney), and transformed human amnion cells, whereas variola virus produces rounding up of cells in these cell lines (Bedson, 1972; Baxby, 1974).

Restriction Endonuclease Mapping of Camelpox Virus DNA. Esposito and Knight (1985) included camelpox virus DNA in their comprehensive comparison of orthopoxvirus DNAs (see Fig. 1-1). It is a typical orthopoxvirus DNA of 196 kbp, with a distinctive restriction enzyme cleavage pattern (Plate 10-2).

Prevention and Control

Leese (1909) noted that in the Punjab and Rajputana, camel breeders practised the inoculation of young camels (the equivalent of variolation) so that they were infected before the onset of rains, thus assuring a mild attack and lifelong immunity. Usually only one or two animals were inoculated and the disease spread to the susceptible animals in the herd. Vaccination with vaccinia virus would also be effective in protecting camels, but care would need to be exercised in choosing a strain of vaccinia virus that did not cause a generalized disease in young camels and spread from one camel to another, or possibly to man.

Summary

Camelpox virus is a distinct species of *Orthopoxvirus* which causes a generalized pox disease in dromedary camels in most parts of the natural geographic range of that animal. It appears to be enzootic in camels and is not known to affect any other animal. If financial considerations permitted, it could be prevented by vaccination of camels with a suitable preparation of vaccinia virus.

RACCOON POXVIRUS

During 1961–1962, 281 wild mammals, belonging to 6 species, were trapped in a forest and swamp area in Chesapeake Bay, Maryland, United States, and examined for a variety of disease agents (Herman, 1964; Alexander *et al.*, 1972). Four viral isolations were made—two poxviruses and two adenoviruses. Raccoon poxvirus was recovered from upper respiratory tissues of each of two raccoons. It produced pocks on the chorioallantoic membrance, and hemagglutination by extracts of infected membranes was neutralized by antiserum to vaccinia virus. Twenty-two out of 92 raccoon sera tested (23.8%) contained antibodies that inhibited hemagglutination by raccoon poxvirus hemagglutinin. No other isolations have been reported.

Disease Production in Raccoons

The captured raccoons that yielded raccoon poxvirus and antibodies were apparently healthy animals. Two raccoons inoculated with large doses of raccoon poxvirus by intravenous, subcutaneous, and intradermal routes showed no apparent clinical signs, but HI antibodies were produced.

Properties of Raccoon Poxvirus

Only one comprehensive study of the properties of raccoon poxvirus has been made (Thomas *et al.*, 1975). The virion is indistinguishable from that of other orthopoxviruses. Serological tests (complement fixation, gel diffusion, hemagglutination inhibition) revealed cross-reactivity between raccoon poxvirus and vaccinia virus antigens, although homologous reactions were stronger than heterologous.

Sections of infected cells revealed that raccoon poxvirus produces A-type inclusion bodies very similar to those produced by cowpox virus (Patel *et al.*, 1986), which were composed of a 155K protein that cross-reacted serologically with the 160K protein found in cowpox virus A-type inclusion bodies.

Host Range.   Raccoon poxvirus produces small white pocks similar to those of ectromelia virus on the chorioallantoic membrane, but unlike other orthopoxviruses, growth of the virus on the chorioallantoic membrane decreased with passage. On intradermal inoculation in rabbits, raccoonpox virus produced a small pink lesion at the site of inoculation. An unusual response was observed in suckling mice inoculated in the footpad; moderate doses produced flaccid paralysis of both hind limbs and skin lesions on the lower parts of the body and tail, before eventually killing the inoculated mice. Larger doses killed the suckling mice before paralysis occurred. Sections of the brains of the paralysed mice revealed chronic multifocal meningoencephalitis, with mild diffuse gliosis.

Raccoon poxvirus grows well in Vero cells, producing large balloon-like plaques with borders formed by giant cell syncytia.

Restriction Endonuclease Mapping of Raccoon Poxvirus DNA.   Although its biological properties identify raccoon poxvirus as an orthopoxvirus, Esposito and Knight (1985) found 11 of 32 fragments in a

*Hind*III digest of raccoon poxvirus DNA did not cross-hybridize with the vaccinia virus probes they used to map the DNAs of other orthopoxvirus species, and they were unable to map raccoon poxvirus DNA. However, Parsons and Pickup (1987) showed that all *Hind*III fragments of the 215-kbp DNA of raccoon poxvirus shared some nucleotide sequences with cowpox virus DNA, and by using fragments obtained by cleavage with several restriction enzymes, they were able to map the raccoon poxvirus DNA (Fig. 10-1). They noted that most of the *Hind*III fragments of raccoon poxvirus DNA and cowpox virus DNA would have aligned well but for a region within the raccoon poxvirus *Hind*III E fragment, which appeared to lack about 14 kbp of DNA that was present in the *Hind*III G/F region of cowpox virus DNA. In contrast, the end regions of raccoon poxvirus DNA, which do not appear to encode any viral proteins, are very similar to the end regions of the DNAs of other orthopoxviruses (Esposito and Knight, 1985; Parsons and Pickup, 1987), suggesting that they must provide some functions that are important in viral replication.

Perhaps the pronounced difference between raccoon poxvirus DNA and the DNAs of other orthopoxviruses reflects the fact that raccoon poxvirus is indigenous to the American continent, whereas except for the recently discovered vole poxvirus (see below), all other orthopoxviruses are native to Europe, Asia, or Africa.

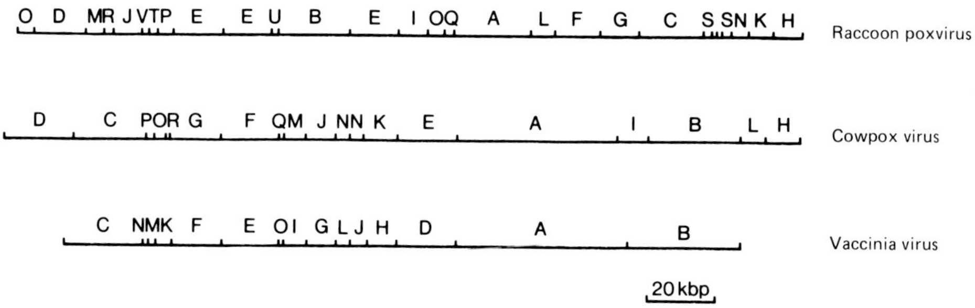

FIG. 10-1. *A comparison of the* Hind*III restriction maps of DNA from raccoon poxvirus, cowpox virus, and vaccinia virus. The maps of cowpox virus DNA (Mackett and Archard, 1979) and vaccinia virus DNA (Mackett and Archard, 1979; De Fillipes, 1982) are aligned with the map of the raccoon poxvirus DNA at the cross-hybridizing regions corresponding to cowpox virus* Hind*III E. vaccinia virus* Hind*III D, and raccoon poxvirus* Hind*III I, O, and Q. (Revised from Parsons and Pickup, 1987, because it now appears that raccoon poxvirus DNA contains a third* Hind*III E fragment (Dr. J. J. Esposito, personal communication, 1988).*

TATERA POXVIRUS

Even less is known of tatera poxvirus than of raccoon poxvirus. It was recovered during a comprehensive program of viral surveillance of small animals in western Africa (Causey and Kemp, 1968) during which more than 7497 samples from at least 101 species were sampled, yielding 83 isolates of 16 different types of virus (Kemp *et al.*, 1974). The tatera poxvirus was obtained from liver–spleen extracts of a wild naked-soled gerbil (*Tatera kempi*) captured in Dahomey (Benin) in April 1968, after intracerebral passage in suckling mice. Most viral isolates from the survey were examined at the Arbovirus Research Unit at Yale University; however, this material was excluded as potential "ectromelia virus" because its infectivity was neutralized by anti-vaccinia virus antibody (see Chapter 9). It was subsequently examined at the Centers for Disease Control, Atlanta, where electron microscopy of the mouse brain material revealed orthopoxvirus virions (Lourie *et al.*, 1975). In several reports this virus is referred to as "gerbilpox virus"; however, we have adopted the name "tatera poxvirus" to avoid confusion with the cowpox virus strain recovered from wild gerbils in Turkmenia (see Chapter 6).

Nothing is known of its potential for disease in its natural hosts, nor even what these hosts may be, other than *Tatera kempi*.

Properties of Tatera Poxvirus

Gispen (1972) compared tatera poxvirus with extromelia virus, and found that tatera poxvirus was less virulent for mice, but could be passaged in the rabbit skin and in RK13 cells. Lourie *et al.* (1975) carried out a more detailed study of tatera poxvirus, and reported that it produced dense white pocks on the chorioallantoic membrane, very similar to those of variola virus; its ceiling temperature was 38°C. It produced reddened, indurated areas 1.0–1.5 cm in diameter at the site of intradermal inoculation in rabbits, and a monkey inoculated by intramuscular and intranasal routes developed a fever and subsequently HI antibodies and resistance to challenge infection with monkeypox virus. The virus grew well in a range of cultured cells derived from monkey, rabbit, and man, producing a cytopathic effect resembling that produced by variola virus, viz., hypertrophic foci which progressed to the formation of syncytial cells and plaques, which appeared 4 days after inoculation.

Tatera poxvirus resembled other orthopoxviruses and differed from variola virus in that its thymidine kinase was not susceptible to inhibition by thymidine triphosphate (Bedson, 1982), and unlike variola virus, it produced cytolytic rather than hyperplastic foci in RK13 cells, with high titers of hemagglutinin (Huq, 1972).

After cleavage with HindIII, tatera poxvirus DNA maps as a typical but distinct species of Orthopoxvirus (Esposito and Knight, 1985; see Fig. 1-1), with a DNA of 183 kbp.

The significance of taterapox virus in unknown; conceivably it may be responsible for some of the orthopoxvirus-positive sera found in screening tests of sera from wild animals in West and Central Africa during investigations of the ecology of monkeypox virus (see Chapter 8).

UASIN GISHU DISEASE VIRUS

Even less work has been carried out with this virus than with those just discussed. It was discovered by Kaminjolo et al. (1974a,b) in skin lesions of horses in Kenya. Since the first recognized cases occurred on the Uasin Gishu plateau, the disease was called "Uasin Gishu skin disease," but it is now known to occur more widely in Kenya (Kaminjolo et al., 1975) and in Zaire and Zambia. Since horses were only introduced into these parts of Africa since European contact, they must be incidental hosts of a virus that is enzootic in wildlife animal species.

The Disease in Horses

The signs of the disease in horses have been described by Kaminjolo et al. (1975). Lesions resembling papillomas (Plate 10-3) are found on the head, neck, back, flanks and quarters, chest, and abdomen. Early lesions are marked by small tufts of hair which project above the level of the skin. On palpation, firm nodules 2–5 mm in diameter can be felt in the skin, with white scabs which leave bald patches that bleed easily when detached. Different stages of development of the lesions can be observed in the same animal, and lesions appear and diappear intermittently. Sometimes they persist for years. Sections of the skin lesions show that the lesions consisted of a swollen and hyperplastic epidermis, the edges of the lesions being sharply demarcated. Affected cells in the stratum spinosum are enlarged and contain B-type inclusion bodies (Kaminjolo and Winquist, 1975), in which poxvirus particles can be seen by electron microscopy.

Previously, pathogenic fungi had been reported as the probable causal agents of Uasin Gishu disease, but Kaminjolo et al. (1975) could not recover pathogenic fungi from the lesions; however, they demonstrated the presence of poxviruses in the lesions of each of nine diseased horses examined.

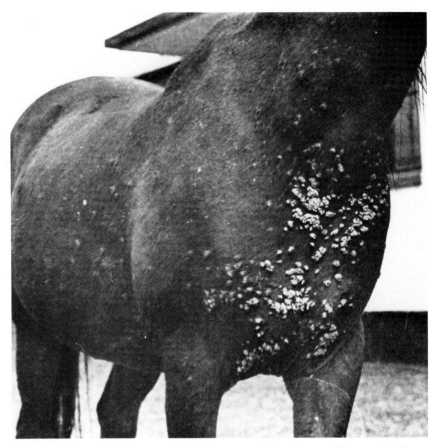

PLATE 10-3.   *Generalized lesions in Uasin Gishu desease of horses in Kenya.*
(Courtesy Dr. J. S. Kaminjolo, Jr.)

Properties of Uasin Gishu Disease Virus

Kaminjolo *et al.* (1974b) have provided the only report so far on the
biological properties of Uasin Gishu disease virus. Virus isolation from
lesion material was achieved in calf kidney cells, but not on the
chorioallantoic membrane, nor in calf testis, bovine fetal skin, or chick
embryo fibroblast cells. However, the latter cells were susceptible to
virus that had been passaged in calf kidney cells. After two blind
passages, fluid from a calf kidney cell culture produced pocks on the
chorioallantoic membrane, which varied in size up to 4 mm in diameter,
with a hemorrhagic periphery. Extracts of infected chorioallantoic

membranes produced typical orthopoxvirus hemagglutination, which was inhibited by antisera to vaccinia and cowpox viruses. Inoculations of rabbit, guinea pig, hamster, and weanling or adult mice were negative, but local lesions developed in calves inoculated intradermally, and after follicle inoculation of 3-day-old chicks. Extensive pustules developed on the bodies of suckling mice inoculated with virus that had been adapted to grow on the chorioallantoic membrane.

Uasin Gishu skin disease of horses thus appears to be caused by an orthopoxvirus that has a rather narrow host range in experimental animals. It is presumably caused by infection of horses from a wildlife reservoir, but nothing is known of its epidemiology in horses nor of its ecology.

VOLE POXVIRUS

Regnery (1987) recently reported the recovery of an orthopoxvirus from a scab on the hind foot of a California vole (*Microtus californicus*). It produced small pocks on the chorioallantoic membrane, extracts of which produced characteristic orthopoxvirus hemagglutination. Sera from naturally and experimentally infected voles inhibited vole poxvirus hemagglutinin and to a lesser extent the hemagglutinins of vaccinia virus and raccoon poxvirus. One-day-old mice died after subcutaneous inoculation, and intradermal inoculation of rabbits produced a slightly raised, pink lesion about 15 mm in diameter. Infection could be transferred by pin prick through a lesion on an infected vole, lesions developing at the pinprick site 6–9 days later and hemagglutinin-inhibiting antibodies about a week later.

Serological surveys in the San Francisco Bay area showed that the virus occured in several geographically separated vole populations, and in some populations for 2 or more years.

It would be of considerable interest to compare the restriction map of the DNA of vole poxvirus with the DNAs of other orthopoxviruses, especially that of the other species native to the Americas, raccoon poxvirus.

CHAPTER 11

The Global Spread, Control, and Eradication of Smallpox

This book is primarily concerned with the orthopoxviruses rather than any particular disease, but we believe that it is essential to include a chapter on smallpox, the human disease that made the orthopoxviruses so important a group in the history of virology, and whose eradication by 1977 represents one of the greatest triumphs of preventive medicine.

Historical aspects of smallpox and of vaccination have been the subjects of many books, most recently "Princes and Peasants. Smallpox in History" (Hopkins, 1983) and the WHO account of the eradication of smallpox, "Smallpox and Its Eradication" (Fenner et al., 1988). This chapter derives mainly from the latter work, which should be consulted by those interested in a fuller account of the subject.

CLINICAL FEATURES OF SMALLPOX

Difficult through it has proved to be, it is easier to make a tentative diagnosis of "smallpox" from descriptions of "plagues" in ancient times than of most other diseases, because its clinical features were in most cases so distinctive. During the twentieth century two varieties of smallpox were recognized, of very different severity: variola major, with case–fatality rates in the unvaccinated of 20–30 percent, and variola minor, with a case–fatality rate of about 1 percent. As described in Chapter 7, the viruses that caused these two varieties of smallpox could sometimes be differentiated by laboratory tests. In the field or the hospital, single cases could not be diagnosed with confidence as either variola major or variola minor. Diagnosis of variola minor depended upon the observation of outbreaks of smallpox with a low case–fatality rate, and in general much less constitutional disturbance and a more rapidly evolving rash than in variola major.

The Rash

The characteristic feature of smallpox was the rash, the lesions of which passed through the stages of macule, vesicle, pustule, and scab before falling off and leaving a pigmented or depigmented spot and, on the face of must sufferers from variola major, a pitted scar, the facial pockmark. The temperature chart and evolution of the rash of a typical case of variola major is illustrated in Fig. 11-1.

The incubation period of 10–12 days ended with the sudden onset of fever, headache, and backache. The fever often fell on the second or third day, as the rash appeared, and rose again as the rash became pustular. The rash appeared first on the face, then on the arms, and later

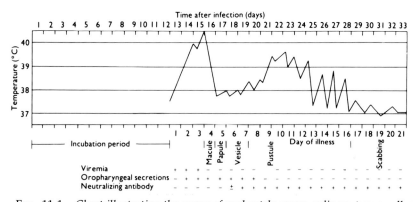

FIG. 11-1. *Chart illustrating the course of moderately severe ordinary-type smallpox in an unvaccinated patient; length of incubation period, temperature chart, development of the rash, the presence of virus in the blood and oropharyngeal secretions, and the appearance of neutralizing antibody in the serum. (From Fenner et al., 1988.)*

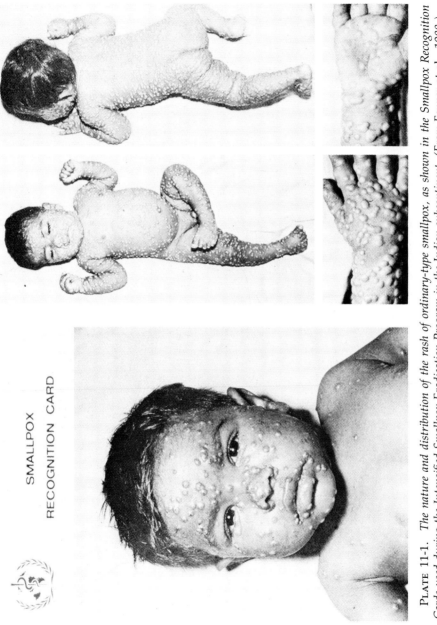

SMALLPOX
RECOGNITION CARD

PLATE 11-1. *The nature and distribution of the rash of ordinary-type smallpox, as shown in the Smallpox Recognition Cards used during the Intensified Smallpox Eradication Program in the Indian subcontinent. (From Fenner et al., 1988.)*

on the lower limbs, and was more profuse on the face and extremities than on the trunk (Plate 11-1). Characteristically there were lesions on the palms of the hands and soles of the feet. The pustules, whose pustular content was due to the action of the virus and not to bacterial infection, were white, hard to the feel, and often loculated. There was an enanthem in the mouth and pharynx, the lesions of which ulcerated and released virus into the oropharyngeal secretions, constituting the principal source of infectivity.

The extent of the pustular rash varied from discrete to confluent. Rarely, it was "flat," or was preceded by a purpuric or hemorrhagic rash. Flat-type smallpox was usually fatal; the hemorrhagic type, to which pregnant women were especially liable, was always fatal. The case–fatality rate in what was called "ordinary-type" smallpox varied from 60 percent in confluent cases to 10 percent in cases with a discrete rash (Table 11-1). In these cold clinical terms, smallpox was clearly an unpleasant and often fatal disease. The reality in endemic countries—almost every country in the world during the nineteenth century and earlier, and the Indian subcontinent up to the mid-1970s—was far worse, as the following description outlines:

Impressions of Smallpox in Bombay in 1958

> The majority of patients had fully developed smallpox in the suppurative stage, with confluent pustules covering the entire body. The head was usually covered by what appeared to be a single pustule; the nose and the lips were glued together. When the tightly filled vesicles burst, the pus soaked through the bedsheet, became smeared on the blanket and formed thick, yellowish scabs and crusts on the skin. When the pulse was taken tags of skin remained stuck to the fingers. . . . when secondary hemorrhage appeared, the affected area of skin formed a single black mass.
>
> All the gravely ill patients were also tortured by mucosal symptoms. The tongue was more or less swollen and misshapen, and hindered breathing through the mouth. The voice was hoarse and faltering. Swallowing was so painful that the patients refused all nourishment and, in spite of agonizing thirst, often also refused all fluids. We saw patients with deep invasion of the respiratory passages . . . wails and groans filled the rooms. The patients were conscious to their last breath.
>
> Some . . . just lay there, dull and unresponsive. They no longer shook off the flies which sat on purulent eyelids, on the openings of mouth and nose, and in swarms on the inflamed areas of the skin. But they were still alive, and with touching gestures they lifted their hands and begged for help. (Translated from A. Herrlich, 1958)

The major factor that mitigated the severity of variola major was vaccination. Successful vaccination for up to 5 years before exposure to smallpox usually prevented disease completely, although serological

TABLE 11-1

The Frequency and Case–Fatality Rates of Different Clinical Types of Variola Major, according to Vaccination Status (Presence of a Scar) in Hospitalized Patients in Madras[a]

Clinical type	Unvaccinated subjects			Vaccinated subjects		
	Number of cases	Percentage of total	Case–fatality rate (%)	Number of cases	Percentage of total	Case–fatality rate (%)
Ordinary type	3147	88.8	30.2	2377	70.0	3.2
Confluent	808	22.8	62.0	156	4.6	26.3
Semiconfluent	847	23.9	37.0	237	7.0	8.4
Discrete	1492	42.1	9.3	1984	58.4	0.7
Modified type	76	2.1	0	861	25.3	0
Flat type	236	6.7	96.5	45	1.3	66.7
Hemorrhagic type	85	2.4	96.4	115	3.4	93.9
Early	25	0.7	100.0	47	1.4	100.0
Late	60	1.7	96.8	68	2.0	89.8
Totals:	3544		35.5	3398		6.3

[a] Based on Rao (1972).

TABLE 11-2

Age and Vaccination Status of Cases of Variola Major Occurring after Importations into Western Countries During the Period 1950–1971[a]

Successfully vaccinated	Age group (years)				Totals	
	0–9	10–49	≥ 50	Unknown	Cases (deaths)	Case–fatality rate (%)
Never	30 (12)[b]	37 (18)	11 (10)	1 (1)	79 (41)	52
Only after exposure	20 (4)	41 (13)	9 (3)	0	70 (20)	29
0–10 years before exposure	18 (0)	48 (1)	5 (0)	1 (0)	72 (1)	1.4
11–20 years before exposure	0	40 (3)	3 (0)	0	43 (3)	7
20 years before exposure	0	187 (8)	96 (25)	14 (0)	297 (33)	11
Unknown	24 (2)	50 (5)	24 (5)	21 (0)	119 (11)	9
Total:	92 (18)	403 (47)	148 (43)	37 (1)	680 (109)	16

[a] Based on Mack (1972).
[b] Number of cases (number of deaths).

studies showed that many such persons suffered subclinical infections (Heiner *et al.*, 1971). As the interval after vaccination increased the likelihood of the development of clinical signs and of a fatal outcome increased (Table 11-2), but the disease was usually much less severe and the signs and symptoms progressed more rapidly. This form of variola major was usually described as "modified smallpox" (see Table 11-1).

The lesions of variola minor were similar to those of variola major in character and distribution, but hemorrhagic-type and flat-type cases were extremely rare, a confluent pustular rash was rare because the individual lesions were smaller than those of variola major, and the lesions usually evolved more rapidly. Toxemia was much less severe, so that cases with an extensive rash were often ambulant, and residual facial pockmarks rarely occurred. Vaccination usually prevented variola minor; in a series of 13,686 cases in the United Kingdom Marsden (1936) recorded on 7 cases in persons who had been vaccinated within the previous 10 years.

Sequelae

The most important sequel to smallpox, which occurred in about 75% of cases of variola major, was the occurrence of facial pockmarks (Plate 11-2). The features important in the retrospective diagnosis of smallpox from descriptions in old writings were the severity of the disease, the character and distribution of the skin lesions, especially their occurrence on the palms and soles, and the residual facial pockmarks. Other rarer sequelae included corneal scarring and blindness, and flail joints.

EPIDEMIOLOGY OF SMALLPOX

Smallpox was a specifically human disease. Monkeys can be infected and transmission can occur between monkeys (Noble and Rich, 1969), but there is no animal reservoir of the virus. The vast majority of infections in unvaccinated persons resulted in disease with obvious clinical manifestations; subclinical attacks sometimes occurred in vaccinated subjects but they did not appear to be infectious. Patients became infectious after the rash had appeared, when the lesions of the enanthem broke down and virus was excreted in the saliva and pharyngeal excretions. Infectivity declined rapidly as the lesions of the enanthem healed and the skin lesions scabbed; although there was a vast amount of virus in the scabs, they were not an important source of infection. Recovery from smallpox was followed by lifelong immunity and recurrences never occurred.

Because the important virus from the point of view of spread was that

PLATE 11-2. *Facial pockmarks and blindness in a woman in Zaire. The former was a frequent sequel of variola major (75% of cases) but rarely occurred after variola minor. Blindness was rare. (From Fenner et al., 1988.)*

released from the lesions of the enanthem, infection was usually by close face-to-face contact, especially between members of the same household. Rarely, infections could be spread by fomites, such as soiled bed linen, and even more rarely, if the index case had a cough early in the course of the disease, infection could be airborne, within the confines of a hospital or a bus.

SMALLPOX IN THE ANCIENT WORLD UP TO 1000 AD

Figure 11-2 illustrates fact and speculation about the spread of smallpox before the year 1000 AD. One unequivocal fact is that smallpox did not occur in the Americas, Australia, or southern Africa during this

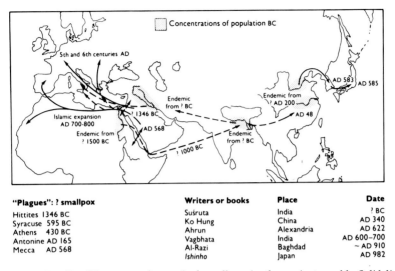

"Plagues": ? smallpox	Writers or books	Place	Date
Hittites 1346 BC	Suśruta	India	? BC
Syracuse 595 BC	Ko Hung	China	AD 340
Athens 430 BC	Ahrun	Alexandria	AD 622
Antonine AD 165	Vagbhata	India	AD 600–700
Mecca AD 568	Al-Razi	Baghdad	~AD 910
	Ishinho	Japan	AD 982

FIG. 11.2. *Possible routes of spread of smallpox in the ancient world. Solid lines indicate known spread; broken lines indicate speculative events or routes of spread. "Plagues" are historical episodes of epidemic disease reported in ancient and classical literature. "Writers or books" refers to the principal descriptions of smallpox before the end of the first millenium A.D. (From Fenner et al., 1988.)*

period. Seeking for the origins of smallpox, it has been speculated that variola virus may have been derived from monkeypox virus and moved with man across the Sahel and Sahara, wetter and more hospitable then than now, some 8000 years ago. Three mummies from Egypt dating from the Eighteenth to the Twentieth Dynasties show pustular lesions which may have been due to smallpox, but no descriptions of a disease like smallpox occur in the Egyptian or Jewish writings of those times.

Historians have suggested that some of the famous "plagues," dating from an epidemic among the Hittites in 1346 BC to the Antonine plague in 165 AD, were due at least in part to smallpox, but the evidence is equivocal. India has long been regarded as the ancient home of smallpox and certainly it was a country with a population on the plain of the Ganges large enough to support endemic smallpox as early as 500 BC. But the ancient writings of the Atharvaveda contain no descriptions of such a disease, which appear only as late as the Susruta, which were gathered together some 2500 years ago but did not reach their final form until about the fourth century AD.

Reliable accounts of smallpox in ancient writings appear from the fourth century onward. Ko Hung, in 340 AD, differentiated smallpox

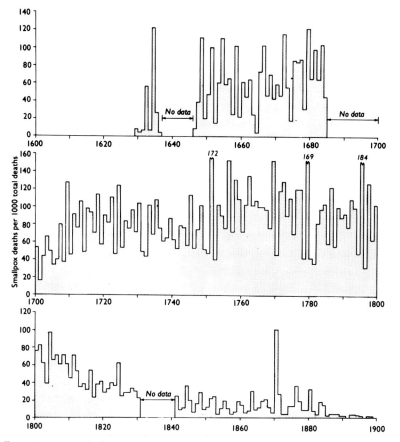

FIG. 11-3. *Deaths from smallpox as a proportion of every 1000 deaths from all causes in London, between 1627 and 1900. Variolation, introduced in 1722, failed to make an impact on the incidence of smallpox deaths. Vaccination, introduced in 1798, permanently reduced the incidence of smallpox deaths, which reached a very low level by the end of the nineteenth century. (From Fenner et al., 1988.)*

from measles and described what was at that time a well-established endemic disease, which was said to have entered China from the west in 48 AD. In the centuries which followed similar descriptions appeared in Alexandria, India, and southwestern Asia. To the north and east, the excellent Chinese records describe the first occurrences of smallpox in Korea in 583 and in Japan in 585. To the west, periodic episodes of smallpox appeared to have occurred in various parts of Europe in the fifth and sixth centuries. However, the major spread of endemic smallpox at this time was that accompanying the great Islamic expansion across north Africa and into the Iberian peninsula in the seventh and eighth centuries. Toward the end of the first millenium AD, al-Razi produced his classical account of the disease and in Japan the book Ishinho described smallpox hospitals and, for the first time, the "red" treatment, which was later practiced throughout Europe and persisted into the twentieth century (Hopkins, 1983). This consisted in draping rooms with red cloth, using red bedclothes, and irradiating patients with red light, a practice supported by Finsen, who was awarded the Nobel Prize in 1903 for his contributions to phototherapy.

By 1000 AD, smallpox was probably endemic in the more densely populated parts of the Eurasian land mass, from Japan to Spain, and the African countries on the southern rim of the Mediterranean. Over the next few centuries, with the constant movements of the Crusades between northern Europe and Southwestern Asia, smallpox became well established in Europe. In Africa, caravans crossing the Sahara to the densely populated kingdoms of West Africa probably carried smallpox with them. The disease was probably repeatedly introduced by Arab traders into the port cities of eastern Africa, but did not spread into the interior.

By the sixteenth century smallpox was common in Europe, but does not appear to have become a really serious problem until the early seventeenth century (Carmichael and Silverstein, 1987). Statistics of smallpox deaths began to be collected at about this time, in Geneva and later in London and Sweden. The London Bills of Mortality (Fig. 11-3) provide concrete evidence of the impact of smallpox in Europe; for most of the seventeenth and eighteenth centuries over 10 percent of all deaths in London each year, and many more in epidemic years, were due to smallpox.

THE SPREAD OF SMALLPOX BY EUROPEAN COLONISTS

With the development of ocean-going ships, the late fifteenth century saw the beginning of the explosion of European explorers and colonists to all the other continents. Over the next three centuries they took smallpox with them; to the Americas, to southwestern, southern, and eastern Africa, and possibly to Australia (Fig. 11-4).

Smallpox played a critical role in the Spanish conquest of Mexico and Peru and in the successful settlement of North America by the English and the French (Stearn and Stearn, 1945). Following an importation from Spain in 1507, smallpox "exterminated whole tribes" of the native

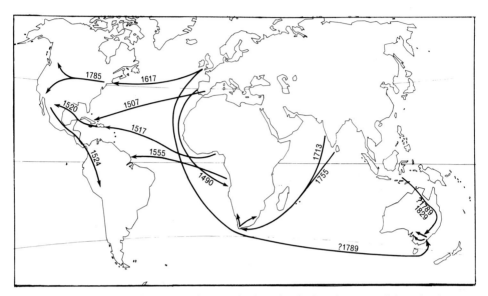

FIG. 11-4. *Map of the world showing the dates for the first (or repeated) introduction of smallpox into new continents following their colonization by Europeans, between the late fifteenth century and the end of the eighteenth century. (From Fenner et al., 1988.)*

inhabitants and was eventually responsible for the destruction of almost the whole of the original population of the Caribbean islands and their replacement by negro slaves and their descendants. Repeated importations occurred into the Caribbean islands from Europe or with slaves from Africa. One of these introductions, in 1520, led to movement of the disease to the mainland and the destruction of Aztec resistance to Cortes. The event was described by a Spanish friar who arrived in Mexico in 1525 (Foster, 1950):

> . . .at the time that Captain Panfilo de Narvaez landed in this country, there was in one of his ships a negro stricken with smallpox, a disease which had never been seen here. At this time New Spain was extremely full of people, and when the smallpox began to attack the Indians it became so great a pestilence among them throughout the land that in most provinces more than half the population died; in others the proportion was little less. For as the Indians did not know the remedy for the disease and were very much in the habit of bathing frequently, whether well or ill, and continued to do so even when suffering from smallpox, they died in heaps, like bedbugs. Many others died of starvation, because, as they were all taken sick at once, they could not care for each other, nor was there anyone to give them bread or anything else. In many places it happened that everyone in a house died, and, as it was impossible to bury the great number of dead they pulled down the houses over them in order

to check the stench that rose from the dead bodies so that their homes became their tombs. This disease was called by the Indians "the great leprosy" because the victims were so covered with pustules that they looked like lepers. Even today one can see obvious evidences of it in some individuals who escaped death, for they were left covered with pockmarks.

Elsewhere in South America, smallpox preceded Pizarro to Peru and was partly responsible for the easy victory of his small band of Conquistadors over the might Inca empire. In Brazil, its effects were carried far into the interior with the proselytizing Christian missionaries. Smallpox and other infectious diseases introduced by the European invaders were in large part responsible for the depopulation of the aboriginal inhabitants of much of South America, which allowed its subsequent repopulation by descendants of the Europeans and their slaves.

The situation was initially less severe among the Indian populations of North America only because they were too sparse to support endemic smallpox, as did the large populations of Mexico and Peru. But there were repeated introductions of smallpox across the Atlantic which devastated the Indians of the east coast. At first they spared the white settlers, who had come from endemic countries in Europe, but as more of the population was born in America the European settlers also began to fear these importations and to take steps to prevent them, like the quarantine of ships with cases of illness on board. Major smallpox epidemics hit the Iroquois confederacy in New York on at least five separate occasions in the seventeenth and eighteenth centuries. Further west, in the eighteenth and nineteenth centuries, smallpox virtually wiped out the Arikara and Mandan tribes and allowed the Teton Sioux to move into the power vaccuum. Similar events occurred throughout the length and breadth of North America, and by the time of the wars between the English and French, and later the War of Independence, smallpox was a factor to be considered by the commanders of the British, French, and American armies. Washington prolonged the siege of Boston in 1775 because of his fear of smallpox there and when the British did leave he ordered "one thousand men who had had the smallpox" to take possession of the city. Further north, smallpox in the Northern Army was so severe as to decide the course of the war there and thus the continued adherence of what is now Canada to the British crown.

In Africa, smallpox was introduced by the Portuguese into Angola, and further south, contaminated bed linen on a ship returning from India via the Cape was responsible for the first importation of smallpox into South Africa in 1713, with disastrous consequences for the Hottentots. There were periodical incursions of smallpox into the interior of Africa with slave traders, but the disease did not become endemic there until the latter part of the nineteenth century.

Even in Australia, months away from Europe by ship, there was an outbreak of smallpox within a year of European settlement in Sydney, from some unknown source. Another outbreak occurred in 1829–1831, and spread widely through the aboriginal population of south eastern Australia. This epidemic may have played an important role in the aboriginal depopulation which preceded and simplified the European occupation of eastern Australia.

METHODS OF CONTROL

Epidemic disease was usually viewed as a visitation by the gods, but various measures were adopted to control it, as well as to treat the victims. Thus the entry of the plague into Europe led to the development of quarantine. Smallpox was even more serious in the seventeenth and eighteenth centuries, because unlike plague and cholera, it was always present and affected the rich and powerful as well as the poor.

Variolation

As long before as the tenth century, intranasal insufflation with powder from smallpox scabs had been introduced as a secret rite in China, to provide protection against "natural" smallpox by causing a less severe disease. It became a more widespread practice there in about the sixteenth century, and at this time or earlier inoculation by cutaneous scarification was practiced in India and spread from there to southwestern Asia and to Egypt. The disease induced by cutaneous inoculation had a shorter incubation period and was usually milder than smallpox acquired by inhalation. European visitors to Constantinople in the early eighteenth century were impressed by the practice of inoculation as practiced by the Turks, and in 1722 the practice was simultaneously introduced into England and Bohemia. At about the same time it was first practiced in the English colony of Boston. Variolation, as inoculation with smallpox material was later called, was practiced on an extensive scale in some parts of Europe, especially in England, in the latter part of the eighteenth century (Razzell, 1977b). It was milder than natural smallpox and carried a mortality of 0.5 to 2 percent instead of about 25 percent. But it was rarely possible to isolate variolated patients, and they frequently transmitted virulent smallpox virus to their uninoculated contacts.

Toward the end of the eighteenth century, Haygarth (1793) in England and Carl (1799) in Bohemia independently suggested elaborate regimes of mass variolation, isolation, and quarantine as a method of eradicating smallpox, but at this time a discovery was made which rapidly supplanted variolation and led eventually to the worldwide eradication of smallpox.

Vaccination

The discovery and development of vaccination, the use of cowpox and vaccinia virus for inoculation instead of variola virus, has been described (Chapter 5). Its virtues were quickly recognized and Carl's Variolation Institute in Prague became a Vaccination Institute, as did Woodville's Smallpox and Inoculation Hospital in London. By the early years of the nineteenth century a few small countries like Denmark had eliminated endemic smallpox,but they faced constant risk of importations from their neighbors. Nevertheless, wherever it was practiced conscientiously, vaccination had a dramatic impact on the incidence and mortality due to smallpox (see Figs. 5-2 and 11-3).

Methods of production and distribution of vaccine were gradually improved and by the end of the nineteenth century the impact of smallpox was reduced in most countries of Europe and North America. Nevertheless, at the beginning of the twentieth century it was still endemic in almost every country in the world (see Fig. 11-7).

Quarantine and Isolation

The idea of quarantining ships, originally introduced by the Venetians to control plague, was extended by colonists in North America, Australia, New Zealand, and South Africa to exclude smallpox. The longer the voyage from a smallpox-endemic country, the more likely was quarantine to succeed and it was a very effective defense against importations of smallpox in Australia (Cumptson, 1914).

The land equivalent of quarantine for ships was the isolation of cases of smallpox, and control of the movement of persons from smallpox-affected places to those free of the disease. Before the advent of vaccination this had little hope of success, but as early as 1667 colonists in the towns on the east coast of North American attempted to prevent the entry of smallpox in this way. The other form of isolation was to nurse smallpox patients in special hospitals, where all attendants were immune to the disease. This was attempted in a few places, partly at the instigation of antivaccinationists, as in the English city of Leicester in the 1870s (Fraser, 1980). Subsequently combined by Millard (1914) with vaccination of contacts, this "Leicester method" anticipated the surveillance–containment method which proved to be of crucial importance in the global smallpox eradication campaign.

THE SPREAD OF VARIOLA MINOR

Outbreaks of mild smallpox, with low case–fatality rates, had been recorded from time to time wherever smallpox occurred, but until the end of the nineteenth century they were always minor incidents in a situation dominated by variola major. At about this time, independently in South Africa and in the United States, observers recorded the existence and the spread of mild smallpox which "bred true" and tended to replace variola major (de Korté, 1904; Chapin, 1913, 1926; Chapin and Smith, 1932). This is the disease we now call variola minor, although it was called alastrim in the Americas.

From a study of the viruses it seems likely that the South African and American strains were mutant strains that developed independently, but shared the properties of low virulence yet effective transmissibility in man (see Chapter 7). Similar mutants must have occurred before, and elsewhere, but the ecological situation was unsuitable for their emergence and dominance. However, by the end of the nineteenth century the response of public health authorities in the United States and South Africa to outbreaks of variola major, by then an uncommon disease, usually imported, was sufficiently vigorous to contain it. Variola minor excited no such concern and conditions were favorable for its spread.

Between the years 1896 and 1900 the American form of variola minor virus spread rapidly throughout the United States and into Canada. Its spread was carefully documented by Chapin (1913). Subsequently it was exported from the United States to many other countries (Fig. 11-5), causing epidemics that were contained in Australia, New Zealand, and the Philippines, but resulted in endemic disease in England and in several countries in South America, especially Brazil.

The situation in Africa is more obscure, but by the 1920s outbreaks of variola minor as well as variola major were being reported in many of the European colonies in Africa; in Kenya, Rhodesia, Tanganyika in the east, in the southern Sudan, and in the Belgian Congo in the west. Later it was recorded in Algeria, whence it spread to Italy during the Second World War, and by 1970 variola minor was the only form of smallpox endemic in Ethiopia. It is not clear whether all variola minor in Africa originated from South Africa, or whether there were also importations from the Americas or Europe, or one or more other independent foci of origin in Africa.

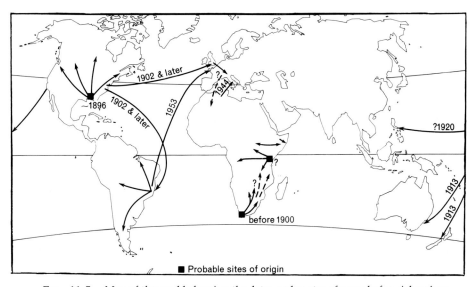

FIG. 11-5. *Map of the world showing the dates and routes of spread of variola minor from foci in the United States and southern Africa during the first quarter of the twentieth century. The original foci in the United States and southern Africa were almost certainly due to independent mutations in variola virus. It is uncertain whether there was more than one focus of origin within Africa. (From Fenner* et al., *1988.)*

CONTROL AND ERADICATION OF SMALLPOX

By the end of the nineteenth century the production of smallpox vaccine had been improved by the application of emerging knowledge of bacteriology and a public health infrastructure had been established in the industrialized countries of Europe and North America. In temperate climates, these tools were adequate to eliminate smallpox by the early 1950s. Encouraged by these results, the World Health organization, which since its establishment in 1948 had urged member states to control smallpox within their own boundaries, in 1959 adopted a resolution calling for the global eradication of smallpox. However, it did not support this initiative with adequate funds, and in 1967 a decision was taken to launch an "Intensified Smallpox Eradication Program," which had the goal of global eradication within 10 years. It is convenient to discuss the control and eradication of smallpox in three time frames: 1900–1958, 1959–1966, and 1967–1980.

National disease statistics for the period 1900–1917 were consolidated by Low (1918), and after the First World War a system of international

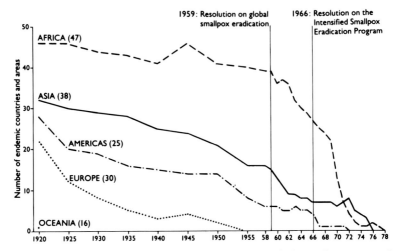

FIG. 11-6. *Numbers of countries and territories in which smallpox was endemic between 1920 and 1978. The figures in brackets indicate the numbers of countries and territories involved. (From Fenner et al., 1988.)*

collection and collation of disease statistics was established by the League of Nations. The global pattern of country-wide elimination and continental eradication of smallpox is illustrated in Fig. 11-6, which show the numbers of countries in which smallpox was endemic, by continents, between 1920 and 1978. The data illustrate the rapid achievement of country-wide elimination in Europe in the period between the two world wars and the much more gradual fall in the number of endemic countries in other parts of the world until the declaration of the goal of global eradication by WHO in 1959. Thereafter there was an acceleration, further enhanced by the initiation of the Intensified Smallpox Eradication Program in 1967.

The Period 1900–1958

The global distribution of smallpox-endemic countries at the beginning and end of this period is illustrated in Fig. 11-7. In 1900 smallpox was endemic in almost every country in the world. Australia, New Zealand, and small islands in the Pacific ocean and elsewhere were free of the disease, because of their isolation and an efficient seaport quarantine and inspection service. So were the Scandinavian countries, thanks to well-conducted vaccination campaigns and containment of outbreaks. But everywhere else smallpox was still endemic, although

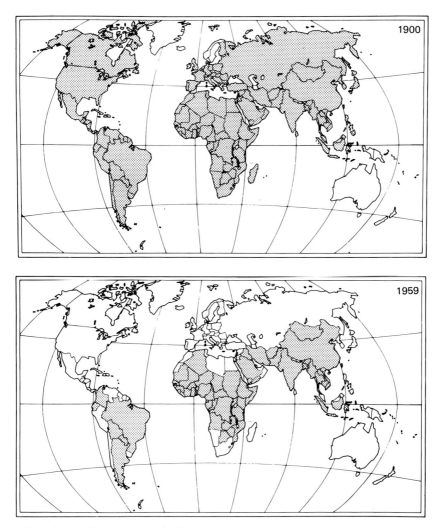

FIG. 11-7. *Countries in which smallpox was endemic in 1900 and 1959. (From Fenner* et al., *1988.)*

much less common in most European countries than it had been during the nineteenth century.

By 1958 Europe, North and Central America, a few countries in South America (Peru, Chile, Venezuela, Uruguay), Africa (Egypt, Libya, Tunisia, Morocco, and Madagascar), and Asia (Japan, Korea, the Philippines, Malaysia, Yemen, and Sri Lanka) were free of endemic smallpox.

TABLE 11-3

Reported Deaths from Smallpox for Various Countries in Europe between 1900 and 1919, by Quinquennia, Arranged in order of Population Size in 1910[a]

Country	Population, 1910 (millions)	Number of deaths from smallpox			
		1900–1904	1905–1909	1910–1914	1915–1919
Russia	134	218,000[b]	221,000[b]	200,000[b]	535,000[b,c]
Germany	65	165	231	136	1,323
France	39	8,448	3,860	825	576
England and Wales	36	4,174	180	65	64
Italy	34	18,590	2,149	8,773	17,453
Austria	28	547	127	350[c]	52,286[c]
Hungary	20	2,672	1,057	284[d]	. . .
Spain	20	24,895	17,083	11,660	13,037
Belgium	7.5	3,391	422	. . .	. . .
Romania	7.0	37	3	38	. . .
Portugal	5.9	2,789[f]	10,510	1,724[g]	15,141
Holland	5.8	51	28	6[g]	9
Sweden	5.5	6	4	3	29
Scotland	4.5	637	12	23	3
Ireland	4.4	60	6	0	0
Switzerland	3.7	75	62	16	0
Finland	3.0	295	155	182	1,605
Denmark	2.8	7	4	3	0
Norway	2.3	0	27	2	0

[a] Data from Low (1918) and Henneberg (1956).
[b] Approximate.
[c] Cases.
[d] To 1912.
[e] . . . , No data.
[f] 1902, 1903, 1904 only.
[g] 1910, 1913, 1914 only.

Elimination of Smallpox from Europe by 1953. Some idea of the incidence of smallpox in Europe before the First World War can be gauged from figures collected by Low and summarized in Table 11-3.

The disruption and mass movements associated with the First World War exacerbated the disease in Russia, where it was particularly severe in Russian Poland, and from Russia it spread to Germany, Austria, and Sweden. Smallpox remained endemic in Russia, Romania, Yugoslavia, Czechoslovakia, and Italy, as well as in Spain and Portugal, and some 300,000 cases occurred in Europe in 1919. For political reasons, Switzerland had sealed it borders throughout the war and this saved it from imported smallpox, which had been a major problem in the Franco–Prussian War in 1870–1871.

In contrast to the relatively well-developed control in most western European countries at this time, the Iberian peninsula continued to suffer from endemic smallpox with periodic severe outbreaks from the turn of the century onward. In the aftermath of the First World War the situation got even worse. During 1918–1920 severe epidemics in Italy killed 36,000 persons, in Portugal 14,000, and in Germany 1500; in Russia 186,000 cases were reported in 1919.

As countries recovered from the ravages of the war and as effective public measures were reinstated, variola major became much less common (Fig. 11-8). However, in the early 1920s variola minor was imported into England from the United States and became established there as an endemic disease, producing over 10,000 reported cases each year between 1926 and 1930. It was finally eliminated in 1934. Although variola minor occurred in a few other European countries, the only other prolonged outbreak was in Switzerland, where over 5000 cases occurred between 1921 and 1926. By the late 1930s endemic smallpox had been eliminated from the USSR and most other countries of Europe, except Portugal, Turkey, and Spain, which experienced a resurgence after the Spanish Civil War, with over 2000 deaths in 1939–1940.

In contrast to the greatly increased incidence of smallpox in Europe after the First World War, only isolated incidents occurred during and after the Second World War. Variola minor was imported into Italy from North Africa in 1944 and had produced over 6000 reported cases before it was eliminated in 1947, while variola major occurred in Turkey during the war (12,000 cases in 1954) and extended from there to Greece.

From 1953 onward Europe was free of endemic smallpox, but countries with colonial possessions in tropical regions (England, France, Portugal, and to some extent the Netherlands and Belgium) were especially liable to importations, rapid aerial travel from endemic regions to Europe greatly increasing the hazard. Small outbreaks, often largely hospital associated, occurred in all these countries.

Eradication as a Declared Goal in the Americas. During the first half of the twentieth century the situation in the United States and Canada closely resembled what happened in Great Britain over the same period. Endemic variola major was eliminated early in the century, although periodic importations continued to occur, especially from Mexico. From 1900, variola minor was endemic throughout both countries and little was done to check it. However, in many states in the United States smallpox vaccination was made a prerequisite for school entry, and in

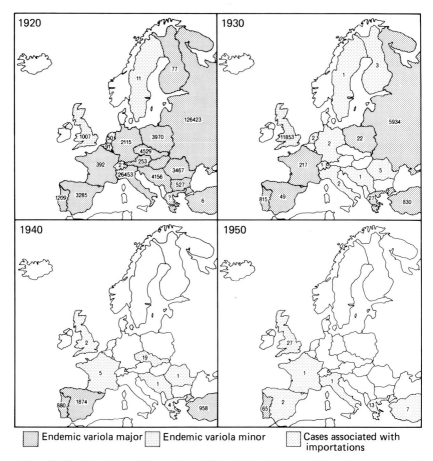

FIG. 11-8. *Countries of Europe in which variola major or variola minor was endemic, or in which imported cases had occurred, in 1920, 1930, 1940, and 1950. Numbers within countries indicate the number of cases reported in the years indicated. (Based on Jezek et al., 1982.)*

these states the incidence of variola minor was reduced to a very low level. The situation gradually improved, and endemic smallpox was eliminated from Canada and United States in the mid-1940s. The six small nations of Central America appear to have achieved elimination of endemic smallpox somewhat earlier than this, but variola major remained a common disease in Mexico until 1951.

The Americas had moved ahead of other groups of countries in establishing a continent-wide public health bureau, the Pan American Sanitary Bureau, in 1902, long before WHO established regional offices in 1948. Initially the Bureau was concerned almost wholly with inter-American quarantine regulations, but the Director of the Bureau just after the Second World War was Dr. F. L. Soper, who was a champion of the concept of eradication. In 1950 the Bureau resolved to eradicate smallpox from the Americas, and the Pan-American Health Organization, which had replaced the Bureau, assisted many countries in South America in the development of laboratories to produce freeze-dried vaccine. Many of these countries undertook mass vaccination programs, and by 1958 endemic smallpox had been eliminated from all but five countries in the Americas: Brazil, Colombia, Ecuador, Argentina, and Bolivia.

Elimination of Smallpox from North Africa. Smallpox was endemic, with periodical epidemics, in all the countries of North Africa during the first quarter of the twentieth century. By the 1930s vaccination programs had been organized by the various colonial powers in the region. There was a severe epidemic in Egypt in 1932–1934, which did not affect the other North African states. At the end of that decade smallpox had been eliminated in all the North African countries except Morocco, but the disturbances associated with the Second World War led to severe outbreaks in all five countries. The contrast to Europe at this time is striking and reflects the relative strengths of the health service infrastructures. The only outbreak of smallpox in Europe after the Second World War (Italy; 1944–1947) was due to an importation from North Africa, where epidemics of variola major were then occurring in Egypt and Libya and of variola minor in Algeria and Morocco. However, from 1948 on improved health services, using liquid vaccine, reduced and finally eliminated smallpox from all the countries of the Mediterranean littoral by the mid-1950s.

Elsewhere in Africa the colonial administrations made little impact on smallpox, their efforts at vaccination being frustrated by the loss of potency of liquid vaccine under field conditions in the tropics. Endemic variola major was eliminated from South Africa by 1958, but variola minor remained endemic among the blacks and was rarely reported. Favored by being an island, Madagascar had eliminated the disease during the First World War.

In most African countries, by the 1920s, smallpox included both variola major and variola minor. In general the public health authorities reacted vigorously to outbreaks of variola major but ignored variola minor. With each of the two world wars there was an exacerbation of smallpox in most countries of Africa, because of increased international and local movements of personnel and a relaxation in vaccination programs. Endemic smallpox remained a major public health problem in 1958.

The Elimination of Smallpox from Some Countries in Asia. Variola major was the only variety of smallpox that occurred in Asia. Smallpox control operated effectively in some of the smaller European colonies in Asia during the period between the wars. Country-wide elimination was achieved in the Philippines in 1931 and in British Malaya and in the Netherlands East Indies by 1938, and control continued to improve in Japan and its colonies, Korea and Formosa.

However, the inadequacy of the public health infrastructures became apparent with the disruption associated with the Second World War, when there were exacerbations of smallpox throughout the region and smallpox again became endemic in Malaysia and Indonesia, although not in the Philippines. Local efforts were successful in eliminating smallpox from Malaysia for the second time, in 1954, and elimination was achieved in Japan in 1951 and in Korea in 1954.

However, the greatest achievement of this period was the elimination of endemic smallpox from the People's Republic of China. This was achieved in all the large towns by 1954, and the last outbreaks in border districts in Yunnan Province and in Tibet occurred as late as 1960. Smallpox had been a virtually universal disease in China for hundreds of years. Before 1949, vaccination was restricted to foreigners and the well to do in the Treaty Ports, whereas the vast majority of China's population were peasants. Immediately after his victory in 1949, Mao Zedong proclaimed the eradication of smallpox in China to be a national objective. It was accomplished by mass vaccination with liquid vaccine, applied on a continental scale from 1951 onward and after preparation of the population by exhortation and example.

Vaccination had reduced the incidence and severity of smallpox in the Indian subcontinent, from levels of about 2000 reported cases per million in 1870 to 300 per million in 1930, but in those populous countries there continued to be millions of cases every year and hundreds of thousands of deaths. Probably only 1 to 2 percent of these were reported.

Halting Progress toward Global Eradication: 1959–1966

As early as 1950 the Third World Health Assembly requested the Director-General of WHO to study ". . .the ways of carrying out a world-wide campaign [against smallpox] including (1) a general programme of work to be implemented by WHO and (2) the estimated costs to the Organization." Each year thereafter the Executive Board and the Assembly urged national member states to control smallpox within their borders, and by 1958 decided that "it [was] opportune to raise the problem of world-wide eradication of smallpox in the near future" (see World Health Organization, 1980a).

At the Twelfth World Health Assembly in 1959, the delegates from the Soviet Union proposed a resolution, which was seconded by the United States, to the effect that world-wide eradication of smallpox was a goal that might be achieved in 4 or 5 years, essentially by vaccinating or revaccinating 80 percent of the population within a period of 4 to 5 years "as had been demonstrated for several countries."

This resolution was accompanied by arrangements for the provision of freeze-dried vaccine to endemic countries, many of which were encouraged to establish national smallpox eradication programs. Standards for smallpox vaccine were laid down (WHO Study Group on Requirements for Smallpox Vaccine, 1959) and a special study group reported on measures to be taken to achieve smallpox eradication (WHO Expert Committee on Smallpox, 1964). However, no special funds were set aside to provide coordination of the program. Country-wide elimination was achieved in several more countries in Asia, Africa, and the Americas (see Fig. 11-6) but by 1966 it was clear that progress on these lines would never achieve world-wide eradication. In consequence, the Nineteenth World Health Assembly in 1966 adopted a resolution which included, *inter alia,* acceptance of the need for coordination of the eradication program of individual countries and the need for WHO finance from the regular budget. This resolution was put into effect the next year, by the establishment of the Intensified Smallpox Eradication Program, to be coordinated by a Smallpox Eradication Unit at WHO Headquarters in Geneva.

The Intensified Smallpox Eradication Program, 1967–1977

The operations of the Intensified Smallpox Eradication Program occupy the greater part of a 1500-page book (Fenner *et al.,* 1988) which we cannot hope even to summarize here. We shall therefore comment briefly on the problems and the strategies used to solve them, summarize the progressive achievement of country-wide elimination and final global eradication, and comment on the features of smallpox that made this feat possible.

The Situation in 1967. The first action of the Smallpox Eradication Unit was to survey the global scene, in terms of the incidence of endemic

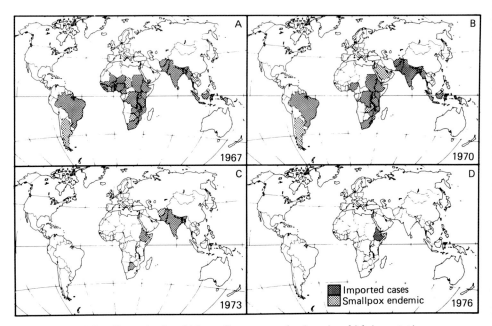

FIG. 11-9.   *Countries in which smallpox was endemic or in which importations were reported, at various periods during the Intensified Smallpox Eradication Program. Global eradication was achieved in 1977. (From World Health Organization, 1980a.)*

smallpox, the quality and quantity of smallpox vaccine, especially in the endemic countries which were mostly in the tropics, and the effectiveness of national smallpox eradication programs in the endemic countries.

Incidence of Smallpox. In 1967 smallpox was endemic in 31 countries and imported cases were reported from another 15 countries (Fig. 11-9A). The incidence of reported cases that year was 131,776; subsequent studies suggested that this probably represented about 1%, or even less, of the real incidence. Thus 170 years after Jenner's discovery of a tool for eradication, there were still some 10–15 million cases and probably over 2 million deaths a year in the world.

Vaccine. Although Collier (1955) at the Lister Institute had developed methods for the large-scale production of a good freeze-dried vaccine several years earlier, many of the endemic countries still used liquid vaccine whoe stability under tropical conditions was such that probably only 15 or 20% of the vaccine used in the field was of acceptable potency. Freeze-dried smallpox vaccine is an exceptionally stable product; indeed the quick test for satisfactory heat stability required that it lose no more than 90% of its titer of over 10^8 pock-forming units per milliliter or higher after being boiled for an hour. The Smallpox Eradication Unit undertook a major campaign to upgrade the quality and increase the quantity of vaccine. Special WHO Collaborating Centers for Smallpox Vaccine were established in Toronto and Utrecht, an international meeting of producers was held and training courses established, and regular control testing of the quality of vaccines was introduced. By 1970 most vaccines in both developed and developing countries had reached WHO standards.

National Smallpox Eradication Campaigns. Following the WHO resolution of 1959, several countries in Asia had mounted national smallpox eradication campaigns, which aimed to achieve eradication by mass vaccination of at least 80% of the population. However, there were many countries in Africa where no special measures were taken to control smallpox. Some of the national eradication campaigns in Asia were effective in controlling smallpox; others, notably in India, appeared to be making no effective headway.

Strategies for Global Eradication. With its limited resources, the Smallpox Eradication Unit decided upon certain regional priorities. A campaign that had already been initiated in western and central Africa with assistance from the United States Agency for International Development was supported by WHO and reached an effective conclusion by 1970. Assistance was given to Brazil to complete the eradication of smallpox from the Americas, which was successful by 1971, and to Indonesia to repeat what it had already accomplished once before in 1937. However, the major attention of the Program was devoted to the ancient strongholds of variola major on the Indian subcontinent and the many countries in Africa where variola major and/or variola minor were still endemic.

Although case finding and ring vaccination had long been recommended as strategies for controlling outbreaks of smallpox following importations into nonendemic countries, the early WHO and national strategy had been to rely on mass vaccination, with the expectation that if at least 80% of the population were vaccinated smallpox transmission would be interrupted. Although apparently successful in some countries, this strategy did not work in highly populous countries like India and Indonesia and it did not work if the 20% of unvaccinated persons were not randomly distributed throughout the population.

The major change in strategy, which occurred early during the course of the Intensified Smallpox Eradication Program, was the elevation of case finding and ring vaccination, now described as surveillance and containment, to preeminence, although still, in most countries, maintaining campaigns for mass primary vaccination.

In parallel with these operational strategies, the Smallpox Eradication Unit undertook important initiatives in four other fields: research, communications, diagnosis, and equipment.

Research. As well as applied research on epidemiology, vaccines and diagnostic methods, links were early established with academic virologists with interests and expertise in poxvirus research. The value of such links were well illustrated when the new disease of human monkeypox was discovered and when suspicions were raised that there might be a wild animal reservoir of variola virus. Several WHO Collaborating Centers for Poxvirus Research were established.

Communications. Information exchange, at all levels, was vital if success was to be achieved in a program of global eradication. Steps were therefore taken to establish liaison with officers of the Intensified Smallpox Eradication Program at the WHO Regional Offices (SEARO, AFRO, and EMRO) which served the areas where most of the problems remained, and to station WHO epidemiologists at national headquarters of as many of the national smallpox eradication programs as possible. Newsletters, correspondence, and publication in the Weekly Epidemiological Record were exploited and regular meetings of headquarters and field staff held in Geneva, at the regional offices, and in the endemic countries. Systems of case reporting were developed and strengthened within each program.

At another level, vigorous publicity was embarked upon to enlist world-wide support for the program and to build up a substantial voluntary fund to supplement the regular WHO budget with meetings, films, special stamp issues, and special articles in "World Health."

Diagnosis. When smallpox was a common endemic disease, diagnosis could safely rest on clinical and epidemiological findings. However, as it became less common and particularly as elimination was in sight, and with suspected importations into nonendemic countries, the diagnosis of individual cases became critical and laboratory confirmation vital. To assist developing countries without their own facilities, WHO Collaborating Centers for Smallpox Diagnosis were established at the Communicable Disease Center at Atlanta and at the Research Institute for Viral Preparations in Moscow. Soon after the establishment of these diagnostic centers, electron microscopy, supplemented by isolation of the virus, had become the major tools used for the laboratory diagnosis of

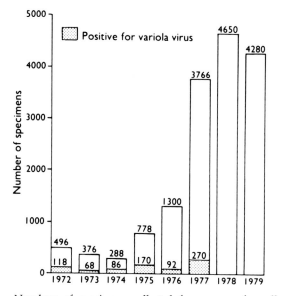

FIG. 11-10. *Numbers of specimens collected from cases of smallpox, suspected smallpox, chickenpox, or suspected monkeypox and tested by one of the Collaborating Centers for Smallpox Diagnosis, in Moscow, U.S.S.R., or Atlanta, Georgia, U.S.A. The large number of negative specimens in 1977–1979 was associated with preparations for the visit of the International Commissions for the Certification of Smallpox Eradication to Ethiopia and Somalia. (From Fenner et al., 1988.)*

smallpox. This service was of the utmost importance, not only in helping in the achievement of eradication but also in the very extensive testing required before eradication could be officially certified (Fig. 11-10).

Equipment. The Smallpox Eradication Unit was involved in the provision of equipment of two kinds. The first involved methods for the inoculation of vaccine, where WHO initially supported use of the jet injector, but efficient use of this instrument required good technical maintenance and the assembly of large numbers of subjects for vaccination, at central collection points. Neither of these requirements could be met in rural areas in many developing countries, and the Unit subsequently concentrated on a very simple and ingenious device, the bifurcated needle. This needle had several advantages: it was cheap, it could be used with a minimum of training, it picked up the requisite quantity of vaccine between its two prongs and batches of a hundred could be readily sterilized after each day's work and used again. It was also well adapted for vaccination in isolated households in remote areas. The second equipment need was means of transportation, a vital and difficult requirement in many developing countries. While often the only method of movement was on foot, a great variety of other methods were used by vaccinators, surveillance staff, and epidemiologists—bicycles, motorbikes, jeeps, boats, and even helicopters.

The Progress of Global Eradication.   The progress of global eradication after the initiation of the Intensified Smallpox Eradication Program in 1967 can be appreciated by examination of Figs. 11-9 and 11-11. During the first 3 years there was little change in the number of endemic countries, but in 1970 the campaign in West and Central Africa was completed and in 1971 smallpox was eliminated from Brazil, the last endemic country in the Americas. The number of endemic countries remained unchanged in Asia until eradication was achieved in Indonesia in 1972, in Afghanistan and Pakistan in 1973 and 1974, respectively, and then in the rest of the Indian subcontinent in 1975. Eradication progressed steadily in Africa until by early 1976 Ethiopia remained the only endemic country in the world, although there were a few imported cases in Somalia and the year before there had been an importation from

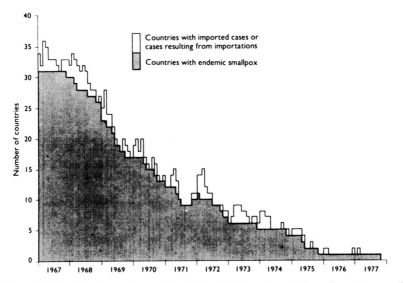

Fig. 11-11. *Numbers of countries with endemic or imported smallpox, by month, 1967–1977. (From Fenner et al., 1988.)*

Ethiopia into Kenya. However, some difficulties still remained. Ethiopia and Somalia were at war, with fighting occurring in the Ogaden desert and hundreds of thousands of refugees pouring into Somalia. The last case in Ethiopia occured in August 1976, but in September an epidemic was recognized in Mogadishu, the capital of Somalia. This smouldered and spread and was not finally eliminated until October 1977, after 947 outbreaks, with 3229 cases, had been found in various parts of Somalia. The last case of endemic smallpox in the world was a hospital cook in the town of Merka, Somalia.

The magnitude of the effort involved in the global eradication of smallpox cannot be appreciated just from these figures and graphs; there is not space to elaborate on the very different problems encountered in the 79 different countries in which WHO engaged in special activities or investigations. The interested reader is referred to the WHO book "Smallpox and Its Eradication" (Fenner *et al.*, 1988). Some of the political and social problems should be mentioned, however. During the eradication campaign in West Africa, there were 23 changes of government in the 18 participating countries and a major civil war in Nigeria, the most populous country in the region. A change of government often meant a change in policy and personnel, creating difficulties in countries and at times when the pool of skilled manpower was perilously small. In Pakistan, in 1970, 3 years after the Intensified Smallpox Eradication Program had started work there, a major governmental reorganization effectively divided the country into six separate provinces in West Pakistan, each of which conducted its own eradication program, and with each of which WHO advisors had to deal separately. Smallpox was eradicated from Bangladesh (then East Pakistan) in August 1970, no mean feat in that former hotbed of the disease, only to return with the 10 million refugees who poured back from India after the 1971 civil war. It took until October 1975, and an enormous effort, to eradicate the disease again. Eradication of variola minor from Ethiopia and Somalia was achieved in spite of the war in the Ogaden, but we might not have been so fortunate in Vietnam and Kampuchea, in Angola, Zaire, Uganda, or Zimbabwe, if eradication had not been achieved before the wars and disruptions that subsequently occurred in those countries.

The Certification of Eradication

The keynotes of the intensified eradication campaign had been quality control and assessment; of the vaccine, the vaccinators and the public health workers and their programs. The final and major assessment had necessarily to be some mechanism for checking the statements by national governments that they had indeed eliminated smallpox; that their "last case" really was the last case. To do this a system of certification of eradication was set up, using outside experts as members of International Commissions for the Certification of Smallpox Eradication. Rules were laid down of the kinds of preparatory work needed to be done by joint WHO and national teams, after the "last case" and before the visit of the Commision. An interval of at least 2 years was set during which there had been no case, and intensive surveillance had been maintained, before the Commission made its inspection. An important feature of these precertification programs was the wide publicity given to a reward for finding smallpox, involving sums of money that moved progressively from $10 to $1000 for finding a case. The scale of the search operations can be gauged from the figures showing deatils of one of several active searches conducted in preparation for the certification of smallpox eradication in India (Table 11-4).

Inspections by the International Commissions were not designed to detect cases of smallpox, but to assess the surveillance activities and decide whether they would have been adequate to detect cases of smallpox if they had occurred. Twenty-two International Commissions, or in a few cases special representatives, visited and certified 79 countries during the period 1973 to 1979 (Fig. 11-12).

Finally, in 1977, a Global Commission for the Certification of Smallpox Eradication was established with the dual tasks of providing evidence that would convince the World Health Assembly that smallpox had indeed been eradicated, and developing a posteradication strategy. During meetings of the full Commission in 1977, 1978, and 1979, and

TABLE 11-4

The Scale of All-India Active Searches Made in Preparation for the Visit of the International Commission for the Certification of Smallpox Eradication in April 1977

In May and in November 1976
  110 million households were visited in
    670,000 villages and
      2,600 urban areas
  by 152,441 health workers
    115,347 to search
     29,046 to supervise
      8,040 to assess
During 1976
  833,412 cases of chickenpox were discovered and verified
  > 500,000 outbreaks of fever-with-rash cases were investigated

subsidiary meetings and visits of international commissions and specialist subcommittees, these goals were achieved. Appropriate recommendations were made in December 1979 (Plate 11-3) and accepted by the World Health Assembly in May 1980.

Factors Contributing to the Success of Smallpox Eradication

There were a number of factors, some biological, some sociopolitical, that made it possible, with a major effort of international cooperation, to eradicate smallpox from the world (Table 11-5).

First, smallpox was such a severe disease that it was clearly worth the effort required for eradication. The next three points relate to the spread of smallpox and were extremely important. Second, as with measles, subclinical infections were virtually unknown. They could be recognized by serological changes in some vaccinated individuals, but such persons excreted very little virus; they were of little epidemiological importance in the presence of endemic smallpox.

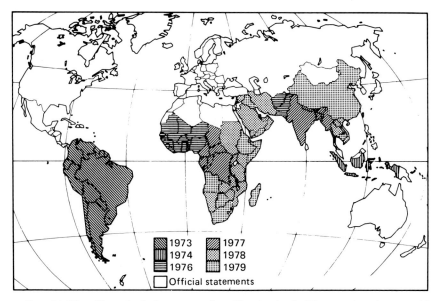

1973	1977
1974	1978
1976	1979
Official statements	

FIG. 11-12. *Chronological progress of certification in the 79 countries where special measures were taken to certify smallpox eradication. All other countries provided an official statement that no case of smallpox had occurred in their country during the preceding 2 years. (From Fenner et al., 1988.)*

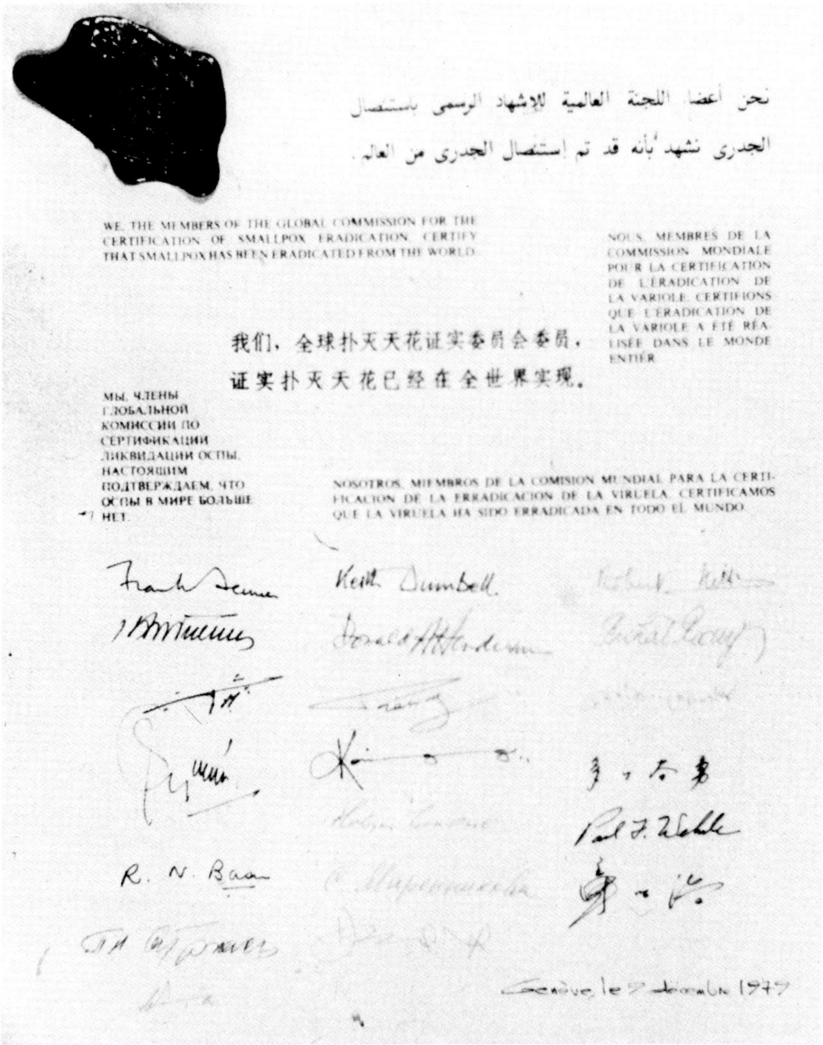

نحن أعضا اللجنة العالمية للإشهاد الرسمي باستئصال
الجدري نشهد بأنه قد تم إستئصال الجدري من العالم.

WE, THE MEMBERS OF THE GLOBAL COMMISSION FOR THE CERTIFICATION OF SMALLPOX ERADICATION, CERTIFY THAT SMALLPOX HAS BEEN ERADICATED FROM THE WORLD.

NOUS, MEMBRES DE LA COMMISSION MONDIALE POUR LA CERTIFICATION DE L'ÉRADICATION DE LA VARIOLE, CERTIFIONS QUE L'ÉRADICATION DE LA VARIOLE A ÉTÉ RÉALISÉE DANS LE MONDE ENTIER.

我们，全球扑灭天花证实委员会委员，
证实扑灭天花已经在全世界实现。

МЫ, ЧЛЕНЫ ГЛОБАЛЬНОЙ КОМИССИИ ПО СЕРТИФИКАЦИИ ЛИКВИДАЦИИ ОСПЫ, НАСТОЯЩИМ ПОДТВЕРЖДАЕМ, ЧТО ОСПЫ В МИРЕ БОЛЬШЕ НЕТ.

NOSOTROS, MIEMBROS DE LA COMISION MUNDIAL PARA LA CERTIFICACION DE LA ERRADICACION DE LA VIRUELA, CERTIFICAMOS QUE LA VIRUELA HA SIDO ERRADICADA EN TODO EL MUNDO.

PLATE 11-3. *The certificate signed by members of the Global Commission for the Certification of Smallpox Eradication on December 9, 1979. The text, in the six official languages of the United Nations, reads: "We, the members of the Global Commission for the Certification of Smallpox Eradication, certify that smallpox has been eradicated from the world." (From Fenner* et al.*, 1988.)*

TABLE 11-5

Features that Favored the Global Eradication of Smallpox

Biological features
1. A severe disease, with high mortality and serious aftereffects
2. Subclinical infections very rare
3. Cases became infectious at the time of onset of the rash
4. Only one serotype
5. An effective heat-stable vaccine available
6. There was no animal reservoir of variola virus

Sociopolitical features
1. Earlier elimination in many countries showed that global eradication was possible
2. There were no social or religious barriers to the recognition of cases
3. The cost of quarantine and vaccination for international travellers provided a strong financial incentive for wealthy countries
4. The Intensified Smallpox Eradication Unit of WHO had inspiring leaders and enlisted devoted health workers

The third feature refers to the period of infectivity of a case. Smallpox patients were not infectious during the incubation period or the preeruptive phase. With the onset of rash, which was usually accompanied by oropharyngeal ulcers, the patient became infectious and remained so until recovery, infectivity being greatest during the first week of illness. However, smallpox was not as highly infectious as influenza, measles, or chickenpox. Spread usually resulted from direct face-to-face contact with patients. If smallpox patients were isolated as soon as the rash was apparent, in a setting in which they had contact only with vaccinated or immune persons, the chain of transmission could be broken.

Fourth, in contrast to diseases like hepatitis, chickenpox, and tuberculosis, neither a prolonged carrier state nor recurrence of clinical illness with associated infectivity ever occurred in smallpox.

The next two points are related. There was only one serotype of variola virus—over centuries of time and all over the world. Indeed, the serological relatedness of all the orthopoxviruses is so great that infection with any one of them provides protection against all the others, although protection wanes with time, especially to heterologous challenge. Since the time of Jenner, we have had an effective live virus vaccine against smallpox in the form of a related orthopoxvirus, initially probably cowpox virus and later vaccinia virus. In the 1950s a freeze-dried vaccine was developed that was stable even under the most adverse conditions; a most important requirement for any vaccine that had to be used in remote areas of tropical countries.

Finally, whatever its origins may have been, variola virus is now a specifically human virus. Chimpanzees, monkeys, and under special conditions a few kinds of laboratory animals can be infected, but there is no animal reservoir of variola virus. The "whitepox" viruses, purportedly recovered from cultured cynomolgus cells and from the organs of wild animals shot in Zaire, were variola virus, but they were probably all laboratory contaminants (see Chapter 7).

The biological features of smallpox were thus especially favorable for eradication, and by 1958 smallpox had been eliminated from all the countries of Europe, North America, and Oceania. If we are thinking of extending the concept of global eradication to other infectious diseases this is an important point to keep in mind; extension to other countries of what has already been accomplished in some clearly provides greater promise than an attack on a disease that is still endemic everywhere. Because smallpox had been eliminated in many countries, it was possible to promote the concept of global eradication as a clear, attainable objective.

The biological features of smallpox, while a necessary precondition for global eradication, were not a sufficient condition. Several sociopolitical factors were equally important (see Table 11-5). A very important attribute of smallpox was that as one of the quarantinable diseases, which had by 1960 been eradicated from the industrialized countries of the world, smallpox still imposed a heavy financial burden on these countries, as well as those where smallpox was endemic. Quite apart from the burden of disease and death due to smallpox itself, the cost of vaccination plus that of maintaining quarantine barriers is calculated to have been about 1 billion dollars (US) a year. This cost would disappear completely if global eradication could be achieved. Further, in countries where smallpox had been eliminated, vaccination caused an unacceptable morbidity and mortality. Thus, if the objective of eradication was demonstrably attainable, it was a sound economic proposition for the industrialized countries of the world to support the campaign financially.

Finally, it was a matter of great importance that the World Health Organization was able to recruit two men who proved to be superb leaders of the program for global eradication, Dr. D. A. Henderson (1967–1976) and Dr. Isao Arita (1976–1985). In addition to being good administrators and managers, both spent a great deal of time in the field as skillful and untiring workers. This was most important in winning and maintaining the respect of the thousands of national and international field staff who did the actual work of eradication.

CHAPTER 12

Vaccinia Virus as a Vector for Vaccine Antigens

Recombination between different strains of vaccinia virus was unequivocally demonstrated by Fenner (1959), and between different species of orthopoxviruses by Woodroofe and Fenner (1960). It was also shown that a poxvirus inactivated by heat or urea treatment could be reactivated by a coinfecting poxvirus (Joklik *et al.*, 1960a,b), even when the later functions of the reactivating virus had been inhibited by treatment with nitrogen mustard (Joklik *et al.*, 1960c), a supraoptimal temperature of incubation (Dumbell and Bedson, 1964), or ultraviolet irradiation (Dunlap and Patt, 1971). This "nongenetic reactivation" was shown to be a convenient way to obtain hybrids between different species of orthopoxviruses (Bedson and Dumbell, 1964a).

Intact molecules of vaccinia virus DNA had been recovered from purified suspensions of viral particles (Joklik, 1962b; Becker and Sarov, 1968), but the isolated DNA was not infectious, since vaccinia virus DNA is transcribed by a virion-associated DNA-dependent RNA polymerase (Kates and McAuslan, 1967b; Munyon *et al.*, 1967). Some years later, Sam and Dumbell (1981) devised experiments to show that the activity of intact orthopoxvirus DNA molecules could be rescued if

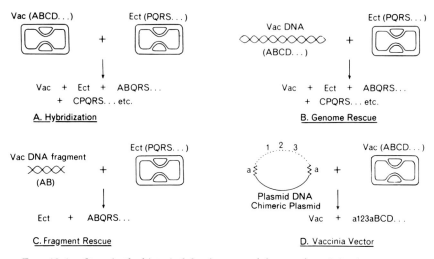

Fig. 12-1. *Steps in the historical development of the use of vaccinia virus as a vector for foreign genes (from Fenner, 1985). (A) Demonstration of hybridization between vaccinia and ectromelia viruses (Woodroofe and Fenner, 1960). (B) Rescue of vaccinia genome and (C) rescue of fragment of vaccinia genome, in cells transfected with vaccinia DNA and infected with ectromelia virus (Sam and Dumbell, 1981). (D) Insertion of foreign gene into vaccinia DNA by homologous recombination. a, Fragment of vaccinia DNA in plasmid; 1, 2, 3, foreign gene (Mackett et al., 1982; Panicali and Paoletti, 1982).*

transfected cells were coinfected with an active orthopoxvirus, and that marker rescue of *ts* mutants could be obtained with particular subgenomic fragments (Fig. 12-1).

Very soon after this, two groups of investigators in the United States (Mackett *et al.*, 1982; Panicali and Paoletti, 1982) showed that a foreign gene (herpesvirus thymidine kinase) could be inserted and expressed in a recombinant vaccinia virus, thus opening the way for the construction of vaccines for the control of infectious diseases. The great potential of this approach for the immunization of children in developing countries was immediately recognized by the World Health Organization, which organized workshops devoted solely to this topic in 1984 (Ada *et al.*, 1985; Quinnan, 1985) and 1985 (World Health Organization, 1985).

During the last few years numerous recombinants have been constructed in many laboratories, and some investigations of immunological aspects of immunization with vaccinia virus recombinant vaccines have been made. We have drawn extensively on the review by Mackett and Smith (1986) for our discussion of various aspects of the construction and use of vaccinia virus vectors.

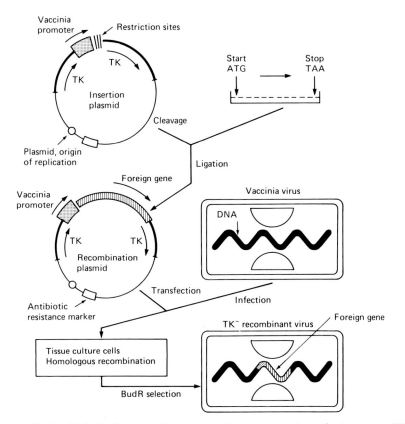

FIG. 12-2. *Method of constructing a vaccinia vector carrying a foreign gene. TK, Thymidine kinase gene of vaccinia virus; BudR, bromodeoxyuridine (Courtesy Dr. B. Moss).*

INSERTION OF FOREIGN GENES INTO VACCINIA VIRUS DNA

Animal viruses, notably SV40, adenoviruses, and retroviruses, have been widely used for obtaining the expression of foreign genes in cells of different types and different animal species (for review, see Rigby, 1983). The usual method of construction of recombinant genomes is by restriction endonuclease cleavage of the viral genome (or cDNA transcribed from it) and religation with the foreign DNA. Cells are transfected with this hybrid DNA and replication is allowed to proceed. This approach is not practicable with orthopoxviruses, since the viral DNA is not infectious and its transcription depends upon unique vaccinia virus promoters and a viral polymerase, rather than cellular polymerases (see Chapter 3). Recombinants are therefore generated by transfecting the foreign gene placed under the control of a viral promoter and flanked by viral sequences into cells infected with vaccinia virus, where recombination occurs between the flanking sequences and the corresponding DNA in the viral genome. This technique has become known as "homologous *in vivo* recombination" (Fig. 12-2). A great advantage of vaccinia virus recombinants is that, in contrast to the packaging constraints which limit the size of the DNA insert in papovaviruses and retroviruses, orthopoxvirus genomes can probably accommodate up to 35 kbp of foreign DNA (see below).

Construction of Plasmid Vectors

An important element in the preparation of vaccinia virus recombinants is to design plasmid vectors that can be used with any one or more of a variety of foreign genes. An essential feature of such plasmids is that they should have a vaccinia virus promoter with multiple unique restriction endonuclease sites just downstream of the RNA start site, translocated within a selected vaccinia virus gene. A series of such "insertion vectors" have been developed (Kieny *et al.*, 1984; Mackett *et al.*, 1984; Boyle *et al.*, 1985; Franke *et al.*, 1985), the most widely used vaccinia gene being the thymidine kinase gene, which is located in the J fragment of *Hind*III digests of vaccinia virus (strain WR) DNA (Weir *et al.*, 1982; see Fig. 2-3).

With correct engineering, authentic foreign proteins are produced and problems associated with incorrect codon phasing and unpredictable properties of fusion proteins are avoided.

Generation, Selection, and Characterization
of Recombinant Viruses

Generation of Recombinant Viruses.   Having constructed an insertion vector containing the foreign gene being studied, the latter is incorporated into vaccinia virus by homologous recombination. To obtain this, cultured cells are infected with the selected strain of vaccinia virus (see below) and transfected with recombinant plasmid DNA. Other methods of generation of recombinants utilize temperature-sensitive mutants of vaccinia virus (Kieny *et al.*, 1984). Cells infected with the *ts* mutant are cotransfected with the insertion vector and intact DNA of wild-type virus and incubated at the nonpermissive temperature. Viral replication can occur only in those cells that have been transfected with the wild-type DNA, which rescues the *ts* mutant. Such cells have normally taken up the insertion plasmid DNA. This procedure reduces the background yield of nonrecombinant virus, but may create a problem with recombinants of unknown biological properties unless the *ts* mutant has been derived from the wild-type virus strain being used.

Selection of Vaccinia Virus Strain. Most research workers in the United States and several other countries (Australia, Switzerland) have used the WR strain; workers in Strasbourg used variants of the Copenhagen strain. When proposals were made for the use of vaccinia recombinants for vaccination of human subjects or for veterinary vaccines, other considerations arose, notably the pathogenicity for man of the virus strain used and the amount of information available about the frequency of complications. The first response to this need was to use the Wyeth (New York City Board of Health) strain that has long been used for smallpox vaccine production in the United States, and has a good record for low pathogenicity and effective immunogenicity (Fenner *et al.*, 1988). Meanwhile, investigations are proceeding on the factors that affect the virulence of vaccinia virus. Early experiments showed that inactivation of the TK gene, a consequence of the most commonly used method of constructing recombinant vaccinia virus, substantially reduces pathogenicity while maintaining immunogenicity (Buller *et al.*, 1985).

Selection of Recombinant Viruses.   Using the standard method, only about 0.1% of the progeny virions are recombinants, so that a selection procedure is necessary to isolate these. Use of the thymidine kinase (TK) gene to provide the flanking vaccinia DNA in the insertion vector provides a convenient method of selection, since recombinants will have a TK$^-$ phenotype and can be selected by plaque production in TK$^-$ cells in the presence of BUdR (Mackett *et al.*, 1984). Spontaneous TK$^-$ mutants can be distinguished by immunological screening (Mackett and Arrand, 1985; Mackett *et al.*, 1985b). An alternate, more generally applicable, selectable marker can be provided in double insertion mutants; e.g., the neomycin resistance gene (Franke *et al.*, 1985). Recombinants expressing this gene, but not wild-type virus, produce plaques in the presence of the synthetic antibiotic G418. This marker has the advantage that recombinants can be generated in a wide range of cell lines and that any nonessential region of the virus genome can be used for insertion of the foreign gene. Other protocols for selecting recombinants utilize double insertion vectors containing the β-galactosidase gene as well as the desired foreign gene (Chakrabarti *et al.*, 1985; Newton *et al.*, 1986).

Recombinant viruses can also be distinguished from parental vaccinia virus by *in situ* plaque DNA:DNA hybridization (Panicali and Paoletti,

1982), or by immunological methods, using ^{125}I-labeled *Staphylococcus* protein A and autoradiography (Panicali *et al.*, 1983) or an enzyme-linked second antibody directed against the primary antibody (Mackett and Arrand, 1985; Newton *et al.*, 1986).

Characterization of Recombinant Viruses. Having obtained a high titer stock of the recombinant virus, its genome DNA is analyzed by restriction endonuclease digestion, Southern blotting, and hybridization with specific DNA probes. In all cases reported so far, the foreign gene has been incorporated without compensatory deletions or other rear-rangements. Very large amounts of foreign DNA can be inserted into vaccinia virus genomes; for example, Smith and Moss (1983) inserted 25 kbp of DNA without compensatory deletions and with no change in growth capacity of the recombinant virus. Further, since viable deletion mutants have been characterized that have lost up to 10 kbp (Moyer and Rothe, 1980) and 18 kbp (Drillien *et al.*, 1981) of their DNA, vaccinia virus recombinants can potentially incorporate at least 35 kbp of foreign DNA. In principle, therefore, one recombinant virus could incorporate genes for a number of foreign antigens, thus providing a polyvalent vaccine in a single dose. To explore this possibility, Perkus *et al.* (1985) inserted genes for hepatitis B virus surface antigen, the herpes simplex glycoprotein D, and influenza virus hemagglutinin into three different nonessential sites in vaccinia virus. Rabbits inoculated either intrave-nously or intradermally with this polyvalent vaccinia virus recombi-nant produced antibodies reactive to all three foreign viral antigens.

Expression of Foreign Genes

Early experiments with the insertion of herpes simplex virus TK gene (Mackett *et al.* 1982) showed that expression of foreign genes depended upon the presence of vaccinia promoters inserted in the correct orienta-tion, although low levels of expression in the absence of correctly engineered promoters have been observed late in infection (Mackett *et al.*, 1984). Since the use of recombinant vaccinia viruses for vaccines or for analyzing gene function depends on high levels of expression of the foreign gene, considerable effort is being devoted to the characterization of additional promoters. One promising source yet to be adequately investigated is the promoter associated with the nonstructural "LS antigen" of vaccinia virus, a very abundant protein which is related to that of the A-type inclusion bodies of cowpox virus (Patel *et al.*, 1986).

Progress in obtaining high levels of expression has recently been made by using two vaccinia virus recombinants (Fuerst *et al.*, 1987). One contains the T7 RNA polymerase gene under the control of a vaccinia virus promoter, and the other the target gene fused to a T7 promoter. When cells are coinfected with the two recombinants, the T7 polymerase is made and transcribes the target gene carried in the second recombinant. This system yields at least 10-fold higher levels of expression than those obtained with more conventional recombinants. In fact, the level of expression seems to be so high that it has not been possible to isolate a single recombinant carrying both genes. The reason for this appears to be that the target gene is so strongly transcribed that such a recombinant may not be able to express its own genes.

The search for active promoters has been simplified by the development of a transient expression system (Cochran *et al.*, 1985a) in which the orthopoxvirus promoter being studied is linked to the gene for an enzyme which can be easily assayed, such as chloramphenicol acetyltransferase (CAT). Such plasmids are transfected into cells infected at high multiplicity and the CAT activity can be assayed directly on infected cell lysates.

Since there is no splicing involved in the formation of vaccinia virus mRNAs (see Chapter 3), it is not possible directly to express foreign genes that contain introns. In such cases, cDNA copies of mature, spliced mRNAs are required.

Detection of Expression of Foreign Genes. Depending on the nature of the foreign gene, its expression can be detected by enzymatic or immunological techniques. Table 12-1 lists five enzymes whose expression has been detected by enzyme assays; all are viral or bacterial enzymes that are absent from eukaryotic cells. As already noted, chloramphenicol acetyltransferase expression is a useful tool for analyzing the temporal control and level of expression obtained with different orthopoxvirus promoters (Mackett *et al.*, 1984; Weir and Moss, 1984; Cochran *et al.*, 1985a,b).

TABLE 12-1

Foreign Enzymes Expressed in Vaccinia Virus-Infected Cells

Gene	Reference
Herpes simplex virus thymidine kinase	Panicali and Paoletti (1982); Mackett *et al.* (1982); Vassef *et al.* (1985)
E. coli β-galactosidase	Chakrabarti *et al.* (1985); Newton *et al.* (1986); Panicali *et al.* (1986)
Chloramphenicol acetyltransferase	Mackett *et al.* (1984); Weir and Moss (1984)
Neomycin resistance	Franke *et al.* (1985)
Influenza virus neuramindase	Gotch *et al.* (1987)

A wide variety of immunological techniques have been used to detect the expression of foreign genes incorporated into recombinant vaccinia viruses: immunoprecipitation, Western blotting, radioimmunoassay, immunofluorescence, and ELISA. Particles consisting of aggregates of hepatitis B surface antigen have been demonstrated with the electron microscope in supernatants of cells infected with vaccinia virus containing the relevant gene (Smith et al., 1983a).

Analysis of foreign gene products on polyacrylamide gels has shown that with correct genetic engineering, the proteins are of the predicted size and have undergone appropriate posttranslational modifications, including glycosylation, proteolytic cleavage, and carboxylation. Experiments with human factor IX have been particularly revealing, since this is an extraordinarily complex protein that undergoes extensive posttranslational modification; a vaccinia virus-factor IX gene recombinant produces fully active factor IX in human hepatoma cells, which normally do not produce this factor (de la Salle et al., 1986).

The foreign gene product is also transported normally in the cell and has been detected on the surface of the cell, or found excreted from infected cells. When a foreign gene like that specifying the HIV envelope glycoprotein is not proteolytically cleaved, the failure in processing may due to lack of the appropriate cell proteases, since the recombinant produced the correctly processed gp 120 in H9 and NIH3T3 cells, but not on HeLa or CV1 cells (Chakrabarti et al., 1986).

ANALYSIS OF VIRAL GENE FUNCTION

Insertion of various genes of a virus in vaccinia virus recombinants provides a powerful way of analyzing gene function and dissecting the immune response of experimental animals to various gene products. The most detailed investigation of the importance of various gene products of a particular virus in the immune response has been made with influenze A virus (Table 12-2). Copy DNA transcribed from each of the genes of influenza virus was inserted into separate vaccinia virus recombinants, and mice were infected with these recombinant viruses by intravenous inoculation. Antibody and cytotoxic T cell (Tc) responses were assayed. From these and other studies (for review, see Ada and Jones, 1986), it has been shown that immunization with the HA recombinant produces a subtype-specific antibody response and subtype-specific Tc cells, and provides protection almost equivalent to that provided by whole virus as antigen, whereas all other viral proteins produce cross-reactive Tc cell responses, but are not protective. The

TABLE 12-2

Immune Responses of Mice to Various Gene Products of Influenza Virus, Studied by Use of Recombinant Vaccinia Virus

Influenza virus protein	Generation of cytotoxic T cells	Protection	Reference
Structural			
Hemagglutinin (HA)	+	+	Boyle *et al.* (1985); Bennink *et al.* (1986)
Neuraminidase (NA)	−	. . .[a]	Gotch *et al.* (1987)
Matrix protein (MI)	+	. . .	Gotch *et al.* (1987)
Nucleoprotein (NP)	+	−	Yewdell *et al.* (1985); Andrew *et al.* (1987a)
Polymerases			
PA	+		
PB1	+	. . .	Bennink *et al.* (1987)
PB2	+		
Nonstructural			
Matrix protein (M2)	+	. . .	Gotch *et al.* (1987)
NSl	+	. . .	Bennink *et al.* (1987)

[a] . . . , Not tested.

nucleoprotein of influenza virus accounts for up to 30% of the total Tc cell response to influenza virus, which is cross-reactive between different types of influenza virus, but bears no relation to protection (Andrew *et al.*, 1986). However, when mice of three different haplotypes (H-2^b, H-2^d, H-2^k) were immunized, a high frequency of low-responder alleles was found. The responses of different congenic mouse strains and of humans to other influenza virus proteins also cosegregate with the MHC haplotype (Gotch *et al.*, 1987). It is possible that MHC antigens that offer favorable or unfavorable responses to critical epitopes on particular viral antigens may affect the responses to vaccination and infection with these viruses.

Vaccinia virus recombinants are also being used to analyze the structural proteins of alphaviruses (Rice *et al.*, 1985). Insertion of the cDNA of the 26S RNA of Sindbis virus into the thymidine kinase gene of vaccinia virus gave rise to the normal precursor proteins and all the mature structural proteins, which were cleaved, glycosylated, and transported in a manner analogous to that seen in authentic Sindbis virus infections. Apart from its potential as a vector for alphavirus vaccines, the vaccinia virus–Sindbis virus recombinants are well suited for site-specific mutagenesis to study the processing, transport, and immunogenicity of alphavirus glycoproteins.

VACCINATION OF EXPERIMENTAL ANIMALS

Having recognized the potential of using genetically engineered vaccinia virus as a method of immunization (Mackett *et al.*, 1982; Panicali and Paoletti, 1982), tests were immediately carried out with model systems, to see whether foreign genes inserted into vaccinia virus would promote an immune response in experimental animals. Smith *et al.* (1983a) and Paoletti *et al.* (1984) showed that mice and rabbits would produce neutralizing antibodies in response to infection with vaccinia virus recombinants containing genes for herpesvirus glycoprotein or hepatitis B virus surface antigen. Subsequently, genes from a wide range of other viruses (Table 12-3) and from certain protozoa, such as *Plasmodium falciparum* (Smith *et al.*, 1984, 1986; Langford *et al.*, 1986), have been inserted in the vaccinia virus genome and tested for expression in cell cultures, and for the production of an immune response in experimental animals.

The results are promising, in that the majority of the inserted genes evoked an immune response in the infected animals, and in the great majority of cases there was evidence of short-term protection. Experiments have yet to be reported on the duration of such protection.

POTENTIAL FOR HUMAN VACCINES

The major incentive for work on vaccinia virus as a vector is its potential as a method of providing vaccines for any infectious agent for which the appropriate protective antigen(s) have been identified and their genes isolated.

Advantages of Vaccinia Virus as a Vector for Vaccines

Smallpox vaccine, as used in the global smallpox eradication campaign, had great advantages over all other live virus vaccines (see Chapters 5 and 11). Briefly, these were as follows: (1) the freeze-dried vaccine is very stable under all ordinary environmental conditions, and after reconstitution is much more stable than other live virus vaccines; (2) production of very large amounts of vaccine was achieved at low cost in developing as well as developed countries; (3) using the bifurcated

TABLE 12-3

Immune Responses to Viral Antigens in Vaccinia Virus Recombinants

Virus species	Antigen	Animal	Neutralizing antibodies	Protection	Reference
Herpes simplex virus	Glycoprotein D	Mouse	+	+	Cremer et al. (1985)
	Glycoprotein B	Mouse	+	+	Cantin et al. (1987)
Epstein–Barr virus	Glycoprotein (gp350)	Rabbit	+	. . .[a]	Mackett and Arrand (1985)
Hepatitis B virus	S protein	Rabbit	+	. . .	Smith et al. (1983)
	MS protein	Rabbit	+	. . .	Cheng and Moss, (1987)
Vesicular stomatitis virus	G protein	Mouse	+	+	Mackett et al. (1985a)
		Cattle	+	+	
Rabies virus	Glycoprotein	Rabbit	+	+	Wiktor et al. (1984)
		Mouse	+	+	
Influenza A virus	Hemagglutinin	Hamster	+	+	Bennink et al. (1986)
		Mouse	+	+	Andrew et al. (1986, 1987a)
		Cattle	+	. . .	Boyle et al. (1986)
		Sheep	+	. . .	
		Poultry	+	. . .	
	Nucleoprotein	Mouse	−	−	Andrew et al. (1986, 1987a); Yewdell et al. (1985)
Respiratory syncytial virus	Glycoprotein G	Cotton rat	+	+	Elango et al. (1986)
	Glycoprotein F	Cotton rat	+	+	Olmsted et al. (1986)
Transmissible gastroenteritis virus of swine	Glycoprotein 195	Mouse	+	. . .	Hu et al. (1985)
Human immunodeficiency virus	Envelope protein	Human	+	. . .	Chakrabarti et al. (1986)
Simian rotavirus	Glycoprotein VP7	Rabbit	+	. . .	Andrew et al. (1987b)

[a] . . . , Not tested.

needle for scarification, the vaccine could be administered without the need for a syringe and needle and the expense and skills necessary to use them; and (4) the local scar provided evidence of successful vaccination in coverage and assessment studies. None of these features was especially advantageous in the industrialized countries; all were important in developing countries. It is clear that the possibility of developing vaccines against other common diseases that have all these advantages is very attractive, especially in the developing countries. Such vaccines would simplify existing global programs like the Expanded Program on Immunization, for example, by eliminating the cumbersome and expensive cold chain, but potentially the method could provide vaccines for diseases like hepatitis B or malaria for which vaccines are either very expensive or not available at all. Indeed, in theory a single polyvalent vaccinia virus recombinant vaccine could be envisaged which contained all the genes necessary for the immunization of persons in particular countries or regions.

Complications of Smallpox Vaccination

With smallpox vaccination there were, of course, some disadvantages. Every successful primary vaccination caused a disease with fever and slight malaise, which was more severe than that found with any other human live virus vaccine. However, if the disease against which protection is sought is common and severe, as was smallpox before its eradication, such effects could be accepted. More important were the occasional severe complications, which affected either the skin or the central nervous system. These complications were acceptable in smallpox-endemic countries, but they were not acceptable to either the public or the health officials in the nonendemic countries when smallpox had ceased to be a threat, in the early 1970s. Since these complications are the major objection to the use of vaccinia virus recombinants as vaccines, the complications observed after vaccination against smallpox, which draw on an experience involving many millions of persons, will be discussed in greater detail here (see also Fenner et al., 1988).

Abnormal Skin Eruptions. This group (Plate 12-1) included four syndromes, each with different predisposing factors and a different prognosis:

Progressive vaccinia (Plate 12-1B) was the most severe and also the rarest complication of vaccination. The local lesion at the vaccination site failed to heal, secondary lesions appeared elsewhere on the body, and all lesions spread progressively until the patient usually died 2 to 5

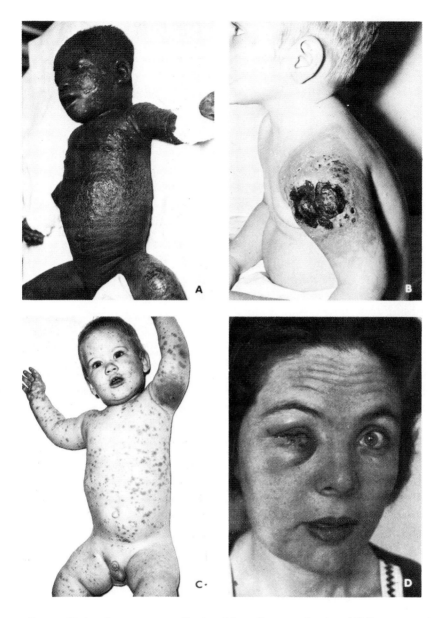

PLATE 12-1. *Cutaneous complications of smallpox vaccination: (A) Eczema vacci-*
natum in the unvaccinated contact of a vaccinated sibling; (B) progressive vaccinia, which
was fatal, in a child with congenital immunodeficiency; (C) generalized vaccinia 10 days
after vaccination. This had a benign course, with no residual scarring; (D) accidental
ocular infection with vaccinia virus (from Fenner et al., 1988; A, courtesy of Dr. I. D.
Ladnyi; B, C, and D, courtesy of Dr. Henry Kempe).

months later. It occurred only among persons with a deficient cell-mediated immune mechanism, either because of a congenital deficiency or a lymphoproliferative disorder, or from immunosuppressive treatment. The case–fatality rate was extremely high, and treatment with vaccinia-immune globulin was rarely effective.

Eczema vaccinatum (Plate 12-1A) was much more common, and occurred among persons with eczema who were vaccinated (although eczema was a contraindication to vaccination in many countries) and also among eczematous contacts of newly vaccinated persons. Either coincident with or shortly after the development of the local vaccinial lesion (or after an incubation period of about 5 days in unvaccinated eczematous contacts), a vaccinial eruption occurred on areas of skin that were eczematous at the time or had previously been so. These areas became intensely inflamed and sometimes the eruption spread to healthy skin. Constitutional symptoms were usually severe. The reported case–fatality rates varied greatly in different series of cases, but according to Kempe (1960), 30 to 40% of such patients died. Treatment with vaccinia-immune globulin reduced the case–fatality rate to 7%.

Generalized vaccinia (Plate 12-1C) followed hematogenous spread of vaccinia virus, with the production of pustular lesions on many parts of the body. The course of the individual skin lesions resembled that of the lesion at the vaccination site, but if the rash was profuse the lesions sometimes varied greatly in size. Generalized vaccinia was not associated with serious immunological deficiencies, and it had a good prognosis. Vaccinia-immune globulin was effective in hastening resolution of the lesions.

Accidental infections (Plate 12-1D) could occur both in vaccinees and contacts, the most serious lesions being those affecting the eyes or the perineum. Vaccinia-immune globulin was helpful in treatment.

Treatment. Several drugs were tried for the treatment of these complications, including cytosine arabinoside, rifampicin, a urea derivative of diphenyl sulfone, and metisazone. None was as effective as vaccinia-immune globulin, which was useful for all skin complications except progressive vaccinia.

Prevention. In countries with well-developed medical services, progressive vaccinia could be reduced by strictly observing the rule that congenital or acquired immunodeficiencies, and immunosuppressive treatment, were absolute contraindications to vaccination. Active eczema or a history of eczema were regarded as contraindications to vaccination in most industrialized countries. Administration of vaccinia-immune globulin at the time of vaccination reduced the frequency of eczema vaccinatum in persons with eczema in whom vaccination was essential (Sharp and Fletcher, 1973). However, if vaccinia vector vaccines are introduced, these measures are unlikely to be effective in developing countries, where vaccinia-immune globulin is not readily available and in general contraindications are not rigorously applied to immunization programs. On the other hand, in such countries children with congenital immunodeficiencies are unlikely to survive other childhood infections.

Accidental or contact vaccinial infections may be a problem if immunization with a recombinant vaccinia virus vaccine is applied only to certain risk groups while the rest of the population remains suceptible. Vaccinial infections among contacts of newly vaccinated military recruits have been reported recently in the United States and Canada.

Diseases Affecting the Central Nervous System (CNS). CNS disease was the most serious complication of smallpox vaccination, in that unlike progressive vaccinia and eczema vaccinatum, its occurrence was unpredictable, and it was associated with a 30% case–fatality rate. In fact, many industrialized countries supported the global smallpox eradication program in the hope that with the eradication of smallpox, the termination of smallpox vaccination programs would eliminate the occurrence of postvaccinial CNS disease.

Weber and Lange (1961) studied autopsy records of 265 cases of postvaccinial CNS disease and demonstrated that the pathology in children less than 2 years of age was usually that of an encephalopathy, with brain edema and hyperemia. In older children and adults encephalomyelitis was found, with demyelinization similar to that seen in other cases of postinfection encephalitis, such as those seen after measles (see Chapter 4). This differentiation of encephalopathy and encephalitis was consistent with earlier observations made by de Vries (1960). Encephalopathy started with convulsions, commonly accompanied by hemiplegia and aphasia. Recovery was incomplete, the patient being left with cerebral impairment and hemiplegia. On the other hand, encephalomyelitis was marked by vomiting, malaise, disorientation, delirum, convulsions, and coma. In nonfatal cases, recovery was usually complete and often rapid. Wilson (1967) suggested that the overall case–fatality rate was about 30%, but encephalopathy appeared to cause more deaths than encephalomyelitis.

There was no effective treatment and no effective preventive measures in either type of CNS disease.

Other Complications of Vaccination. Fetal vaccinia occurred as a very rare complication of pregnancy, and multiple sclerosis and malignant skin tumors were sometimes said to be related to smallpox vaccination, but the latter associations were probably purely coincidental. None of these complications was a public health problem.

Frequency of Complications.   A great deal has been published regarding the occurrence of progressive vaccinia, eczema vaccinatum, generalized vaccinia, and accidental infections with vaccinia virus, but it is difficult ot draw definite conclusions on their frequency since the diagnostic criteria at different times and in different countries varied greatly. Wilson (1967) reviewed 10 different reports from 1931 to 1963 in the United Kingdom, West Germany, and the United States (1964) and found that the frequency of eczema vaccinatum per million primary vaccinations varied from 2.9 in Bavaria (1945–1953) to 185 in England and Wales in 1962, when mass vaccination was carried out because of outbreaks of smallpox. This great variation reflects the care with which eczematous subjects were excluded from vaccination; in other surveys in the United Kingdom the frequency was 3.2 per million.

Two surveys conducted in the United States in 1968 provide the most comprehensive surveys of vaccination complications of all kinds in a large country over a defined period. One was a national survey of smallpox vaccination complications (Lane et al., 1969) and the other a 10-state survey conducted in the same year (Lane et al., 1970) to assess

TABLE 12-4

Frequency of Complications of Smallpox Vaccination in the United States in 1968, as Determined by a National Survey[a]

Complication	Reported cases (deaths) per million vaccinations		
	Primary vaccination	Revaccination	Unvaccinated contacts
	5,594,000[b]	8,547,000[b]	. . .[c]
Encephalopathy (age < 1 year)	6.5 (4.9)	—	—
Encephalitis (age > 1 year)	2.2 (0.2)	—	—
Progressive vaccinia	0.9 (0.4)	0.7 (0.2)	—
Eczema vaccinatum	10.4	0.9	60 cases (1 death)
Generalized vaccinia	23.4	1.2	2 cases
Accidental infection	25.4	0.8	44 cases
Other complications	11.8	1.0	8 cases
Total:	74.7	4.7	114 cases

[a] Based on Lane et al. (1969).
[b] Numbers in group.
[c] . . . , Not known.

the validity of the national survey. The national survey utilized essentially passive methods for collecting data, finding complications through miscellaneous routes such as the distribution of vaccinia-immune globulin and metisazone, death certificates, and reported cases of encephalitis, etc. For the other survey physicians were approached directly. Although the sample numbers were smaller, frequencies of all complications were greater in the 10-state survey, especially for the less severe complications. The results of these surveys are summarized in Table 12-4. All complications were much less frequent after revaccination than after primary vaccination.

Looking at the total incidence for all age groups may not be relevant, since each recombinant vaccinia virus vaccine, if used, is likely to be targeted at a certain age group in the population, which may be different from the target populations of the smallpox vaccination program in the United States in the 1960s.

Despite many studies, it is difficult to assess the true incidence of postvaccinial CNS disease, partly because of confusion with coincidental "background" encephalitis due to other causes. Utilization of a vaccinia vector vaccine would raise the problem again, if vaccine strains similar to those used previously were employed. Data on the frequency of this disease are summarized in Table 12-5.

There is some evidence that the occurrence of CNS disease is related to the strain of vaccinia virus. For example, after the Bern strain, once used in Austria, Switzerland, and West Germany, was replaced by the Lister strain in 1971, the occurrence of complications of the central nervous system declined. Surveys numbered 1 to 5 showed a high incidence of this complication, and the vaccinia strain used in those countries at that time played the major role. If the Lister or the New York City Board of Health strains are utilized for vaccinia vector vaccine, the magnitude of CNS complications might be about that found in surveys number 6 or 7, assuming that the 10-state survey in the United States provides a better index of the frequency of complications than did the national survey.

Just as the frequency of CNS complications varied greatly in different surveys, so did the case–fatality rate. Wilson (1967) cites an average figure of 30%; for the United States surveys there was 1 death (an infant) in 8 cases in the 10-state survey and 4 deaths (3 infants) in the 16 cases recorded in the national survey.

Conclusions. Our past knowledge and experience relating to complications of smallpox vaccination have been summarized to aid in evaluating whether or not the frequency of such complications would be acceptable should vaccinia virus be used as a vector for other vaccines.

TABLE 12-5

Summary of Postvaccinial Complications of the Central Nervous System

Survey number	Country	Investigator	Year investigated	Encephalopathy (age < 2 years)	Encephalomyelitis (age > 2 years)	Vaccinia virus strain
					Cases per million primary vaccinations	
1	Austria	Berger and Puntigam (1954)	1948–1953	103	1211	(?) Bern
2	Hamburg, West Germany	Seelemann (1960)	1939–1958	93	449	(?) Bern
3	Bavaria, West Germany	Herrlich (1954)	1945–1953	56	121	?
4	Holland	Stuart (1947)	1930–1943	50	348	?
5	Holland	van den Berg (1946)	1924–1928	39	232	?
6	United Kingdom	Conybeare (1964)	1951–1960	15[a]	15	Lister
7	United States (10 states)	Lane et al. (1970)	1968	42[a]	9	New York City Board of Health
8	United States (national)	Lane et al. (1969)	1968	7[a]	2	New York City Board of Health

[a] Age less than 1 year.

The complications involving the CNS are the most important since they are severe, unpredictable and therefore cannot be avoided by withholding vaccination from high-risk subjects, and there is no effective treatment. With the least pathogenic strains of vaccinia virus widely used in the smallpox eradication campaign, such as the Lister or the New York City Board of Health strains, between 10 and 40 children under 2 years of age might be expected to suffer in every 1 million primary vaccinees of whom between 3 and 12 might die.

Progressive vaccinia and eczema vaccinatum might be expected at a rate of 10 per million primary vaccinations of children under 2 years old, but the rate would clearly be dependent on the specified contraindications to vaccination and the care with which these were excluded, or in cases of eczema, given vaccinia-immune globulin.

Accidental vaccinia infection among contacts might cause some unpleasant social problems but it is hard to estimate its frequency. All other complications including generalized vaccinia, fever, malaise, lymphadenitis, etc., are grouped together as they are mild and may not be necessary to include in an evaluation of risks and benefits. Perhaps 1000 such cases might occur among each million primary vaccinations of children under 2 years of age. The figures for complications presented in the tables for subjects over 2 years of age are of the same order of magnitude as in infants.

The frequency and severity of complications in the United Kingdom and the United States, which occurred in well-nourished populations with good medical services, may not be directly applicable to estimates of morbidity and deaths in populations in developing countries.

These estimated hazards must be weighed against the public health importance of the target disease, taking into consideration the advantages of recombinant vaccinia virus vaccines, namely, ease of production, ease of administration, low cost, and heat stability. If any further attentuation of virulence of vaccinia virus can be made, without loss of immunogenicity, so much the better, but the field evaluation of the frequency of complications would be difficult, since the incidence of such important complications as CNS disease may well be less that 1 in 100,000 primary vaccinations, which would be difficult to differentiate from the background level.

Development of Vaccinia Virus Vector Vaccines

Further developmental work needs to be focused on four problems: reduction of complications, expression of antigen(s), immune responses important for protection, and potential for revaccination.

Reduction of Complications.   In countries with well-developed health services the incidence of serious skin complications can be reduced by proper screening, so that persons with congenital or acquired immuno-deficiencies are excluded and persons with eczema vaccinated under cover of immune γ-globulin. Although the pathogenesis of postvaccinial encephalitis is not understood, the incidence of this complication appears to be related to the virulence of the orthopoxvirus involved (see Table 5-3). It is therefore clear that the strain of vaccinia virus to be used for new vaccines must be selected with two features in mind; immuno-genicity and safety. The need for effective immunogenicity excludes potential "safe" vaccines like the attenuated CVI strain (Galasso *et al.*, 1977). As a first approach, it is logical to use one of the strains with a good record in the Intensified Smallpox Eradication Program, i.e., either the Lister strain or the New York City Board of Health (Wyeth) strain. However, increased knowledge of the molecular biology of vaccinia virus suggests that even better (i.e., safer but immunogenic) strains of virus could be developed. Initial tests with recombinant vaccines with foreign gene insertions in the TK gene have given promising results (Buller *et al.*, 1985). Tests in mice revealed that compared with wild-type WR or Wyeth strains, TK⁻ vectored vaccines were immunogenic but less virulent (Table 12-6). There are other possibilities for reducing the potential of vaccinia virus to spread through the body without affecting its capacity to replicate to high titer in the skin. One obvious approach is to reduce the potential for the production of enveloped forms of vaccinia, which spread through the body more effectively than nonen-veloped forms (Payne, 1980).

TABLE 12-6

Virulence in Mice of Vaccinia Virus TK⁻ Recombinants Compared with Wild-Type Vaccinia Virus[a]

Virus	Intracerebral inoculation (log LD_{50})	Spleen titer after intraperitoneal inoculation (day)		
		1	2	6
WR	1.0	9.5	9.4	7
WR-Infl-HA	6.6	5.5	6.1	< 3
WR-Herpes-gp	5.4	5.0	4.0	< 3
Wyeth	6.5	6.3	4.2	< 3
Wyeth-HBsAg	7.9	5.0	3.0	< 3

[a] Based on Buller *et al.*, 1985. Titers expressed as reciprocal $\log_{10}$ titers.

Expression of Antigens. Much of the effort of molecular biologists, described above, has been devoted to efforts to increase the expression of inserted genes by the use of appropriately chosen promoters. In addition, steps may need to be taken to genetically engineer inserted genes so that the relevant antigens are expressed in the most suitable way to evoke a protective immune response, e.g., it may be desirable that the gene product should be excreted from the cell in a soluble form, or that it should be incorporated in cell membranes, or available in whatever form is optimum for promoting cell-mediated immunity.

Immune Responses Important for Protection. For many diseases it is still not clear what component of the immune response and what antigen or antigens of a virus are most important for protection. For diseases in which there is a viremic phase with virions circulating free in plasma, e.g., yellow fever, measles, and poliomyelitis, it is clear that circulating antibody to the coat or envelope proteins is of prime importance in preventing generalization, although mucosal immunity may play a role in preventing the establishment of infection (e.g., in the gut in poliomyelitis). In other infections, notably those of the respiratory and alimentary tracts, mucosal immunity may be of primary importance. Cell-mediated immunity plays a major role in recovery in many infections (Mims and White, 1984); its precise role in preventing infection is less clear. However, there are several situations in which the aim of immunization may be to allow a harmless, abortive infection to occur rather than to completely prevent infection, and cell-mediated immunity may play an important role in such situations. Finally, it is clear that more work is needed on defining exactly which antigens are important in promoting protective immunity. Investigations into the pathogenesis and immunology of infection need to be carried out with every agent for which vaccine development is planned—first in experimental animals and then in human subjects. Even if recombinant vaccinia virus vaccines never come into widespread use, the use of recombinant vaccinia virus offers great potential for defining the role of different antigens in eliciting protective immunity.

Potential for Revaccination. It is likely that as knowledge develops more and more antigens may be incorporated into a single vaccinia virus vector. The question therefore arises as to when a vaccinia-vectored vaccine should start to be used, if such a course would preclude the later use of more effective vaccines.

In rabbits, Perkus *et al.* (1985) have shown that revaccination of rabbits with either the same or a new foreign antigen a year after their first vaccination will provoke appropriate immune responses. One rabbit that had been vaccinated with a vaccinia virus–hepatitis B virus surface antigen recombinant responded to revaccination with the same recombinant with increases in the titer of anti-HBs-Ag and anti-vaccinia virus antibodies. Another that had initially been vaccinated with a vaccinia virus HBs-Ag recombinant was revaccinated with a vaccine virus-infl-HA recombinant; it produced a high titer of anti-infl-HA antibody.

Successful vaccination of humans with vaccinia virus does not prevent effective revaccination (i.e., with replication of vaccinia virus) a few years later. For persons exposed to smallpox, such as hospital employees, it was recommended that revaccination should be carried out at intervals of 3 years. If a potent vaccine was used, revaccination at such intervals produced an accelerated reaction, accompanied by a rise in antibody level, i.e., viral replication had occurred. Indeed Henderson and Arita (1985) note that "most individuals to whom a high-titered vaccinia virus was administered experienced a so-called 'major reaction,' i.e., erythema and induration site at the vaccination site at the sixth to eighth day, even when as little as 6 to 12 months had elapsed since their previous vaccination." More effective responses occurred if revaccination was carried out on the ventral surface of the forearm, following primary vaccination over the deltoid muscle. Revaccination against smallpox was designed to maintain a high level of immunity to this disease; in the event of the general use of a range of vaccinia virus vectors it will be important to measure the response of vaccinated persons to revaccination with virus containing another insert at various intervals after the previous vaccination.

If an effective recombinant vaccinia virus vaccine is developed against a widespread and important pathogen like hepatitis B virus, its immediate widespread use would be fully justified in countries of Asia and Africa, irrespective of the possible effect on the later use of another preparation that might contain, for instance, an effective combination of malaria parasite genes. Given the time taken to achieve such developments, most of the population initially vaccinated wuld be susceptible to revaccination with the new vaccine.

POTENTIAL FOR VETERINARY VACCINES

Infectious diseases are the most important health-related constraint to profitable livestock production, and immunization is the most generally applicable way of preventing many of these diseases. There are important differences between immunization practices in humans and animals. Except in developing countries, economic constraints are of little importance in human medicine, but very important in all areas of veterinary practice except, perhaps, for companion animals. Also, far more latitude is allowed in the manufacture and use of veterinary vaccines than has ever been allowed by regulatory agencies for human vaccines. Finally, even a very small number of vaccine-associated illnesses or deaths constitute a major objection to the use of a vaccine in humans whereas in animals mild illness (and even, in the past, occasional deaths) can be tolerated if the vaccine is effective and the potential cost of failure to control the disease is sufficiently high.

The usage of veterinary vaccines in countries like the United States is very large; e.g., in 1982 the total number of all vaccine formulations was over 21 billion doses. However, this very large volume supports a biologics industry with annual sales of only $300 million (Murphy, 1985). Virus-vector vaccines have a chance of entry into this market only if they are effective and cheap. In contrast to human vaccination, for which vaccinia virus appears to be the most attractive of the poxvirus-vectored vaccines, orf virus (a *Parapoxvirus*) and fowlpox virus (an *Avipoxvirus*) have potential in the sheep and poultry industries, respectively.

In spite of these negative features, vaccinia virus-vectored vaccines would be attractive if polyvalent vaccines tailored for covering the range of diseases found in particular countries or regions were available. Effective vectored vaccines might also be used in two areas that involve public responsibility for animal disease prevention and control; for stockpiling as part of the defense against exotic livestock diseases like foot-and-mouth disease, rinderpest, African swine fever, Rift Valley fever, etc., and for the control of rabies in dogs in the developing countries and in wild life in the industrialized countries.

Although there is more latitude in terms of rare complications than with human vaccines, decisions on the use of vaccinia-vectored vaccines in livestock and companion animals must take into account the great diversity in the animals in which they are to be used—diversity in species and breed, immunological maturity at birth, placental or colostral transfer of antibodies, etc. Ideally, investigations are required for each disease–host species combination, preferable with a range of potential vectors, including viruses belonging to other poxvirus genera, herpesviruses, adenoviruses, etc. For vaccinia-vectored vaccines, especially in developing countries, the advantages already listed for such vaccines for human use also apply: ease of administration, ease of manufacture, and stability. Other important considerations for veterinary vaccines include a duration of immunity—lifetime immunity is not important for livestock because such animals do not live very long. On the other hand, even year-long immunity from a vectored foot-and-mouth vaccine would be attractive, since some current vaccine protocols call for three vaccinations a year.

Priority in research for new veterinary vaccines will vary in each country, depending upon the relative importance of its livestock industries and the prevalent diseases. For developing countries, a 1982 workshop on "Priorities in Biotechnology Research for International Development," sponsored by the United States National Academy of Sciences, United States National Research Council, United States for International Development, the World Bank, and the Rockefeller Foundation, identified the following animal diseases as having highest priority internationally: neonatal diarrheas of all species (including rotavirus infections), bovine respiratory disease complex, foot-and-mouth disease in cattle, African swine fever, hemotropic protozoa (African trypanosomiasis, theileriosis, babesiosis, and anaplasmosis), animal tuberculosis, rabies, Rift Valley fever of ruminants, Newcastle disease of fowl, rinderpest of cattle and bluetongue of sheep, hog cholera, African horse sickness, equine encephalitis, pulmonary adenomatosis and other retrovirus infections of sheep, pseudorabies of swine, and vesicular stomatitis of cattle and horses. The workshop was pessimistic about the possibilities of transfer of recombinant DNA technology to developing countries. However, vaccinia-vectored vaccines developed in industrialized countries could be readily transferred in final form for production in developing countries by methods similar to those used during the smallpox eradication campaign.

Bibliography

Ada, G. L., and Jones, P. D. (1986). The immune response to influenza infection. *Curr. Top. Microbiol. Immunol.* **128**, 1.

Ada, G. L., Jackson, D. C., Blanden, R. V., Tha Hla, R., and Bowern, N. W. (1976). Changes in the surface of virus-infected cells recognized by T cells. I. Minimal requirements for lysis of ectromelia-infected P-815 cells. *Scand. J. Haematol.* **5**, 23.

Ada, G. L., Leung, K.-N., and Ertl, H. (1981). An analysis of effector T cell generation in mice exposed to influenza A or Sendai viruses. *Immunol. Rev.* **58**, 5.

Ada, G. L. Brown, F., and Schild, G.C. (rapporteurs). (1985). Recombinant vaccinia viruses as live virus vectors for vaccine antigens: Memorandum for a WHO/USPHS/NIBSC meeting. *Bull. W.H.O.* **63**, 471.

Aldershoff, H., and Broers, C. M. (1906). Contribution à l'étude des corps intra-epitheliaux de Guarnieri. *Ann. Inst. Pasteur, Paris* **20**, 779.

Alekseeva, A. K., and Akopova, I. I. (1966). [A study of the properties of the virus causing rabbit smallpox isolated from a culture of kidney tissue from *Macaca rhesus* monkeys.] *Vopr. Virusol.* **11**, 532 (in Russian).

Alexander, A. D., Flyger, V., Herman, Y. F., McDonnell, S. J., Rothstein, N., and Yager, R. H. (1972). Survey of wild mammals in a Chesapeake Bay area for selected zoonoses. *J. Wildl. Dis.* **8**, 119.

Al Falluji, M. M., Tantawi, H. H., and Shony, O. (1979). Isolation, identification and characterization of camelpox virus in Iraq. *J. Hyg.* **83**, 267.

Allen, A. M., Clarke, G. L., Ganaway, J. R., Lock, A., and Werner, R. M. (1981). Pathology and disgonosis of mousepox. *Lab. Anim. Sci.* **31**, 599.

Allsop, P. J. (1958). A study of cell tropisms of cowpox virus and its variant in tissue culture. M.Sc. Thesis, University of Liverpool.

Allsop, P. J., Downie, A. W., and Dumbell, K. R. (1958). Studies on a variant of cowpox virus. *Abstr., Int. Cong. Microbiol., 7th, 1958,* p. 263.

Al-Razi, Abu Bakr Muhammad ibn Zakarija (Rhazes) (1766). "De Variolis et Morbillis Commentarius." G. Bowyer, London [Engl. Transl.: *Med. Classics* **4**, 22 (1939)].

Amano, H., and Tagaya, I. (1981). Isolation of cowpoxvirus clones deficient in the production of type A inclusions: Relationship to the production of diffusible LS antigen. *J. Gen. Virol.* **54**, 203.

Amano, H., Ueda, Y., and Tagaya, I. (1979). Orthopoxvirus strains deficient in surface antigen induction. *J. Gen. Virol.* **44**, 265.

Anderson, A. (1861). "Study of Fever." Churchill, London.

Anderson, R., and Dales, S. (1978). Biogenesis of poxviruses: Glycolipid metabolism in vaccinia-infected cells. *Virology* **84**, 108.

Anderson, R. M., and May, R. M. (1979). Population biology of infectious diseases. Part 1. *Nature (London)* **280**, 361.

Andrew, M. E., Coupar, B. E. H., Ada, G. L., and Boyle, D. B. (1986). Cell-mediated immune responses to influenza virus antigens expressed by vaccinia virus recombinants. *Microbiol. Pathog.* **1**, 443.

Andrew, M. E., Coupar, B. E. H., Boyle, D. B., and Ada, G. L. (1987a). The roles of influenza virus haemagglutinin and nucleoprotein in protection: Analysis using vaccinia virus recombinants. *Scand. J. Immunol.* **25**, 21.

Andrew, M. E., Boyle, D. B., Coupar, B. E. H., Whitfield, P. L., Both, G. W., and Bellamy, A. R. (1987b). Vaccinia virus recombinants expressing the SA11 rotavirus VP7 glycoprotein gene induce serotype-specific neutralizing antibodies. *J. Virol.* **61**, 1054.

Andrewes, C. H. (1951). Viruses and Linnaeus. *Acta Pathol. Microbiol. Scand.* **28**, 211.

Andrewes, C. H. (1952). Classification and nomenclature of viruses. *Annu. Rev. Microbiol.* **6**, 119.

Andrewes, C. H. (1953). The Rio Congress decisions with regard to the study of selected groups of viruses. *Ann. N. Y. Acad. Sci.* **56**, 428.

Andrewes, C. H., and Elford, W. J. (1947). Infectious ectromelia: Experiments on interference and immunization. *Br. J. Exp. Pathol.* **28**, 278.

Anslow, R. A., Ewald, B. H., Pakes, S. P., Small, J. D., and Whitney, R. A. (1975). Institutional outbreak of ectromelia. *Lab. Anim. Sci.* **25**, 532.

Appleyard, G., and Westwood, J. C. N. (1964a). The growth of rabbitpox virus in tissue culture. *J. Gen. Microbiol.* **37**, 391.

Appleyard, G., and Westwood, J. C. N. (1964b). A protective antigen from the poxviruses. II. Immunization of animals. *Br. J. Exp. Pathol.* **45**, 162.

Appleyard, G., Hapel, A. J., and Boulter, E. A. (1971). An antigenic difference between intracellular and extracellular rabbitpox virus. *J. Gen. Virol.* **13**, 9.

Aragão, H. de B. (1927). Myxoma dos coelhos. *Mem. Inst. Oswaldo Cruz* **20**, 225.

Archard, L. C. (1979). *De novo* synthesis of two classes of DNA induced by vaccinia virus infection of HeLa Cells. *J. Gen Virol.* **42**, 223.

Archard, L. C. (1983). Synthesis of full-length, virus genomic DNA by nuclei of vaccinia-infected HeLa cells. *J. Gen. Virol.* **64**, 2561.

Archard, L. C., and Mackett, M. (1979). Restriction endonuclease analysis of red cowpox virus and its white pock variant. *J. Gen. Virol.* **45**, 51.

Archard, L. C., Mackett, M., Barnes, D. E., and Dumbell, K. R. (1984). The genome structure of cowpox virus white pock variants. *J. Gen. Virol.* **65**, 875.

Archard, L. C., Johnson, K., and Malcolm, A. D. B. (1985). Specific transcription of orthopoxvirus DNA by HeLa cell RNA polymerase II. *FEBS Lett.* **192**, 53.

Arita, I. (1979). Virological evidence for the success of the smallpox eradication campaign. *Nature (London)* **279**, 293.

Arita, I., and Henderson, D. A. (1968). Smallpox and monkeypox in non-human primates. *Bull. W. H. O.* **39**, 277.

Arita, I., Shafa, E., and Kader, M. A. (1970). Role of hospital in smallpox outbreak in Kuwait. *Am. J. Public Health* **60**, 1960.

Arita, I., Gispen, R., Kalter, S. S., Wah, L. T., Marennikova, S. S., Netter, R., and Tagaya, I. (1972). Outbreaks of monkeypox and serological surveys in non-human primates. *Bull. W. H. O.* **46**, 625.

Arita, M., and Tagaya, I. (1980). Virion polypeptides of poxviruses. *Arch. Virol.* **63**, 209

Arzoglou, P., Drillien, R., and Kirn, A. (1978). Evidence for an alkaline protease in vaccinia virus. *Virology* **95**, 211

Bajszar, G., Wittek, R., Weir, J. P., and Moss, B. (1983). Vaccinia virus thymidine kinase and neighboring genes: mRNAs and polypeptides of wild-type virus and putative nonsense mutants. *J. Virol.* **45**, 62.

Baltazard, M., Boué, A., and Siadat, H. (1958). Étude du comportement du virus de la variole en cultures du tissus. *Ann. Inst. Pasteur, Paris* **94**, 560.

Baltimore, D. (1971). Expression of animal virus genomes. *Bacteriol. Rev.* **35**, 235.

Barbosa, E., and Moss, B. (1978a). mRNA (nucleoside-2'-)-methyltransferase from vaccinia virus. Purification and physical properties. *J. Biol. Chem.* **253**, 7692.

Barbosa, E., and Moss, B. (1978b). mRNA (nucleoside-2'-)-methyltransferase from vaccinia virus. Characteristics and substrate specificity. *J. Biol. Chem.* **253**, 7698.

Barker, L. F. (1969). Further attenuated vaccinia virus: A possible alternative for primary immunization. *6 Annu. Immun. Conf., 1969*, p. 55.

Barnard, J. E., and Elford, W. J. (1931). The causative organism in infectious ectromelia. *Proc. R. Soc. London, Ser. B* **109**, 360.

Baroudy, B. M., and Moss, B. (1980). Purification and characterization of a DNA-dependent RNA polymerase from vaccinia virions. *J. Biol. Chem.* **255**, 4372.

Baroudy, B. M., and Moss, B. (1982). Sequence homologies of diverse length tandem repetitions near ends of vaccinia virus genome suggest unequal crossing over. *Nucleic Acids Res.* **10**, 5673.

Baroudy, B. M., Venkatesan, S., and Moss B. (1982). Incompletely base-paired flip-flop terminal loops link the two DNA strands of the vaccinia virus genome into one uninterrupted polynucleotide chain. *Cell (Cambridge, Mass.)* **28**, 315.

Baroudy, B. M., Venkatesan, S., and Moss, B. (1983). Structure and replication of vaccinia virus telomeres. *Cold Spring Harbor Symp. Quant. Biol.* **47**, 723.

Barry, F. W. (1889). "Report on an Epidemic of Smallpox in Sheffield, 1887–8." H. M. Stationery Office, London.

Basse, A., Freundt, E. A., and Hansen, J. F. (1964). *Int. Symp. Dis. Zoo Anim., 6th*, p. 55.

Bateman, A. J. (1975). Simplification of palindromic telomere theory. *Nature (London)* **253**, 379.

Bauer, D. J., and Sadler, P. W. (1960). New antiviral chemotherapeutic agent active against smallpox infection. *Lancet* **1**, 1110.

Bauer, D. J., Dumbell, K. R., Fox-Hulme, P., and Sadler, P. W. (1962). The chemotherapy of variola major infection. *Bull. W. H. O.* **26**, 727.

Bauer, W. R., Ressner, E. C., Kates, J., and Patzke, J. U. (1977). A DNA nicking–closing enzyme encapsidated in vaccinia virus: Partial purification and properties. *Proc. Natl. Acad. Sci. U.S.A.* **74**, 1841.

Bawden, F. C., Pirie, N. W., Bernal, J. D., and Fankuchen, I. (1936). Liquid crystalline substances from virus-infected plants. *Nature (London)* **138**, 1051.

Baxby, D. (1972). Smallpox-like viruses from camels in Iran. *Lancet* **2**, 1063.

Baxby, D. (1974). Differentiation of smallpox and camelpox viruses in cultures of human and monkey cells. *J. Hyg.* **72**, 251.

Baxby, D. (1975). Laboratory characteristics of British and Dutch strains of cowpox virus. *Zentralbl. Veterinaermed., Reihe B* **22**, 480.

Baxby, D. (1977a). Is cowpox misnamed? A review of 10 human cases. *Br. Med. J.* **1**, 1379.

Baxby, D. (1977b). The origins of vaccinia virus. *J. Infect. Dis.* **136**, 453.

Baxby, D. (1981). "Jenner's Smallpox Vaccine. The Riddle of the Origin of Vaccinia Virus." Heinemann, London.

Baxby, D. (1982a). The natural history of cowpox. *Bristol Med.-Chir. J.*, p. 12.

Baxby, D. (1982b). The surface antigens of orthopoxviruses detected by cross-neutralization tests on cross-absorbed sera. *J. Gen. Virol.* **58**, 251.

Baxby, D. (1985). The genesis of Edward Jenner's *Inquiry* of 1798: A comparison of the two unpublished manuscripts and the published version. *Med. Hist.* **29**, 193.

Baxby, D., and Ghaboosi, B. (1977). Laboratory characteristics of poxviruses isolated from captive elephants in Germany. *J. Gen. Virol.* **37**, 407

Baxby, D., and Hill, B. J. (1971). Characteristics of a new poxvirus isolated from Indian buffaloes. *Arch. Gesamte Virusforsch.* **35**, 70.

Baxby, D., and Osborne, A. D. (1979). Antibody studies in natural bovine cowpox. *J. Hyg.* **83**, 425.

Baxby, D., and Rondle, C. J. M. (1967). The relative sensitivity of chick and rabbit tissues for the titration of vaccinia and cowpox viruses. *Arch. Gesamte Virusforsch.* **20**, 263.

Baxby, D., Ramyar, H., Hessami, M., and Ghaboosi, B. (1975). Response of camels to intradermal inoculation with smallpox and camelpox viruses. *Infect. Immun.* **11**, 617.

Baxby, D., Ashton, D. G., Jones, D., Thomsett, L. R., and Denham, E. M. (1979a). Cowpox virus infection in unusual hosts. *Vet. Rec.* **109**, 175.

Baxby, D., Shackleton, W. B., Wheeler, J., and Turner, A. (1979b). Comparison of cowpox-like viruses isolated from European zoos. *Arch. Virol.* **61**, 337.

Baxby, D., Ashton, D. G., Jones, D. M., and Thomsett, L. R. (1982). An outbreak of cowpox in captive cheetahs: Virological and epidemiological studies. *J. Hyg.* **89**, 365.

Beaud, G., Kirn, A., and Gros, F. (1972). In vitro protein synthesis directed by RNA transcribed from vaccinia DNA. *Biochem. Biophys. Res. Commun.* **49**, 1459.

Becker, Y., and Sarov, I. (1968). Electron microscopy of vaccinia virus DNA. *J. Mol. Biol.* **34**, 655.

Béclère, A., Chambon, and Menard (1898). Études sur l'immunité vaccinale. II. L'immunité consecutive à l'inoculation sous-cutanée du vaccin. *Ann. Inst. Pasteur, Paris* **12**, 837.

Béclère, G., Chambon, and Menard (1899). Études sur l'immunité vaccinale. III. Le pouvoir antivirulent du serum de l'homme et des animaux immunisés contre l'infection vaccinale ou variolique. *Ann. Inst. Pasteur, Paris* **13**, 81.

Bedson, H. S. (1964). The ceiling temperature of pox viruses. M.D. Thesis, University of London.

Bedson, H. S. (1972). Camelpox and smallpox. *Lancet* **2**, 1253.

Bedson, H. S. (1982). Enzyme studies for the characterization of some orthopoxvirus isolates. *Bull. W. H. O.* **60**, 377.

Bedson, H. S., and Duckworth, M. J. (1963). Rabbit pox: An experimental study of the pathways of infection in rabbits. *J. Pathol. Bacteriol.* **85**, 1.

Bedson, H. S., and Dumbell, K. R. (1961). The effect of temperature on the growth of pox viruses in the chick embryo. *J. Hyg.* **59**, 457.

Bedson, H. S., and Dumbell, K. R. (1964a). Hybrids derived from the viruses of alastrim and rabbit pox. *J. Hyg.* **62**, 141.

Bedson, H. S., and Dumbell, K. R. (1964b). Hybrids derived from the viruses of variola major and cowpox. *J. Hyg.* **62**, 147.

Bedson, H. S., Dumbell, K. R., and Thomas, W. R. G. (1963). Variola in Tanganyika. *Lancet* **2**, 1085.

Beijerinck, M. W. (1899). Über ein Contagium vivum fluidum als Ursache der Fleckenkrankheit der Tabaksblatter. *Zentralbl. Bakteriol., Parasitenkd., Infektionskr. Hyg., Abt. 2, Naturwiss.: Allg., Landwirtsch. Tech. Mitrobioli.* **5**, 27.

Belle Isle, H., Venkatesan, S., and Moss, B. (1981). Cell-free translation of early and late mRNAs selected by hybridization to cloned DNA fragments derived from the left 14 million to 72 million daltons of the vaccinia virus genome. *Virology* **112**, 306.

Benison, S. (1967). "Tom Rivers. Reflections on a Life in Medicine and Science." MIT Press, Cambridge, Massachusetts.

Bennett, M., Gaskell, C. J., Gaskell, R. M., Baxby, D. and Gruffydd-Jones, T. J. (1986).

Poxvirus infection in the domestic cat: Some clinical and epidemiological observations. *Vet. Rec.* **118**, 387.

Bennink, J. R., Yewdell, J. W., Smith, G. L., and Moss, B. (1986). Recognition of cloned influenza virus hemagglutinin gene products by cytotoxic T lymphocytes. *J. Virol.* **57**, 786.

Bennink, J. R., Yewdell, J. W., Smith, G. L., and Moss, B. (1987). Anti-influenza virus cytotoxic lymphocytes recognize the three viral polymerases and a nonstructural protein: Responsiveness to individual viral antigens is major histocompatibility complex controlled. *J. Virol.* **61**, 1098.

Berg, T. O., and Stevens, D. A. (1971). "Catalogue of Viruses, Rickettsiae, Chlamydiae." 4th ed. American Type Culture Collection, Rockville, Maryland.

Berger, K., and Puntigam, F. (1954). Ueber die Erkrankungshaufigkeit verschiedener Altersklassen von Erstimpflingen an postvakzinaler Enzephalitis nach subcutaner Pockenschutzimpfung. *Wien. Med. Wochenschr.* **104**, 487.

Berns, K. I., and Silverman, C. (1970). Natural occurrence of cross-linked vaccinia virus deoxyribonucleic acid. *J. Virol.* **5**, 299.

Berry, G. P., and Dedrick, H. M. (1936). A method for changing the virus of rabbit fibroma (Shope) into that of infectious myxomatosis (Sanarelli). *J. Bacteriol.* **31**, 50.

Bertholet, C., Drillien, R., and Wittek, R. (1985). One hundred base pairs of 5' flanking sequence of a vaccinia virus late gene are sufficient to temporally regulate late transcription. *Proc. Natl. Acad. Sci. U.S.A.* **82**, 2096.

Bertholet, C., Stocco, P., Van Meir, E., and Wittek, R. (1986). Functional analysis of the 5' flanking sequence of a vaccinia virus late gene. *EMBO J.* **5**, 1951.

Bertholet, C., Van Meir, E., ten Heggeler-Bordier, B., and Wittek, R. (1987). Vaccinia virus produces late mRNAs by discontinuous synthesis. *Cell (Cambridge, Mass.)* **50**, 153.

Bhatt, P. N., and Jacoby, R. O. (1987a). Mousepox in inbred mice innately resistant or susceptible to lethal infection with ectromelia virus. I. Clinical responses. *Lab. Anim. Sci.* **37**, 11.

Bhatt, P. N., and Jacoby, R. O. (1987b). Mousepox in inbred mice innately resistant or susceptible to lethal infection with ectromelia virus. III. Experimental transmission of infection and derivation of virus-free progeny from previously infected dams. *Lab. Anim. Sci.* **37**, 23.

Bhatt, P. N., and Jacoby, R. O. (1987c). Stability of ectromelia virus strain NIH-79 under various laboratory conditions. *Lab. Anim. Sci.* **37**, 33.

Bhatt, P. N., and Jacoby, R. O. (1987d). Effect of vaccination on the clinical response, pathogenesis and transmission of mousepox. *Lab. Anim. Sci.* **37**, 610.

Bhatt, P. N., Oruns, W. G., Buckley, S. M., Casals, J., Shope, R. E., and Jonas, A. M. (1981). Mousepox epizootic in an experimental and a barrier mouse colony at Yale University. *Lab. Anim. Sci.* **31**, 560.

Bhatt, P. N., Jacoby, R. O., Paturxo, F. X., and Gras, L. (1985). Transmission of mousepox strain NIH-79 under various conditions. *Lab. Anim. Sci.* **35**, 523.

Birnstiel, M. L., Busslinger, M., and Strub, K. (1985). Transcription termination and 3' processing: The end is in sight. *Cell (Cambridge, Mass.)* **41**, 349.

Blackman, K. E., and Bubel, H. C. (1972). Origin of the vaccinia virus hemagglutinin. *J. Virol.* **9**, 290.

Blanden, R. V. (1970). Mechanisms of recovery from a generalized viral infection: Mousepox. I. The effects of antithymocyte serum. *J. Exp. Med.* **132**, 1035.

Blanden, R. V. (1971a). Mechanisms of recovery from a generalized viral infection: Mousepox. II. Passive transfer of recovery mechanisms with immune lymphoid cells. *J. Exp. Med.* **133**, 1074.

Blanden, R. V. (1971b). Mechanisms of recovery from a generalized viral infection: Mousepox. III. Regression of infectious foci. *J. Exp. Med.* **133**, 1090.

Blanden, R. V. (1974). T cell response to viral and bacterial infection. *Transplant. Rev.* **19**, 56.

Blaxall, F. R. (1930). Cowpox. *In* "A System of Bacteriology in Relation to Medicine," Vol. 7, p. 150. H. M. Stationery Office, London.

Bleyer, J. G. (1922). Ueber Auftreten von Variola unter Affen der genera *Mycetes* und *Cebus* bei Vordringen einer Pockenepidemie im Urwaldgebiete an den Nebenflüssen des Alto Uruguay in Sübrasilien. *Muench. Med. Wochenschr.* **69**, 1009.

Blomquist, M. C., Hunt, L. T., and Barker, W. C. (1984). Vaccinia virus 19-kilodalton protein: Relationship to several mammalian proteins, including two growth factors. *Proc. Natl. Acad. Sci. U.S.A.* **81**, 7363.

Boone, R. F., and Moss, B. (1978). Sequence complexity and relative abundance of vaccinia virus mRNA's synthesized *in vivo* and *in vitro*. *J. Virol.* **26**, 554.

Boone, R. F., Parr, R. P., and Moss, B. (1979). Intermolecular duplexes formed from polyadenylated vaccinia virus RNA. *J. Virol.* **30**, 365.

Borries, B., Ruska, R., and Ruska, H. (1938). Bakterie und Virus in übermikroscopischer Aufnahme (mit einer Einfuhrung in die Technik des Übermikroskops). *Klin. Wochenschr.* **17**, 921.

Bossart, W., Nuss, D. L., and Paoletti, E. (1978). Effect of UV-irradiation on the expression of vaccinia virus gene products sythesized in a cell-free system coupling transcription and translation. *J. Virol.* **26**, 673.

Boulter, E. A. (1969). Protection against poxviruses. *Proc. R. Soc. Med.* **62**, 295.

Boulter, E. A., and Appleyard, G. (1973). Differences between extracelluelar and intracellular forms of poxvirus and their implications. *Prog. Med. Virol.* **16**, 86.

Boulter, E. A., Maber, H. B., and Bowen, E. T. W. (1961a). Studies on the physiological disturbances occurring in experimental rabbit pox: An approach to rational therapy. *Br. J. Exp. Pathol.* **42**, 433.

Boulter, E. A., Westwood, J. C. N., and Maber, H. B. (1961b). Value of serotherapy in a virus disease (rabbitpox). *Lancet* **2**, 1012.

Boulter, E. A., Zwartouw, H. T., Titmuss, D. H. I., and Maber, H. B. (1971). The nature of the immune state produced by inactivated vaccinia virus in rabbits. *Am. J. Epidemiol.* **94**, 612.

Bourke, A. T. C., and Dumbell, K. R. (1972). An unusual poxvirus from Nigeria. *Bull. W.H.O.* **46**, 621.

Boyle, D. B., Coupar, B. E. H., and Both, G. W. (1985). Multiple-cloning-site plasmids for the rapid construction of recombinant poxviruses. *Gene* **35**, 169.

Boyle, D. B., Coupar, B. E. H., Parsonson, I. M., Bagust, T. J., and Both, G. W. (1986). Responses of cattle, sheep and poultry to a recombinant vaccinia virus expressing a swine influenza hemagglutinin. *Res. Vet. Sci.* **41**, 40.

Boyle, D. B., Coupar, B. E. H., Gibbs, A. J., Seigman, L. J., and Both, G. W. (1987). Fowlpox virus thymidine kinase: Nucleotide sequence and relationships to other thymidine kinases. *Virology* **156**, 355.

Bradshaw, H. D. and Deininger, P. L. (1984). Human thymidine kinase gene; Molecular cloning and nucleotide sequence of a cDNA expressible in mammalian cells. *Mol. Cell. Biol.* **4**, 2316.

Bras, G. (1952a). The morbid anatomy of smallpox. *Doc. Med. Geogr. Trop.* **4**, 303.

Bras, G. (1952b). Observations on the formation of smallpox scars. *Arch. Pathol.* **54**, 149.

Breathnach, R., and Chambon, P. (1981). Organization and expression of eucaryotic split genes coding for proteins. *Annu. Rev. Biochem.* **50**, 349.

Breman, J. G., Bernadou, J., and Nakano, J. H. (1977). Poxvirus in West African nonhuman primates: Serological survey results. *Bull. W.H.O.* **55**, 165.

Breman, J. G., Kalisa-Ruti, Steniowski, M. V., Zanotto, E., Gromyko, A. I., and Arita, I. (1980). Human monkeypox, 1970–79. *Bull. W. H. O.* **58**, 165.

Brinckerhoff, W. R., and Tyzzer, E. E. (1906). Studies upon experimental variola and vaccinia in Quadrumana. *J. Med. Res.* **14**, 223.

Briody, B. A. (1955). Mouse pox (ectromelia) in the United States. *Lab. Anim. Sci.* **6**, 1.

Briody, B. A. (1959). Response of mice to ectromelia and vaccinia viruses. *Bacteriol. Rev.* **23**, 61.

Briody, B. A. (1966). The natural history of mousepox. *Natl. Cancer Inst. Monogr.* **20**, 105.

Briody, B. A., Hauschka, T. S., and Mirand, E. A. (1956). The role of genotype in resistance to an epizootic of mouse pox (ectromelia). *Am. J. Hyg.* **63**, 59.

Bronson, L. H., and Parker, R. F. (1941). Neutralization of vaccine virus by immune serum. Titration by means of intracerebral inoculation of mice. *J. Immunol.* **41**, 269.

Brown, A., Elsner, V., and Officer, J. E. (1960). Growth and passage of variola virus in mouse brain. *Proc. Soc. Exp. Biol. Med.* **104**, 605.

Brown, J. P., Twardzik, D. R., Marquardt, H., and Todaro, G. J. (1985). Vaccinia virus encodes a polypeptide homologous to epidermal growth factor and transforming growth factor. *Nature (London)* **313**, 491.

Brown, K., Briles, W. E., and Brown, E. R. (1973). Vaccinia virus hemagglutinin receptor in chickens determined by K isoantigen locus. *Proc. Soc. Exp. Biol. Med.* **142**, 16.

Broyles, S. S., and Moss, B. (1986). Homology between RNA polymerases of poxviruses, prokaryotes, and eukaryotes: Nucleotide sequence and transcriptional analysis of vaccinia virus genes encoding 147-KDa and 22-KDa subunits. *Proc. Natl. Acad. Sci. U.S.A.* **83**, 3141.

Broyles, S. S., and Moss, B. (1987). Sedimentation of an RNA polymerase complex from vaccinia virus that specifically initiates and terminates transcription. *Mol. Cell. Biol.* **7**, 7.

Buddingh, G. J. (1938). Infection of chorio-allantois of chick embryo as diagnostic test for variola. *Am. J. Hyg.* **28**, 130.

Bugbee, L. M., Like, A. A., and Stewart, R. B. (1960). The effects of cortisone on intradermally induced vaccinia infection in rabbits. *J. Infect. Dis.* **106**, 166.

Buist, J. B. (1886). The life-history of the micro-organisms associated with variola and vaccinia. *Proc. R. Soc. Edinburgh* **13**, 603.

Buller, R. M. L. (1986). Answer. *In* "Viral and Mycoplasmal Infections of Laboratory Rodents. Effects on Biomedical Research" (P. N. Bhatt, R. O. Jacoby, H. C. Morse, III, and A. E. New, eds.), p. 551. Academic Press, Orlando, Florida.

Buller, R. M. L., and Wallace, G. D. (1985). Reexamination of the efficacy of vaccination against mousepox. *Lab. Anim. Sci.* **35**, 473.

Buller, R. M. L., Bhatt, P. N., and Wallace, G. D. (1983). Evaluation of an enzyme-linked immunosorbent assay for the detection of ectromelia (mousepox) antibody. *J. Clin. Microbiol.* **18**, 1220.

Buller, R. M. L., Smith, G. L., Cremer, K., Notkins, A. L., and Moss, B. (1985). Decreased virulence of recombinant vaccinia virus expression vector is associated with a thymidine kinase-negative phenotype. *Nature (London)* **317**, 813.

Buller, R. M. L., Potter, M., and Wallace, G. D. (1986). Variable resistance to ectromelia (mousepox) virus among genera of *Mus. Curr. Top. Microbiol. Immunol.* **127**, 319.

Buller, R. M. L., Chakrabarti, S., Cooper, J. A., Twardzik, D. R., and Moss, B. (1987a). Deletion of the vaccinia virus growth factor gene reduces virulence. *J. Virol.* **62**, 866.

Buller, R. M. L., Chakrabarti, S., Moss, B., and Fredrickson, T. (1987b). Cell proliferative response to vaccinia virus is mediated by VGF *in vitro* and *in vivo*. *Virology* **164**, 182.

Buller, R. M. L., Weinblatt, A. C., Hamburger, A. W., and Wallace, G. D. (1987c). Observations on the replication of ectromelia virus in mouse-derived cell lines: Implications for the epidemiology of mousepox. *Lab. Anim Sci.* **37**, 28.

Buller, R. M. L., Holmes, K. L., Hugin, A., Fredrickson, R. N., and Morse H. C., III (1987d). Induction of cytotoxic T-cell responses *in vivo* in the absence of CD4 helper cells. *Nature (London)* **327**, 77.

Buller, R. M. L., Yetter, R. A., Frederickson, T. N., and Morse H. C., III (1987e). Abrogation of resistance to severe mousepox in C57BL/6 mice infected with LP-BM5 murine leukemia viruses. *J. Virol.* **61**, 383.

Burnet, F. M. (1936). The use of the developing egg in virus research. *Med. Res. Counc. G. B. Spec. Rep. Ser.* **SRS-220**.

Burnet, F. M. (1945). An unsuspected relationship between the viruses of vaccinia and infectious ectromelia of mice. *Nature (London)* **155**, 543.

Burnet, F. M. (1946). Vaccinia haemagglutinin. *Nature (London)* **158**, 199.

Burnet, F. M., and Boake, W. C. (1946). The relationship between the virus of infectious ectromelia of mice and vaccinia virus. *J. Immunol.* **53**, 1.

Burnet, F. M., and Lush, D. (1936a). Inapparent (subclinical) infection of the rat with the virus of infectious ectromelia of mice. *J. Pathol. Bacteriol.* **42**, 469.

Burnet, F. M., and Lush, D. (1936b). The propagation of the virus of infectious ectromelia of mice in the developing egg. *J. Pathol. Bacteriol.* **43**, 105.

Burnet, F. M., and Stone, J. D. (1946). The haemagglutinins of vaccinia and ectromelia viruses. *Aust. J. Exp. Biol. Med. Sci.* **24**, 1.

Burnet, F. M., Keogh, E. V., and Lush, D. (1937). The immunological reactions of filterable viruses. *Aust. J. Exp. Biol. Med. Sci.* **15**, 227.

Cairns, J. (1960). The initiation of vaccinia infection. *Virology* **11**, 603.

Calmette, A., and Guérin, C. (1901). Recherches sur la vaccine expérimentale. *Ann. Inst. Pasteur, Paris* **15**, 161.

Camus, L. (1917). Influence de la vasodilatation sur la localisation des pustules vaccinales spontanées. *Bull. Acad. Med., Paris* **77** (Ser. 3), 111.

Cantin, E. M., Eberle, E., Baldick, J. L., Moss, B., Willey, D. E., Notkins, A. L., and Openshaw, H. (1987). Expression of herpes simplex virus 1 glycoprotein B by a recombinant vaccinia virus and protection of mice against lethal herpes simplex virus 1 infection. *Proc. Natl. Acad. Sci. U.S.A.* **84**, 5908.

Carl, A. (1799). "Darstellung des dritten Jahrgangs der zu Brunn gestifteten Blatternimpfanstalt." Brno.

Carmichael, A. G., and Silverstein, A. M. (1987). Smallpox in Europe before the seventeenth century: Virulent killer or benign disease? *J. Hist. Med. Allied Sci.* **42**, 147.

Carra, L., and Dumbell, K. R. (1987). Characterization of poxviruses from sporadic human infections. *S. Afr. Med. J.* **72**, 846.

Carrasco, L., and Bravo, R. (1986). Specific proteins synthesized during the viral lytic cycle in vaccinia virus-infected HeLa cells: Analysis by high resolution, two-dimensional gel electrophoresis. *J. Virol.* **58**, 569.

Carthew, P., Hill, A. C., and Verstraete, A. P. (1977). Some observations on the diagnosis of an outbreak of ectromelia in 1976. *Vet. Rec.* **100**, 293.

Cassel, W. A. (1957). Multiplication of vaccinia virus in the Ehrlich ascites carcinoma. *Virology* **3**, 514.

Causey, O. R., and Kemp, G. E. (1968). Surveillance and study of viral infections of vertebrates in Nigeria. *Niger. J. Sci.* **2**, 131.

Cavalier-Smith, T. (1974). Palindromic base sequences and replication of eukaryote chromosome ends. *Nature (London)* **250**, 467.

Ceely, R. (1842). Further observations on the variolae vaccinae. *Trans. Prov. Med. Surg. Assoc.* **10**, 209.

Chakrabarti, S., Brechling, K., and Moss, B. (1985). Vaccinia virus expression vector: Co-expression of β-galactosidase gene provides visual screening of recombinant virus plaques. *Mol. Cell. Biol.* **5**, 3403.

Chakrabarti, S., Robert-Guroff, M., Wong-Staal, F., Gallo, R. C., and Moss B. (1986). Expression of HTLV-III enveloped gene by recombinant vaccinia virus. *Nature (London)* **320**, 535.

Challberg, M. D., and Englund, P. T. (1979). Purification and properties of the deoxyribonucleic acid polymerase induced by vaccinia virus. *J. Biol. Chem.* **254**, 7812.

Chandra, R. K. (1979). Nutritional deficiency and susceptibility to infection. *Bull. W.H.O.* **57**, 167.

Chang, A., and Metz, D. H. (1976). Further investigations on the mode of entry of vaccinia virus into cells. *J. Gen. Virol.* **32**, 275.

Chapin, C. V. (1913). Variation in type of infectious disease as shown by the history of smallpox in the United States 1895–1912. *J. Infect. Dis.* **13**, 171.

Chapin, C. V. (1926). Changes in type of contagious disease with special reference to smallpox and scarlet fever. *J. Prev. Med.* **1**, 1.

Chapin, C. V., and Smith, J. (1932). Permanency of the mild type of smallpox. *J. Prev. Med.* **6**, 273.

Chen, H. R., and Barker, W. C. (1985). Similarity of vaccinia 28K, *v-erb-B* and EGF receptors. *Nature (London)* **316**, 219.

Cheng, K.-C., and Moss, B. (1987). Selective synthesis and secretion of particles composed of the hepatitis B virus middle surface protein directed by a recombinant vaccinia virus: Induction of antibodies to epitopes to pre-S and S epitopes. *J. Virol.* **61**, 1286.

Cho, C. T., and Wenner, H. A. (1973). Monkeypox virus. *Bacteriol. Rev.* **37**, 1.

Christensen, L. R., Bond, E., and Matanic, B. (1967). "Pock-less" rabbit pox. *Lab. Anim. Care* **17**, 281.

Chu, C. M. (1948). Studies on vaccinia haemagglutinin. I. Some physicochemical properties. *J. Hyg.* **46**, 42.

Cochran, M. A., Mackett, M., and Moss, B. (1985a). Eukaryotic transient expression system dependent on transcription factor and regulatory DNA sequences of vaccinia virus. *Proc. Natl. Acad. Sci. U.S.A.* **82**, 19.

Cochran, M. A., Puckett, C., and Moss, B. (1985b). *In vitro* mutagenesis of the promoter region of a vaccinia virus gene: Evidence for tandem early and late regulatory signals. *J. Virol.* **54**, 30.

Cohen, G. H., and Wilcox, W. C. (1968). Soluble antigens of vaccinia-infected mammalian cells. III. Relation of "early" and "late" proteins to virus structure. *J. Virol.* **2**, 449.

Colby, C., and Duesberg, P. H. (1969). Double-stranded RNA in vaccinia virus infected cells. *Nature (London)* **222**, 940.

Colby, C., Jurale, C., and Kates, J. R. (1971). Mechanism of synthesis of vaccinia virus double-stranded ribonucleic acid *in vivo* and *in vitro*. *J. Virol.* **7**, 71.

Cole, G. A., and Blanden, R. V. (1982). Immunology of poxvirus infections. *Compr. Immunol.* **9** (Part 2), 1.

Collier, L. H. (1954). The preservation of vaccinia virus. *Bacteriol. Rev.* **18**, 74.

Collier, L. H. (1955). The development of a stable smallpox vaccine. *J. Hyg.* **53**, 76.

Collins, M. J., Jr, Peters, R. L., and Parker, J. C. (1981). Serological detection of ectromelia virus antibody. *Lab. Anim. Sci.* **31**, 595.

Conybeare, E. T. (1964). Illnesses attributed to smallpox vaccination, 1951–1960. Part II—Illnesses reported as affecting the central nervous system. *Monthly Bulletin of the Ministry of Health and the Public Health Laboratory Service* **23**, 150.

Cooper, J. A., and Moss, B. (1978). Transciption of vaccinia virus mRNA coupled to translation *in vitro. Virology* **88**, 149.

Cooper, J. A., and Moss, B. (1979). Translation of specific vaccinia virus RNAs purified as RNA–DNA hybrids on potassium iodide gradients. *Nucleic Acids Res.* **6**, 3599.

Cooper, J. A., Wittek, R., and Moss, B. (1981a). Hybridization selection and cell-free translation of mRNAs encoded within the inverted terminal repetition of the vaccinia virus genome. *J. Virol.* **37**, 284.

Cooper, J. A., Wittek, R., and Moss, B. (1981b). Extension of the transcriptional and translational map of the left end of the vaccinia virus genome to 21 kilobase pairs. *J. Virol.* **39**, 733.

Copeman, S. M. (1892). The bacteriology of vaccine lymph. *Trans. Int. Congr. Hyg. Demog., 7th, 1891,* Vol. 2, p. 319.

Copeman, S. M. (1894). Variola and vaccinia, their manifestations and inter-relations in the lower animals: A comparative study. *J. Pathol. Bacteriol.* **2**, 407.

Craigie, J. (1932). The nature of the vaccinia flocculation reaction and observations on the elementary bodies of vaccinia. *Br. J. Exp. Pathol.* **13**, 259.

Craigie, J., and Wishart, F. O. (1936a). Studies on the soluble precipitating substances of vaccinia. I. The dissociation *in vitro* of soluble precipitable substances from elementary bodies of vaccinia. *J. Exp. Med.* **64**, 803.

Craigie, J., and Wishart, F. O. (1936b). Studies on the soluble precipitating substances of vaccinia. II. The soluble precipitable substances of dermal vaccine. *J. Exp. Med.* **64**, 819.

Creighton, C. (1889). "Jenner and Vaccination." Sonnenschein, London.

Cremer, K., Mackett, M., Wohlenberg, C., Notkins, A. L., and Moss, B. (1985). Vaccinia virus recombinants expressing herpes simplex type 1 glycoprotein D prevents latent herpes in mice. *Science* **228**, 737.

Crookshank, E. M. (1889). "History and Pathology of Vaccination." Lewis, London.

Cumpston, J. H. L. (1914). "The History of Small-pox in Australia, 1788–1908," Publ. No. 3. Commonwealth of Australia, Quarantine Service, Government Printer, Melbourne.

Dahl, R., and Kates, J. (1970). Intracellular structure containing vaccinia DNA: Isolation and partial characterization. *Virology* **42**, 453.

Dales, S. (1963). The uptake and development of vaccinia virus in strain L cells followed with labeled viral deoxyribonucleic acid. *J. Cell Biol.* **18**, 51.

Dales, S. (1965). Penetration of animal viruses into cells. *Prog. Med Virol.* **7**, 1.

Dales, S., and Mosbach, E. H. (1968). Vaccinia as a model for membrane biogenesis. *Virology* **35**, 564.

Dales, S., and Pogo, B. T. (1981). The poxviruses. *Virol. Monogr.* **18**, 1.

Dales, S., Milovanovich, V., Pogo, B. G. T., Weintraub, S. B., Huima, T., Wilton, S., and McFadden, G. (1978). Biogenesis of vaccinia: Isolation of conditional lethal mutants and electron microscopic characterization of their phenotypically expressed defects. *Virology* **84**, 403.

Dalldorf, G., and Gifford, R. (1955). Recognition of mouse ectromelia. *Proc. Soc. Exp. Biol. Med.* **88**, 290.

Dalling, T., Robertson, A., Boddie, G. F., and Spruell, J. S. A. (1966). "International Encyclopaedia of Veterinary Medicine," Vol. 4, p. 2387. Sweet & Maxwell, London.

Daniell, E. (1976). Genome structure of imcomplete particles of adenovirus. *J. Virol.* **19**, 685.

Davies, F. G., Mungat, J. N., and Shaw, T. (1975). Characteristics of a Kenyan camelpox virus. *J. Hyg.* **75**, 318.

Davies, J. H. T., Janes, L. R., and Downie, A. W. (1938). Cowpox infection in farm workers. *Lancet* **2**, 1534.

Dawson, I. M., and McFarlane, A. S. (1948). Structure of an animal virus. *Nature (London)* **161**, 464.

De Filippes, F. M. (1982). Restriction enzyme mapping of vaccinia virus DNA. *J. Virol.* **43**, 136.

Dekking, F. (1964). Cowpox and vaccinia. *In* "Zoonoses" (J. van der Hoeden, ed.), p. 411. Elsevier, Amsterdam.

de Korté, W. E. (1904). Amaas, or kaffir milk-pox. *Lancet* **1**, 1273.

DeLange, A. M., and McFadden, G. (1986). Sequence-nonspecific replication of transfected plasmid in poxvirus-infected cells. *Proc. Natl. Acad. Sci. U.S.A.* **83**, 614.

DeLange, A. M., Reddy, M., Scraba, D., Upton, C., and McFadden, G. (1986). Replication and resolution of cloned poxvirus telomeres *in vivo* generates linear microchromosomes with intact viral hairpin termini. *J. Virol.* **59**, 249.

de la Salle, H., Altenburger, W., Elkaim, R., Dott, K., Dieterk, A., Drillien, R., Cuzenave, J. P., Tolstoshev, P., and Lecocq, J. P. (1986). Active γ-carboxylated human factor is expressed using recombinant DNA techniques. *Nature (London)* **316**, 268.

de Vries, E. (1960). "Post Vaccinial Perivenous Encephalitis." Elsevier, Amsterdam.

Dinger, J. E. (1956). Differences in persistence of smallpox and alastrim virus on the chorioallantois. *Docum. Med. Geogr. Trop.* **8**, 202.

Dixon, C. W. (1962). "Smallpox." Churchill, London.

Dorst, J., and Dandelot, P. (1969). "A Field Guide to the Larger Mammals of Africa." Houghton Mifflin, Boston.

Downie, A. W. (1930a). Study of the lesions produced experimentally by cowpox virus. *J. Pathol. Bacteriol.* **48**, 361.

Downie, A. W. (1939b). Immunological relationship of virus of spontaneous cowpox to vaccinia virus. *Br. J. Exp. Pathol.* **20**, 158.

Downie, A. W. (1965). Poxvirus group. *In* "Viral and Rickettsial Infections of Man" (F. L. Horsfall and I. Tamm, eds.), 4th ed., p. 932. Lippincott, Philadelphia, Pennsylvania.

Downie, A. W., and Dumbell, K. R. (1947). The isolation and cultivation of variola virus on the chorioallantois of the chick embryo. *J. Pathol. Bacteriol.* **59**, 189.

Downie, A. W., and Dumbell, K. R. (1956). Pox viruses. *Annu. Rev. Microbiol.* **10**, 237.

Downie, A. W., and Haddock, D. W. (1952). A variant of cowpox virus. *Lancet* **1**, 1049.

Downie, A. W., and McCarthy, K. (1950). The viruses of variola, vaccinia, cowpox and ectromelia—neutralization tests on the chorioallantois with unabsorbed and absorbed immune sera. *Br. J. Exp. Pathol.* **31**, 789.

Downie, A. W., and McCarthy, K. (1958). The antibody response in man following infection with viruses of the pox group. III. Antibody response in smallpox. *J. Hyg.* **56**, 479.

Downie, A. W., and McCarthy, K., and Macdonald, A., MacCallum, F. O., and Macrae, A. D. (1953). Virus and virus antigen in the blood of smallpox patients. *Lancet* **2**, 164.

Downie, A. W., Dumbell, K. R., Ayroza Galvao, P. A., and Zatz, I. (1963). Alastrim in Brazil. *Trop. Geogr. Med.* **15**, 25.

Downie, A. W., Fedson, D. S., St Vincent, L., Rao, A. R., and Kempe, C. H. (1969). Haemorrhagic smallpox. *J. Hyg.* **67**, 619.

Driessen, J. H., and Greenham, L. W. (1959). Haemadsorption in vaccinia-infected tube tissue cultures. *Arch. Gesamte Virusforsch.* **9**, 45.

Drillien, R., Koehren, F., and Kirn, A. (1981). Host range deletion mutant of vaccinia virus defective in human cells. *Virology* **111**, 488.

Dubbs, D. R., and Kit, S. (1964). Isolation and properties of vaccinia mutants deficient in thymidine kinase-inducing activity. *Virology* **22**, 214.

Dudgeon, J. A. (1963). Development of smallpox vaccine in England in the eighteenth and nineteenth centuries. *Br. Med. J.* **1**, 1367.

Duesberg, P. H., and Colby, C. (1969). On the biosynthesis and structure of double-stranded RNA in vaccinia virus-infected cells. *Proc. Natl. Acad. Sci. U.S.A.* **64**, 393.

Dulbecco, R. (1952). Production of plaques in monolayer tissue cultures by single particles of an animal virus. *Proc. Natl. Acad. Sci. U.S.A.* **38**, 747.

Dumbell, K. R. (1974). "Wild white" viruses and smallpox. *Lancet* **2**, 585.

Dumbell, K. R., and Archard, L. C. (1980). Comparison of white pock (h) mutants of monkeypox virus with parental monkeypox and with variola-like viruses isolated from animals. *Nature (London)* **286**, 29.

Dumbell, K. R., and Bedson, H. S. (1964). The use of ceiling temperature and reactivation in the isolation of pox virus hybrids. *J. Hyg.* **62**, 133.

Dumbell, K. R., and Bedson, H. S. (1966). Adaptation of variola virus to growth in the rabbit. *J. Pathol. Bacteriol.* **91**, 459.

Dumbell, K. R., and Huq, F. (1975). Epidemiological implications of the typing of variola isolates. *Trans. R. Soc. Trop. Med. Hyg.* **69**, 303.

Dumbell, K. R., and Huq, F. (1986). The virology of variola minor. Correlation of laboratory tests with the geographic distribution and human virulence of variola isolates. *Am. J. Epidemiol.* **123**, 403.

Dumbell, K. R., and Kapsenberg, J. G. (1982). Laboratory investigation of two "whitepox" viruses and comparison with two variola strains from southern India. *Bull. W.H.O.* **60**, 381.

Dumbell, K. R., and Wells, D. G. T. (1982). The pathogenicity of variola virus. A comparison of the growth of standard strains of variola major and variola minor in cell cultures from human embryos. *J. Hyg.* **89**, 389.

Dumbell, K. R., Downie, A. W., and Valentine, R. C. (1957). The ratio of the number of virus particles to infective titer of cowpox and vaccinia virus suspensions. *Virology* **4**, 467.

Dumbell, K. R., Bedson, H. S., and Rossier, E. (1961). The laboratory differentiation between variola major and variola minor. *Bull. W.H.O.* **25**, 73.

Dumbell, K. R., Bedson, H. S., and Nizamuddin, M. (1967). Thermo-efficient strains of variola major virus. *J. Gen. Virol.* **1**, 379.

Dunlap, R. C., and Patt, J. K. (1971). Reactivation of heated vaccinia virus by UV-irradiated vaccinia virus. *Proc. Soc. Exp. Biol. Med.* **136**, 1.

Duran-Reynals, F. (1931). On the spontaneous immunization of rabbits to vaccine virus. *J. Immunol.* **20**, 389.

Earl, P. L., Jones, E. V., and Moss, B. (1986). Homology between DNA polymerases of poxviruses, herpesviruses, and adenoviruses: Nucleotide sequence of the vaccinia virus DNA polymerase gene. *Proc. Natl. Acad. Sci. U.S.A.* **83**, 3659.

Easterbrook, K. B. (1966). Controlled degradation of vaccinia virions *in vitro*: An electron-microscopic study. *J. Ultrastruct. Res.* **14**, 484.

Edgar, R. S. (1966). Conditional lethals. *In* "Phage and the Origins of Molecular Biology" (J. Cairns, G. S. Stent, and J. D. Watson, eds.), p. 166. Cold Spring Harbor Lab., Cold Spring Harbor, New York.

Edward, D. G. F., Elford, W. J., and Laidlaw, P. P. (1943). Airborne virus infections. I.

Experimental technique and preliminary observations on influenza and infectious ectromelia. *J. Hyg.* **43**, 1.

Elango, N., Prince, G. A., Murphy, B. R., Venkatesan, S., Chanock, R. M., and Moss, B. (1986). Resistance to human respiratory syncytial virus (RSV) infection induced by immunization of cotton rats with a recombinant vaccinia virus expressing RSV G glycoprotein. *Proc. Natl. Acad. Sci. U.S.A.* **83**, 1906.

El Dahaby, H., El Sabbagh, A., Nassar, M., Kamell, M., and Iskander, M. (1966). Investigations of an outbreak of cowpox with special reference to the disease in Egypt, UAR. *J. Arab Vet. Med. Assoc.* **26**, 11.

Epstein, D. A., Marsh, Y. V., Schreiber, A. B., Newman, S. R., Todaro, G. J., and Nestor, J. J. (1985). Epidermal growth factor receptor occupancy inhibits vaccinia virus infection. *Nature (London)* **318**, 663.

Ermolaeva, S. N., Blandova, Z. K., and Dushkin, V. A. (1974). Genetic study on susceptibility of different mouse lines to ectromelia virus. *Sov. Genet. (Engl. Transl.)* **8**, 681.

Espmark, J. A., and Magnusson, B. (1964). A non-specific inhibitor to haemagglutination in post mortem human sera. *Acta Pathol. Microbiol. Scand.* **62**, 595.

Esposito, J. J., and Knight, J. C. (1984). Nucleotide sequence of the thymidine kinase gene region of monkeypox and variola viruses. *Virology* **135**, 561.

Esposito, J. J., and Knight, J. C. (1985). Orthopoxvirus DNA: A comparison of restriction profiles and maps. *Virology* **143**, 230.

Esposito, J. J., Obijeski, J. F., and Nakano, J. H. (1977a). Serological relatedness of monkeypox, variola and vaccinia viruses. *J. Med. Virol.* **1**, 35.

Esposito, J. J., Obijeski, J. F., and Nakano, J. H. (1977b). The virion and soluble antigen proteins of variola, monkeypox and vaccinia viruses. *J. Med. Virol.* **1**, 95.

Esposito, J. J., Obijeski, and Nakano, J. H. (1978). Orthopoxvirus DNA: Strain differentiation by electrophoresis of restriction endonuclease fragmented virion DNA. *Virology* **89**, 53.

Esposito, J. J., Cabradilla, C. D., Nakano, J. H., and Obijeski, J. F. (1981). Intragenomic sequence transposition in monkeypox virus. *Virology* **109**, 231.

Esposito, J. J., Nakano, J. H., and Obijeski, J. F. (1985). Can variola-like viruses be derived from monkeypox virus? An investigation based on DNA mapping. *Bull. W.H.O.* **63**, 695.

Essani, K., and Dales, S. (1979). Biogenesis of vaccinia: Evidence for more than 100 polypeptides in the virion. *Virology* **95**, 385.

Essani, K., Dugre, R., and Dales, S. (1982). Biogenesis of vaccinia: Involvement of spicules of the envelope during virion assembly examined by means of conditional lethal mutants and serology. *Virology* **118**, 279.

Esteban, M., and Holowczak, J. A. (1977a). Replication of vaccinia DNA in mouse L cells. I. *In vivo* DNA synthesis. *Virology* **78**, 57.

Esteban, M., and Holowczak, J. A. (1977b). Replication of vaccinia DNA in mouse L cells. III. Intracellular forms of viral DNA. *Virology* **82**, 308.

Esteban, M., Flores, L., and Holowczak, J. A. (1977). Model for vaccinia virus DNA replication. *Virology* **83**, 467.

Fairbrother, R. W., and Hoyle, L. (1937). Observations on the aetiology of influenza. *J. Pathol. Bacteriol.* **44**, 213.

Feller, A. E., Enders, J. F., and Weller, T. H. (1940). The prolonged existence of vacinia virus in roller tube cultures of chick embryonic tissues. *J. Exp. Med.* **72**, 367.

Fenner, F. (1947a). Studies in infectious ectromelia of mice. I. Immunization of mice against ectromelia with living vaccinia virus. *Aust. J. Exp. Biol. Med. Sci.* **25**, 257.

Fenner, F. (1947b). Studies in infectious ectromelia of mice. II. Natural transmission: The portal of entry of the virus. *Aust. J. Exp. Biol. Med. Sci.* **25**, 275.

Fenner, F. (1948a). The clinical features of mouse-pox (infectious ectromelia of mice) and the pathogenesis of the disease. *J. Pathol. Bacteriol.* **60**, 529.

Fenner, F. (1948b). The pathogenesis of the acute exanthems. *Lancet* **2**, 915.

Fenner, F. (1948c). The epizootic behaviour of mouse-pox (infectious ectromelia of mice). *Br. J. Exp. Pathol.* **29**, 69.

Fenner, F. (1948d). The epizootic behaviour of mousepox (infectious ectromelia of mice). II. The course of events in long-continued epidemics. *J. Hyg.* **46**, 383.

Fenner, F. (1949a). Studies in mousepox (infectious ectromelia of mice). IV. Quantitative investigations on the spread of virus through the host in actively and passively immunized animals. *Aust. J. Exp. Biol. Med. Sci.* **27**, 1.

Fenner, F. (1949b). Studies in mousepox (infectious ectromelia of mice). VI. A comparison of the virulence and infectivity of three strains of ectromelia virus. *Aust. J. Exp. Biol. Med. Sci.* **27**, 31.

Fenner, F. (1949c). Studies in mousepox (infectious ectromelia of mice). VII. The effect of the age of the host on the response to infection. *Aust. J. Exp. Biol. Med. Sci.* **27**, 45.

Fenner, F. (1949d). Mousepox (infectious ectromelia of mice): A review. *J. Immunol.* **63**, 341.

Fenner, F. (1958). The biological characteristics of several strains of vaccinia, cowpox and rabbitpox virus. *Virology* **5**, 502.

Fenner, F. (1959). Genetic studies with mammalian poxviruses. II. Recombination between two strains of vaccinia virus in single HeLa cells. *Virology* **8**, 499.

Fenner, F. (1976). Classification and nomenclature of viruses. Second report of the International Committee on Taxonomy of Viruses. *Intervirology* **7**, 1.

Fenner, F. (1979). Portraits of viruses: The poxviruses. *Intervirology* **11**, 137.

Fenner, F. (1982). Mousepox. *In* "The Mouse in Biomedical Research" (H. L. Foster, J. D. Small, and J. G. Fox, eds.), Vol. 2, p. 209. Academic Press, New York.

Fenner, F. (1985). Vaccination, its birth, death and resurrection. *Aust. J. Exp. Biol. Med. Sci.* **63**, 607.

Fenner, F. (1988). *Poxviridae. In* "Viral Infections of Vertebrates" Viral Infections of Rodents and Lagomorphs" (A. D. M. E. Osterhaus, ed.), Vol. 6. Elsevier, Amsterdam. (In press.)

Fenner, F., and Burnet, F. M. (1957). A short description of the poxvirus group (vaccinia and related viruses). *Virology* **4**, 305.

Fenner, F., and Comben, B. M. (1958). Genetic studies with mammalian poxviruses I. Demonstration of recombination between two strains of vaccinia virus. *Virology* **5**, 530.

Fenner, F., and Fenner, E. M. B. (1949). Studies in mouse-pox (infectious ectromelia of mice). V. Closed epidemics in herds of normal and vaccinated mice. *Aust. J. Exp. Biol. Med. Sci.* **27**, 19.

Fenner, F., and Nakano, J. H. (1988). *Poxviridae. In* "Laboratory diagnosis of Infectious Diseases: Principles and Practice" (E. H. Lennette, P. Halonen, and F. A. Murphy, eds.), Vol. 2. Springer-Verlag, New York. (In press.)

Fenner, F., and Ratcliffe, F. N. (1965). "Myxomatosis." Cambridge Univ. Press, London and New York.

Fenner, F., and Sambrook, J. F. (1966). Conditional lethal mutants of rabbitpox virus. II. Mutants (*p*) that fail to multiply in PK-2a cells. *Virology* **28**, 600.

Fenner, F., and Woodroofe, G. M. (1960). Reactivation of poxviruses. II. Range of reactivating viruses. *Virology* **11**, 185.

Fenner, F., Holmes, I. H., Joklik, W. K., and Woodroofe, G. M. (1959). The reactivation of heat-inactivated poxviruses: A general phenomenon which includes the fibroma–myxoma virus transformation of Berry and Dedrick. *Nature (London)* **183**, 1340.

Fenner, F., McAuslan, B. R., Mims, C. A., Sambrook, J. F., and White, D. O. (1974). "The Biology of Animal Viruses," 2nd ed. Academic Press, New York.

Fenner, F., Henderson, D. A., Arita, I., Jezek, A., and Ladnyi, I. D. (1988). "Smallpox and Its Eradication." World Health Organization, Geneva.

Fleming, G. (1880a). Human and animal variolae: A study in comparative pathology. *Lancet* **1**, 164, 246, 443, 484, 832.

Fleming, G. (1880b). Human and animal variolae: A study in comparative pathology. *Lancet* **2**, 374, 453.

Flynn, R. J. (1963). The diagnosis and control of ectromelia infection of mice. *Lab. Anim. Care* **13**, 130.

Flynn, R. J., and Briody, B. A. (1962). Relative susceptibilities of vaccinated and non-vaccinated Syrian hamsters and mice to ectromelia virus. *Proc. Anim. Care Panel* **12**, 263.

Foglesong, P. D. (1985). *In vitro* transcription of a cloned vaccinia virus gene by a soluble extract prepared from vaccinia virus-infected HeLa cells. *J. Virol.* **53**, 822.

Food and Agriculture Organization (1981). "Forest Resources of Tropical Africa: Tropical Forest Resources Assessment Project, 1981," UN 32.6 1301-78-04, Tech. Rep. 2. FAO/United Nations, Rome

Foster, E. A., transl. and ed. (1950). "Motolinía's History of the Indians of New Spain." The Cortés Society, Berkeley, California (reprinted by Greenwood Press, Westport, Connecticut, 1973).

Foster, S. O., Brink, E. W., Hutchins, D. L., Pifer, J. M., Lourie, B., Moser, C. R., Cummings, E. C., Kuteyi, O. E. K., Eke, R. E. A., Titus, J. B., Smith, E. A., Hicks, J. W., and Foege, W. H. (1972). Human monkeypox. *Bull. W.H.O.* **46**, 569

Fournier, F., Tovell, D. R., Esteban, M., Metz, D. H., Ball, L. A., and Kerr, I. M. (1973). The translation of vaccinia virus messenger RNA in animal cell-free systems. *FEBS Lett.* **30**, 268.

Franke, C. A., Rice, C. M., Strauss, J. H., and Hruby, D. E. (1985). Neomycin resistance as a dominant selectable marker for selection and isolation of vaccinia virus recombinants. *Mol. Cell. Biol.* **5**, 1918.

Fraser, S. M. F. (1980). Leicester and smallpox: The Leicester method. *Med. Hist.* **24**, 315.

Freed, E. R., Richard, J. D., and Escobar, M. R. (1972). Vaccinia necrosum and its relationship to impaired immunologic responsiveness. *Am. J. Med.* **52**, 420.

Fuerst, T. R., Earl, P. L., and Moss, B. (1987). Use of a hybrid vaccinia virus-T7 RNA polymerase system for expression of target genes. *Mol. Cell. Biol.* **7**, 2538.

Fulginiti, V. A., Kempe, C. H., Hathaway, W. E., Pearlman, D. S., Serber, O. F., Jr., Ellen, J. J., Joyner, J. J., and Robertson, A. (1968). Progressive vaccinia in immunologically deficient individuals. *In* "Immunological Deficiency Diseases in Man" (D. Bergsma and R. A. Good, eds.), p. 129. National Foundation, New York.

Gafford, L. G., and Randall, C. C. (1976). Virus-specific RNA and DNA in nuclei of cells infected with fowlpox virus. *Virology* **69**, 1.

Galasso, G. J. (1970). Report of a conference. Project studies on immunization against smallpox. *J. Infect. Dis.* **121**, 575.

Galasso, G. J., Mattheis, M. J., Cheny, J. D., Connor, J. D., McIntosh, K., Benenson, A. S., and Alling, D. W. (1977). Summary. *J. Infect. Dis.* **135**, 183.

Gangemi, J. D., and Sharp, D. G. (1976). Use of a restriction endonuclease in analyzing the genomes from two different strains of vaccinia virus. *J. Virol.* **20**, 319.

Ganiev, M. K., and Ferzaliev, I. A. (1964). [Pox in buffaloes from contact with vaccinated human beings.] *Veterinariya* **41**, 31 (in Russian); abstract in English: *Vet. Bull. (London)* (1965) **35**, 25.

Garon, C. F., Barbosa, E., and Moss, B. (1978). Visualization of an inverted terminal repetition in vaccinia virus DNA. *Proc. Natl. Acad. Sci. U.S.A.* **75**, 4863.

Gehring, H., Mahnel, H., and Mayer, H. (1972). Elefanten pocken. *Zentralbl. Veterinaermed., Riehe B* **19**, 258.

Gell, P. G. H., Coombes R. A. A., and Lachmann, P. J. (1975). "Clinical Aspects of Immunology." Blackwell, Oxford.

Gemmell, A., and Cairns, J. (1959). Linkage in the genome of an animal virus. *Virology* **8**, 381.

Gemmell, A., and Fenner, F. (1960). Genetic studies with mammalian poxviruses. III. White (u) mutants of rabbitpox virus. *Virology* **11**, 219.

Geshelin, P., and Berns, K. I. (1974). Characterization and localization of the naturally occurring crosslinks in vaccinia virus DNA. *J. Mol. Biol.* **88**, 785.

Ghendon, Y. Z., and Chernos, V. I. (1964). Comparative study of genetic markers of some poxvirus strains. *Acta Virol. (Engl. Ed.)* **8**, 359.

Ghosh, T. K., Arora, R. R., Sehgal, C. L., Ray, S. N., and Wattal, B. L. (1977). An investigation of buffalopox outbreak in animals and human beings in Dhulia District (Maharashtra State). 2. Epidemiological studies. *J. Commun. Dis.* **9**, 93.

Gibbs, A., and Fenner, F. (1984). Methods for comparing sequence data such as restriction endonuclease maps or nucleotide sequences of viral nucleic acid molecules. *J. Virol. Methods* **9**, 317.

Gibbs, E. P. J., and Osborne, A. D. (1974). Observations on the epidemiology of pseudocowpox in south-west England and south Wales. *Br. Vet. J.* **130**, 150.

Gibbs, E. P. J., Johnson, R. H., and Osborne, A. D. (1970). The differential diagnosis of viral skin infections of the bovine teat. *Vet. Rec.* **87**, 602.

Gibbs, E. P. J., Johnson, R. H., and Collings, D. F. (1973). Cowpox in a dairy herd in the United Kindgom. *Vet. Rec.* **92**, 56.

Gillard, S., Spehner, D., Drillien, R., and Kirn, A. (1986). Localization and sequence of a vaccinia virus gene required for multiplication in human cells. *Proc. Natl. Acad. Sci. U.S.A.* **83**, 5573.

Gillespie, J. H., and Timoney, J. F. (1981). "Hagan and Bruner's Infectious Diseases of Domestic Animals," 7th ed., pp. 283, 534. Cornell Univ. Press, Ithaca, New York.

Ginsberg, A. H., and Johnson, K. P. (1977). The effect of cyclophosphamide on intracerebral vaccinia virus infection in Balb/c mice. *Exp. Mol. Pathol.* **27**, 285.

Gispen, R. (1949). De herbesmetting van Indonesie met pokken. *Ned. Tijdschr. Geneenskd.* **93**, 3686.

Gispen, R. (1952). Silver impregnation of smallpox elementary bodies after treatment with xylol. *Antonie van Leeuwenhoek* **18**, 107.

Gispen, R. (1955). Analysis of pox-virus antigens by means of double diffusion. A method for direct serological differentiation of cowpox. *J. Immunol.* **74**, 134.

Gispen, R. (1970). Witte monkeypox-virusstammen. *Volksgezond. Versl. Rapp.* **20**, 151.

Gispen, R. (1972). Onderscheiding van een pokvirus afkomstig uit een woestijnmuis. *Volksgezond. Versl. Rapp.* **24**, 164.

Gispen, R., and Brand-Saathof, B. (1972). "White" poxvirus strains from monkeys. *Bull. W. H. O.* **46**, 585.

Gispen, R., and Brand-Saathof, B. (1974). Three specific antigens produced in vaccinia, variola and monkeypox infections. *J. Infect. Dis.* **129**, 289.

Gispen, R., and Kapsenberg, J. G. (1966). Monkeypox virus-infectie in cultures van apeniercellen zonder duidelijk epizootisch verband met pokken, en in een kolonie van apen lijdende aan pokken. *Versl. Meded. Volksgezond.* **12**, 140.

Gispen, R., Verlinde, J. D., and Zwart, P. (1967). Histopathological and virological studies on monkeypox. *Arch. Gesamte Virusforsch.* **21**, 205.

Gispen, R., Huisman, J., Brand-Saathof, B., and Hekker, A. C. (1974). Immunofluorescence test for persistent poxvirus antibodies. *Arch. Gesamte Virusforsch.* **44**, 391.

Gispen, R., Brand-Saathof, B., and Hekker, A. C. (1976). Monkeypox-specific antibodies in human and simian sera from the Ivory Coast and Nigeria. *Bull. W.H.O.* **53**, 355.

Gledhill, A. W. (1962a). Latent ectromelia. *Nature (London)* **196**, 298.

Gledhill, A. W. (1962b). Viral diseases in laboratory animals. *In* "The Problems of Laboratory Animal Disease" (R. J. C. Harris, ed.), p. 99. Academic Press, London.

Gold, P., and Dales, S. (1968). Localization of nucleotide phosphohydrolase within vaccinia virus. *Proc. Natl. Acad. Sci. U.S.A.* **60**, 845.

Golini, F., and Kates, J. R. (1985). A soluble transcription system derived from purified vaccinia virions. *J. Virol.* **53**, 205.

Goodpasture, E. W. (1933). Borreliotoses. Fowlpox, molluscum contagiosum, variola-vaccinia. *Science* **77**, 119.

Goodpasture, E. W., Woodruff, A. M., and Buddingh, G. J. (1932). Vaccinal infection of the chorioallantoic membrane of the chick embryo. *Am. J. Pathol.* **8**, 271.

Gordon, M. H. (1925). Studies of the viruses of vaccinia and variola. *Med. Res. Counc. G. B., Spec. Rep. Ser.* **SPS-98**.

Gordon, M. H. (1937). Virus bodies. John Buist and the elementary bodies of vaccinia. *Edinburgh Med. J.* **44**, 65.

Gotch, F., McMichael, A., Smith, G., and Moss, B. (1987). Identification of viral molecules recognized by influenza-specific human cytotoxic T lymphocytes. *J. Exp. Med.* **165**, 408.

Grady, L. J., and Paoletti, E. (1977). Molecular complexity of vaccinia DNA and the presence of reiterated sequences in the genome. *Virology* **79**, 337.

Greene, H. S. N. (1933). A pandemic of rabbit-pox. *Proc. Soc. Exp. Biol. Med.* **30**, 892.

Greene, H. S. N. (1934a). Rabbit pox. I. Clinical manifestations and cause of disease. *J. Exp. Med.* **60**, 427.

Greene, H. S. N. (1934b). Rabbit pox. II. Pathology of the epidemic disease. *J. Exp. Med.* **60**, 441.

Greene, H. S. N. (1935). Rabbit pox. III. Report of an epidemic with especial reference to epidemiological factors. *J. Exp. Med.* **61**, 807.

Greenwood, M., Hill, A. B., Topley, W. W. C., and Wilson, J. (1936). Experimental epidemiology. *Med. Res. Counc. G. B. Spec. Rep. Ser.* **SPS-209**.

Griffith, F. (1928). The significance of pneumococcal types. *J. Hyg. Camb.* **27**, 113.

Grimley, P. M., Rosenblum, E. N., Mims, S. J., and Moss, B. (1970). Interruption by rifampin of an early stage in vaccinia virus morphogenesis: Accumulation of membranes which are precursors of virus envelopes. *J. Virol.* **6**, 519.

Gröppel, K.-H. (1962). Über das Vorkommen von Ektromelie (Mäusepocken) unter Wildmäusen. *Arch. Exp. Veterinaermed.* **16**, 243.

Guarnieri, G. (1892). Ricerche sulla patogenesi ed etiologia dell'infezione vaccinica e variolosa. *Arch. Sci. Med.* **16**, 403.

Guillon, J.-C. (1970). Recherche sur la transmission du virus de l'ectromélie de la souris par un acarien hématophage: *Ornithonyssus bacoti. Exp. Anim.* **3**, 177.

Guillon, J.-C. (1975). Prophylaxie sanitaire et médicale de l'ectromélie de la souris. *Recl. Med. Vet.* **151**, 19.

Gurvich, E. B., and Marennikova, S. S. (1964). Laboratory diagnosis of smallpox and similar viral diseases by means of tissue culture methods. III. Additional modes of differentiating viruses of the pox group in tissue culture. *Acta Virol. (Engl. Ed.)* **8**, 435.

Hahon, N. (1958). Cytopathogenicity and propagation of variola virus in tissue culture. *J. Immunol.* **81**, 426.

Hahon, N. (1961). Smallpox and related poxvirus infections in the simian host. *Bacteriol. Rev.* **25**, 459.

Hahon, N., and Wilson, B. J. (1960). Pathogenesis of variola in *Macaca iris* monkeys. *Am. J. Hyg.* **71**, 69.

Hamilton, A., Kinchington, D., Greenaway, P. J., and Dumbell, K. (1985). Recombinant bacterial plasmids containing inserts of variola DNA. *Lancet* **2**, 1356.

Hanafusa, T., Hanafusa, H., and Kamahora, J. (1959a). Transformation of ectromelia into vaccinia virus in tissue culture. *Virology* **8**, 525.

Hanafusa, T., Hanafusa, H., and Kamahora, J. (1959b). Transformation phenomena in the pox group viruses. II. Transformation between several members of the pox group. *Biken J.* **2**, 85.

Hänggi, M., Bannwarth, W., and Stunnenberg, H. G. (1986). Conserved TAAAT motif in vaccinia virus late promoters: Overlapping TATA box and site of transcription initiation. *EMBO J.* **5**, 1071.

Hansson, O., Johansson, S. G., and Vahlquist, B. (1966). Vaccinia gangrenosa with normal humoral antibodies. A case possibly due to deficient cellular immunity treated with N-methylisatin beta-thiosemicarbazone (compound 33T57, Marboran). *Acta Paediatr. Scand.* **55**, 264.

Harford, C. G., Hamlin, A., and Riders, E. (1966). Electron microscopic autoradiography of DNA synthesis in cells infected with vaccinia virus. *Exp. Cell Res.* **42**, 50.

Harper, J. M., Parsonage, M. T., Pelham, H. R. B., and Darby, G. (1978). Heat inactivation of vaccinia virus particle-associated functions: Properties of heated particles *in vivo* and *in vitro*. *J. Virol.* **26**, 646.

Harper, L., Bedson, H. S., and Buchan, A. (1979). Identification of orthopoxviruses by polyacrylamide gel electrophoresis of intracellular polypeptides. I. Four major groupings. *Virology* **93**, 435.

Hashizume, S. (1975). [A new attenuated strain of vaccinia virus, LC16m8: Basic information.] *J. Clin. Virol.* **3**, 229 (in Japanese).

Hashizume, S., Morita, T., Yoshizawa H., Suzuki, K., Arita, M., Komatsu, T., Amario, H., and Tagaya, I. (1973). Intracerebral inoculation of monkeys with several vaccinia strains: An approach to the comparison of different strains. *Symp. Ser. Immunobiol. Stand.* **19**, 325.

Haygarth, K. J. (1785). "An Inquiry How to Prevent the Smallpox." Johnson, London.

Haygarth, J. (1793). "Sketch of a Plan to Exterminate the Casual Small-Pox from Great Britain and to Introduce General Inoculation." Johnson, London.

Heberling, R. L., Kalter, S. S., and Rodriquez, A. R. (1976). Poxvirus infection of the baboon *(Papio cynocephalus)*. *Bull, W.H.O.* **54**, 285.

Heiner, G. C., Fatima, N., Daniel, R. W., Cole, J. L., Anthony, R. L., and McCrumb, F. R., Jr. (1971). A study of inapparent infection in smallpox. *Am. J. Epidemiol.* **94**, 252.

Helbert, D. (1957). Smallpox and alastrim. Use of the chick embryo to distinguish between the viruses of variola major and variola minor. *Lancet* **1**, 1012.

Henderson, D. A., and Arita, I. (1985). Utilization of vaccine in the global eradication of smallpox. *In* "Vaccinia Viruses as Vectors for Vaccine Antigens" (G. V. Quinnan, Jr., ed.), p. 61. Am. Elsevier, New York.

Henneberg, G. (1956). The distribution of smallpox in Europe 1919–1948. *In* "World-Atlas of Epidemic Diseases," Part II, p. 69. Falk, Hamburg.

Herman, Y. F. (1964). Isolation and characterization of a naturally occurring pox virus of raccoons. *Bacteriol. Proc.*, p. 117.

Herrlich, A. (1954). Probleme der Pocke und Pockenschutzimpfung. *Muench. Med. Wochenschr.* **96**, 529.

Herrlich, A. (1958). Variola. Eindrücke von einer epidemie in Bombay im jahre 1958. *Dtsch. Med. Wochenschr.* **83**, 1426.

Herrlich, A., Mayr, A., Mahnel, H., and Munz, E. (1963). Experimental studies on the transformation of the variola virus into the vaccinia virus. *Arch. Gesamte Virusforsch.* **12**, 579.

Hiller, G., and Weber, K. (1985). Golgi-derived membranes that contain an acylated viral polypeptide are used for vaccinia virus envelopment. *J. Virol.* **55**, 651.

Hiller, G., Eibl, H., and Weber, K. (1981a). Characterization of intracellular and extracellular vaccinia virus variants: N_1-isonicotinoyl-N_2-3-methyl-4-chlorobenzoylhydrazine interferes with cytoplasmic virus dissemination and release. *J. Virol.* **39**, 903.

Hiller, G., Jungwirth, C., and Weber, K. (1981b). Fluorescence microscopical analysis of the life cycle of vaccinia virus in chicken embryo fibroblasts. Virus cytoskeleton interactions. *Exp. Cell Res.* **132**, 82.

Hime, T. W. (1896). Animal vaccination. *Br. Med. J.* **1**, 1279.

Hirst, G. K. (1941). The agglutination of red cells by allantoic fluid of chick embryos infected with influenza virus. *Science* **94**, 22.

Hirt, P., Hiller, G., and Wittek, R. (1986). Localization and fine structure of a vaccinia virus gene encoding an envelope antigen. *J. Virol.* **58**, 757.

Hoagland, G. L., Smadel, J. E., and Rivers, T. M. (1940a). Constituents of elementary bodies of vaccinia. I. Certain basic analyses and observations on lipid components of the virus. *J. Exp. Med.* **71**, 737.

Hoagland, G. L., Lavin, G. I., Smadel, J. E., and Rivers, T. M. (1940b). Constituents of elementary bodies of vaccinia. II. Properties of nucleic acid obtained from vaccinia virus. *J. Exp. Med.* **72**, 139.

Hoagland, G. L., Ward, S. M., Smadel, J. E., and Rivers, T. M. (1942). Constituents of elementary bodies of vaccinia. VI. Studies on the nature of the enzymes associated with the purified virus. *J. Exp. Med* **76**, 163.

Hochstein-Mintzel, V., Hauichen, T., Huber, H. C., and Stickl, H. (1975). Vaccinia- und variolaprotektive Wirkung des modifizierten Vaccinia-Stammes MVA bei Intramuskularer Immunisierung. *Zentralbl. Bakteriol., Parasitenkd., Infektionsk. Hyg., Abt. 1:, Orig., Reihe A* **230**, 283.

Holmes, F. O. (1948). Filterable viruses. *In* "Bergeys Manual of Determinative Bacteriology" (R. S. Breed, E. D. C. Murray, and A. P. Hitchins, eds.), 6th ed., p. 1127. Williams & Wilkins, Baltimore, Maryland.

Holowczak, J. A., and Joklik, W. K. (1967a). Studies on the structural proteins of vaccinia virus. I. Structural proteins of virions and cores. *Virology* **33**, 717.

Holowczak, J. A., and Joklik, W. K. (1967b). Studies on the structural proteins of vaccinia virus. II. Kinetics of the synthesis of individual groups of structural proteins. *Virology* **33**, 726.

Hopkins, D. R. (1983). "Princes and Peasants. Smallpox in History." Univ. of Chicago Press, Chicago, Illinois.

Horgan, E. S., and Haseeb, M. A. (1939). Cross immunity experiments in monkeys between variola, alastrim and vaccinia. *J. Hyg.* **39**, 615.

Horgan, E. S., and Haseeb, M. A. (1944). Some observations on accidental vaccination on the hands of workers in a vaccine lymph institute. *J. Hyg.* **43**, 273.

Howard, W. T., and Perkins, R. G. (1905). Studies on the etiology and pathology of vaccinia in the rabbit and in man. *J. Med. Res.* **14**, 51.

Hruby, D. E., and Ball, L. A. (1981). Control of expression of the vaccinia virus thymidine kinase genome. *J. Virol.* **40**, 456.

Hruby, D. E., and Ball, L. A. (1982). Mapping and indentification of the vaccinia thymidine kinase gene. *J. Virol.* **43**, 403.

Hruby, D. E., Lynn, D. L., and Kates, J. R. (1979). Vaccinia virus replication requires active participation of the host cell transcriptional apparatus. *Proc. Natl. Acad. Sci. U.S.A.* **76**, 1887.

Hruby, D. E., Maki, R. A., Miller, D. B., and Ball, L. A. (1983). Fine structure analysis and nucleotide sequence of the vaccinia virus thymidine kinase gene. *Proc. Natl. Acad. Sci. U.S.A.* **80**, 3411.

Hu, S.-K., Brussewski, J., and Smalling, R. (1985). Infectious vaccinia virus recombinant that expresses the surface antigen of porcine transmissible gastroenteritis virus (TGEV). *In* "Vaccinia Viruses as Vectors for Vaccine Antigens" (G. V. Quinnan, Jr., ed.), p. 201. Am. Elsevier, New York.

Hu, S.-K., Kosowski, S. G., and Dalrymple, J. M. (1986). Expression of AIDS virus envelope gene in recombinant vaccinia viruses. *Nature (London)* **320**, 527.

Huq, F. (1972). Studies on variants of variola virus. Ph.D. Thesis, University of London.

Huq, F. (1976). Effect of temperature and relative humidity on variola virus in crusts. *Bull. W.H.O.* **54**, 710.

Hutchinson, H. D., Ziegler, D. W., Wells, D. E., and Nakano, J. H. (1977). Differentiation of variola, monkeypox and vaccinia antisera by radioimmunoassay. *Bull. W.H.O.* **55**, 613.

Ichihashi, Y., and Dales, S. (1971). Biogenesis of poxviruses: Interrelationship between hemagglutinin production and polykaryocytosis. *Virology* **46**, 533.

Ichihashi, Y., and Dales, S. (1973). Biogenesis of poxviruses: Relation between a translation complex and formation of A-type inclusions. *Virology* **51**, 297.

Ichihashi, Y., and Matsumoto, S. (1966). Studies on the nature of Marchal bodies (A-type inclusion) during ectromelia virus infection. *Virology* **29**, 264.

Ichihashi, Y., and Matsumoto, S. (1968a). The relationship between poxvirus and A-type inclusion body during double infection. *Virology* **36**, 262.

Ichihashi, Y., and Matsumoto, S. (1968b). Genetic character of poxvirus A-type inclusion. *Virus (Kyoto)* **18**, 237.

Ichihashi, Y., Matsumoto, S., and Dales, S. (1971). Biogenesis of poxviruses: Role of A-type inclusions and host cell membranes in virus dissemination. *Virology* **46**, 507.

Iftimovici, R., Iacobescu, V., Mutiu, A., and Puca, D. (1976). Enzootic with ectromelia symptomatology in Sprague-Dowley rats. *Rev. Roum. Med., Virol.* **27**, 65.

Ikuta, K., Miyamoto, H., and Kato, S. (1979). Serologically cross-reactive polypeptides in vaccinia, cowpox and Shope fibroma viruses. *J. Gen. Virol.* **44**, 557.

Ipsen, J. (1945). Quantitative Studien über mausepathogene Fleckfieberrickettsien. II. Das Wachstum der Infektion und ihre Ausbreitung im Mauseorganismus. *Schweiz. Z. Pathol. Bakteriol.* **8**, 57.

Isaacs, A., and Westwood, M. A. (1959). Inhibition by interferon of the growth of vaccinia virus in the rabbit skin. *Lancet* **2**, 324.

Iwad, F. I., Saber, M. S., Amin, M. M., and Yousef, H. M. (1981). Some epizootiological studies of buffalopox in Egypt. *Acta Vet. (Belgrade)* **31**, 41.

Jackson, D. C., Ada, G. L., and Tha Hla, R. (1976). Cytotoxic T cells recognize very early minor changes in ectromelia virus-infected target cells. *Aust. J. Exp. Biol. Med. Sci.* **54**, 349.

Jackson, T. M., Zaman, S. N., and Huq, F. (1977). T and B rosetting lymphocytes in the blood of smallpox patients. *Am. J. Trop. Med. Hyg.* **26**, 517.

Jacoby, R. O., and Bhatt, P. N. (1987). Mousepox in inbred mice innately resistant or susceptible to lethal infection with ectromelia virus. II. Pathogenesis. *Lab. Anim. Sci.* **37**, 14.

Jansen, J. (1941). Tödlich Infektionen von Kaninchen durch ein filtrierbares Virus. *Zentralbl. Bakteriol., Parasitenkd., Infektionskr. Hyg., Abt. 1:Orig.* **148**, 65.

Jansen, J. (1946). Immunity in rabbit plague—immunological relationship with cowpox. *Antonie van Leeuwenhoek* **11**, 139.

Jansen, J. (1962). Rabbit virus diseases. *In* "The Problems of Laboratory Animal Disease" (R. J. C. Harris, ed.), p. 185. Academic Press, London.

Japan, Ministry of Health (1975). Report of Committee on Smallpox Vaccination. Investigation of treatment of complications caused by smallpox vaccination. *J. Clin. Virol.* **3**, 269 (in Japanese).

Jaureguiberry, G., Ben-Hamida, F., Chapeville, F., and Beaud, G. (1975). Messenger activity of RNA transcribed *in vitro* by DNA–RNA polymerase associated to vaccinia virus cores. *J. Virol.* **15**, 1467.

Jenner, E. (1798). "An Inquiry into the Cause and Effects of the Variolae Vaccinae, a Disease Discovered in Some of the Western Counties of England, particularly Gloucestershire, and Known by the Name of Cowpox," London. (Reprinted in C. N. B. Camac, ed., "Classics of Medicine and Surgery," p. 213. Dover, New York, 1959.)

Jenner, E. (1799). "Further Observations on the Variolae Vaccinae or Cowpox." (Reprinted in C. N. B. Camac, ed., "Classics of Medicine and Surgery," p. 241. Dover, New York, 1959.)

Jezek, Z., and Fenner, F. (1988). Human monkeypox. *Monogr. Virol.* **17**, 1.

Jezek, Z., Basu, R. N., and Arya, Z. S. (1978). Problem of persistence of facial pockmarks among smallpox patients. *Indian J. Public Health* **22**, 95.

Jezek, Z., Sery V., Zikmund, V., and Slonim, D. (1982). ["Smallpox and its Eradication."] Avicenum, Praha (in Czech).

Jezek, Z., Kriz, B., and Rothbauer, V. (1983). Camelpox and its risk to the human population. *J. Hyg., Epidemiol., Microbiol., Immunol.* **27**, 29.

Jezek, Z., Arita, I., Mutombo, M., Dunn, C., Nakano, J. H., and Szczeniowski, M. (1986a). Four generations of probable person-to-person transmission of human monkeypox. *Am. J. Epidemiol.* **123**, 1004.

Jezek, Z., Marennikova, S. S., Mutombo, M., Nakano, J. H., Paluku, K. M., and Szczeniowski, M. (1986b). Human monkeypox: A study of 2,510 contacts of 214 patients. *J. Infect. Dis.* **154**, 551.

Jezek, Z., Nakano, J. H., Arita, I., Mutombo, M., Szczeniowski, M., and Dunn, C. (1987a). Serological survey for human monkeypox infections in a selected population in Zaire. *J. Trop. Med. Hyg.* **90**, 31.

Jezek, Z., Grab, B., and Dixon, H. (1987b). Stochastic model for interhuman spread of monkeypox. *Am. J. Epidemiol.* **126**, 1082.

Jezek, Z., Sczceniowski, M., Paluku, K. M., and Mutombo, M. (1987c). Human monkeypox: Clinical features of 282 patients. *J. Infect. Dis.* **156**, 293.

Joklik, W. K. (1962a). The purification of four strains of poxvirus. *Virology* **18**, 9.

Joklik, W. K. (1962b). Some properties of poxvirus DNA. *J. Mol. Biol.* **5**, 265.

Joklik, W. K. (1964). The intracellular uncoating of poxvirus DNA. I. The fate of radioactively-labeled rabbitpox virus. *J. Mol. Biol.* **8**, 263.

Joklik, W. K. (1966). The poxviruses. *Bacteriol. Rev.* **30**, 33.

Joklik, W. K., Woodroofe, G. M., Holmes, I. H., and Fenner, F. (1960a). The reactivation of poxviruses. I. Demonstration of the phenomenon and techniques of assay. *Virology* **11**, 168.

Joklik, W. K., Holmes, I. H., and Briggs, M. J. (1960b). The reactivation of poxviruses. III. Properties of reactivable particles. *Virology* **11**, 202.

Joklik, W. K., Abel, P., and Holmes, I. H. (1960c). Reactivation of poxviruses by a non-genetic mechanism. *Nature (London)* **186**, 992.

Jolly, J. (1901). "Indian Medicine." (translated from the German by C. G. Kashikar, Munshiram Manoharlal, New Delhi, 1977.)

Jones, E. V., and Moss, B. (1984). Mapping of the vaccinia virus DNA polymerase gene by marker rescue and cell-free translation of selected RNA. *J. Virol.* **49**, 72.

Jones, E. V., and Moss, B. (1985). Transcriptional mapping of the vaccinia virus DNA polymerase gene. *J. Virol.* **53**, 312.

Jones, E. V., Puckett, C., and Moss, B. (1987). DNA-dependent RNA polymerase subunits encoded within the vaccinia virus genome. *J. Virol.* **61**, 1765.

Jungwirth, C., and Joklik, W. K. (1965). Studies on "early" enzymes in HeLa cells infected with vaccinia virus. *Virology* **27**, 80.

Kalter, S. S., and Heberling, R. L. (1971). Comparative virology of primates. *Bacteriol. Rev.* **35**, 310.

Kalter, S. S., Rodriguez, A. R., Cummins, L. B., Heberling, R. L., and Foster, S. O. (1979). Experimental smallpox in chimpanzees. *Bull. W.H.O.* **57**, 637.

Kaminjolo, J. S., Jr., and Winquist, G. (1975). Histopathology of skin lesions in Uasin Gishu skin disease of horses. *J. Comp. Pathol.* **85**, 391.

Kaminjolo, J. S., Jr., Johnson, L. W., Frank, H., and Gicho, J. N. (1974a). Vaccinia-like poxvirus indentified in a horse with a skin disease. *Zentralbl. Veterinaermed., Reihe B* **21**, 202.

Kaminjolo, J. S., Jr., Nyaga, P. N., and Gicho, J. N. (1947b). Isolation, cultivation and characterization of a poxvirus from some horses in Kenya. *Zentralbl. Veterinaermed., Reihe B* **21**, 592.

Kaminjolo, J. S., Jr., Johnson, L. W., Muhammed, S. I., and Berger, J. (1975). Uasin Gishu skin disease of horses in Kenya. *Bull. Anim. Health Prod. Afr.* **23**, 225.

Kaplan, C. (1969). Immunization against smallpox. *Br. Med. Bull.* **25**, 131.

Kaplan, C., Healing T. D., Evans, N., Healing, L., and Prior, A. (1980). Evidence of infection by viruses in small British field rodents. *J. Hyg.* **84**, 285.

Kassavetis, G. A., Zentner, P. G., and Geiduschek, E. P. (1986). Transcription at bacteriophage T4 variant late promoters. An application of a newly devised promoter-mapping method involving RNA chain retraction. *J. Biol. Chem.* **261**, 14256.

Kataria, R. S., and Singh, I. P. (1970). Serological relationship of buffalopox virus to vaccinia and cowpox viruses. *Acta Virol. (Engl. Ed.)* **14**, 307.

Kates, J. R., and Beeson, J. (1970a). Ribonucleic acid synthesis in vaccinia virus. II. Synthesis of polyriboadenylic acid. *J. Mol. Biol.* **50**, 19.

Kates, J. R., and Beeson, J. (1970b). Ribonucleic acid synthesis in vaccinia virus. I. The mechanism of synthesis and release of RNA in vaccinia cores. *J. Mol. Biol.* **50**, 1.

Kates, J. R., and McAuslan, B. R. (1967a). Messenger RNA synthesis by a "coated" viral genome. *Proc. Natl. Acad. Sci. U.S.A.* **57**, 314.

Kates, J. R., and McAuslan, B. R. (1967b). Poxvirus DNA-dependent RNA polymerase. *Proc. Natl. Acad. Sci. U.S.A.* **58**, 134.

Kato, S., Takahashi, M., Kameyama, S., and Kamahora, J. (1959). A study on the morphological and cyto-immunological relationship between the inclusions of variola, cowpox, rabbitpox, vaccinia (variola origin) and vaccinia IHD and a consideration of the term "Guarnieri body." *Biken J.* **2**, 353.

Kato, S., Kameyama, S., and Kamahora, J. (1960). Autoradiography with tritium-labelled thymidine of pox virus and human amnion cell system in tissue culture. *Biken J.* **3**, 135.

Katz, E., and Moss, B. (1970a). Formation of a vaccinia virus structural polypeptide from a high molecular weight precursor: Inhibition by rifampicin. *Proc. Natl. Acad. Sci. U.S.A.* **66**, 677.

Katz, E., and Moss, B. (1970b). Vaccinia virus structural polypeptide derived from a high-molecular-weight precursor: Formation and integration into the virus particle. *J. Virol.* **6**, 717.

Kaverin, N. V., Varich, N. L., Surgay, V. V., and Chernos, V. I. (1975). A quantitative estimation of poxvirus genome fraction transcribed as "early" and "late" mRNA. *Virology* **65**, 112.

Keidan, S. E., McCarthy, K., and Haworth, J. C. (1953). Fatal generalized vaccinia with failure of antibody production and absence of serum gamma globulin. *Arch. Dis. Child.* **28**, 110.

Kelsch, Teissier, and Camus (1909). De la variole-vaccine. Recherche expérimentelle presentée à l'Académie de Médicine. *Bull. Acad. Méd. (Paris)* **62**, 13.

Kemp, G. E., Causey, O. R., Setzer, H. W., and Moore, D. L. (1974). Isolation of viruses from wild mammals in West Africa 1966–1970. *J. Wildl. Dis.* **10**, 279.

Kempe, C. H. (1960). Studies on smallpox and complications of smallpox vaccination. *Pediatrics* **26**, 176.

Kempe, C. H. (1980). Acceptance of the Howland award. *Pediatr. Res.* **14**, 1155.

Kempe, C. H., Fulginiti, V., Mitamitami, M., and Shinefield, H. (1968). Smallpox vaccination of eczema patients with a strain of attenuated live vaccinia (CVI-78). *Pediatrics* **42**, 980.

Keogh, E. V. (1936). Titration of vaccinia virus on chorioallantoic membrane of chick embryo and its application to immunological studies of neuro-vaccinia. *J. Pathol. Bacteriol.* **43**, 441.

Khodakevich, L., Widy-Wirski, R., Arita, I., Marennikova, S. S., Nakano, J., and Meunier, D. (1985). Orthopoxvirose simiene de l'homme en République Centrafricaine. *Bull. Soc. Pathol. Exot.* **78**, 311.

Khodakevich, L., Jezek, Z., and Kinzanza, K. (1986). Isolation of monkeypox virus from a wild squirrel infected in nature. *Lancet* **1**, 98.

Khodakevich, L., Sczceniowski, M., Mambu-ma-Disu, Jezek, Z., Marennikova, S., Nakano, J., and Meier, F. (1987a). Monkeypox virus in relation to the ecological features surrounding human settlements in Bumba zone, Zaire. *Trop. Geogr. Med.* **39**, 56.

Khodakevich, L., Szczeniowski, M., Mambu-ma-Disu, Jezek, Z., Marennikova, S., Nakano, J., and Messinger, D. (1987b). The role of squirrels in sustaining monkeypox virus transmission. *Trop. Geogr. Med.* **39**, 115.

Kieny, M. P., Lathe, R., Drillien, R., Spehner, D., Skory, S., Schmitt, D., Wiktor, T., Koprowski, H., and Lecocq, J. P. (1984). Expression of rabies virus glycoprotein from a recombinant vaccinia virus. *Nature (London)* **312**, 163.

Kikuth, W., and Gönnert, R. (1940). Erzeugung von Ektromelie durch Provokation. *Arch. Gesamte Virusvorsch.* **1**, 295.

Kinchington, D., Dollery, A., Greenaway, P. J., and Dumbell, K. R. (1984). The detection of subtle differences between different orthopoxvirus genomes by heteroduplex analysis. *Virus Res.* **1**, 351.

Kirn, A., and Braunwald, J. (1964). Selection par passages à basse température d'un variant "froid" à virulence atténuée du virus vaccinial. *Ann. Inst. Pasteur, Paris* **106**, 427.

Kirtland, G., and Gold, C. (1802–1806). "Plates of the Small Pox and Cow Pox Drawn from Nature." G. Kirtland, London.

Kit, S., Jorgensen, G. N., Liau, A., and Zaslavsky, V. (1977). Purification of vaccinia virus induced thymidine kinase activity from (^{35}S) methionine-labelled cells. *Virology* **77**, 661.

Kitamoto, N., Goto, E., Tanimoto, S., Tanaka, T., Miyamoto, H., Wakamiya, N., Ikuta, K., Ueda, S., and Kato, S. (1984). Cross-reactivity between cowpox and vaccinia viruses with monoclonal antibodies. *Arch. Virol.* **82**, 129.

Kitamoto, N., Tanimoto, S., Hiroi, K., Ozaki, M., Miyamoto, H., Wakamiya, N., Ikuta, K., Ueda, S., and Kato, S. (1987). Monoclonal antibodies to cowpox virus: Polypeptide analysis of several major antigens. *J. Gen. Virol.* **68**, 239.

Kitamura, T. (1968). Studies on the formation of hyperplastic focus by variola virus in human cell cultures. I. *In vitro* quantitation of variola virus by focus counting in HeLa and FL cell cultures. *Virology* **36**, 174.

Kitamura, T., and Shinjo, N. (1972). Assay of neutralizing antibody against variola virus by the degree of focus reduction on HeLa cell cultures and its application to revaccination with smallpox vaccines of various potencies. *Bull. W.H.O.* **46**, 15.

Kitamura, T., and Tanaka, Y. (1973). Differential diagnosis of variola viruses by microfocus assay. *Bull. W.H.O.* **48**, 495.

Kleiman, J. H., and Moss, B. (1975a). Purification of a protein kinase and two phosphate acceptor proteins from vaccinia virions. *J. Biol. Chem.* **250**, 2420.

Kleiman, J. H., and Moss, B. (1975b). Characterization of a protein kinase and two phosphate acceptor proteins from vaccinia virions. *J. Biol. Chem.* **250**, 2430.

Kling, C., and Packalén, T. (1947). An experimental study on a derivative of Laigret-Durand's strain of murine rickettsiae for the preparation of living typhus vaccine. I. Serological and immunological analysis. *Acta Pathol. Microbiol. Scand.* **24**, 371.

Kozak, M. (1986). Bifunctional messenger RNAs in eukaryotes. *Cell (Cambridge Mass.)* **47**, 481.

Kraft, L. M., D'Amelio, E. D., and D'Amelio, F. E. (1982). Morphological evidence for natural poxvirus infection in rats. *Lab. Anim. Sci.* **32**, 648.

Krikun, V. A. (1974). [Isolation of pneumotropic virus from rats during the outbreak of disease of unknown etiology in rats nursery.] *Lab. Anim. Med. Stud.*, p. 96. (in Russian.)

Kristensson, K., Lundh, B., Norrby, E., Payne, L., and Orvell, C. (1984). Asymmetric budding of viruses in ependymal and choroid plexus epithelial cells. *Neuropathol. Appl. Neurobiol.* **10**, 209.

Kriz, B. (1982). A study of camelpox in Somalia. *J. Comp. Pathol.* **92**, 1.

Krupenko, S. S. (1972). [Camelpox caused by vaccinia.] *Veterinariya (Moscow)* **8**, 61 (in Russian).

Kulikova, O. S., and Lobanova, Z. I. (1980). [Spontaneous pox in laboratory rats.] *Veterinariya (Moscow)* **6**, 70 (in Russian).

Kumar, L., Abdul Salam, N. M., Datta, U., and Walia, B. N. S. (1977). Cell-mediated immuno-deficiency with normal immunoglobulins (Nezelof's syndrome) with progressive vaccinia. *Indian Pediatr.* **14**, 69.

Laboratory Animal Science (1981). Ectromelia (mousepox) in the United States. *Lab. Anim. Sci.* **31**, 549.

LaColla, P., and Weissbach, A. (1975). Vaccinia virus infection of HeLa cells. I. Synthesis of vaccinia DNA in host cell nuclei. *J. Virol.* **15**, 305.

Ladnyi, I. D., Ziegler, P., and Kims, A. (1972). A human infection caused by monkeypox virus in Basankusu Territory, Democratic Republic of the Congo. *Bull. W.H.O.* **46**, 593.

Ladnyi, I. D., Ogorodinkova, Z. I., Shelukhina, E. M., Gerasimenko, R. T., and Voronin,

U. S. (1975). [Poxvirus circulation among wild rodents in Turkemenia.] *Probl. Quarant. Dis.* **3-4**, 165 (in Russian).

Laigret, J., and Durand, R. (1939). La vaccination contre le typhus exanthématique. Nouvelle technique de préparation du vaccin: Emploi des cervaux de souris. *Bull. Soc. Pathol. Exot.* **32**, 735.

Laigret, J., and Durand, R. (1941). Précisions techniques sur le vaccin vivant en enrobé contre le typhus exanthématique. *Bull. Soc. Pathol. Exot.* **34**, 139.

Lake, J. R., and Cooper, P. D. (1980). Deletions of the terminal sequences in the genomes of the white pock (*u*) and host-restricted (*p*) mutants of rabbitpox virus. *J. Gen. Virol.* **48**, 135.

Lakritz, N., Foglesong, P. D., Reddy, M., Baum, S., Hurwitz, J., and Bauer, W. R. (1985). A vaccinia virus DNase preparation which cross-links superhelical DNA. *J. Virol.* **53**, 935.

Lal, S. M., and Singh, I. P. (1973). Serological characterization of buffalopox virus. *Arch. Gesamte Virusforsch.* **43**, 393.

Lal, S. M., and Singh, I. P. (1977). Buffalopox—a review. *Trop. Anim. Health Prod.* **9**, 107.

Lane, J. M., Ruben, F. L., Neff, J. M., and Millar, J. D. (1969). Complications of smallpox vaccination, 1968. National surveillance in the United States. *N. Engl. J. Med.* **281**, 1201.

Lane, J. M., Ruben, F. L., Neff, J. M., and Millar, J. D. (1970). Complications of smallpox vaccination, 1968: Results of ten statewide surveys. *J. Infect. Dis.* **122**, 303.

Langford, C. J., Edwards, S. J., Smith, G. L., Mitchell, G. F., Moss, B., Kemp, D. J., and Anders, R. F. (1986). Anchoring a secreted *Plasmodium* antigen on the surface of recombinant vaccinia virus-infected cells increases its immunogenicity. *Mol. Cell. Biol.* **6**, 3191.

Leake, J. P. (1927). Questions and answers on smallpox and vaccination. *Public Health Rep.* **60**, 221.

Ledingham, J. C. G. (1931). The aetiological importance of the elementary bodies in vaccinia and fowlpox. *Lancet* **2**, 525.

Lees, D. N., and Stephen, J. (1985). Ectromelia virus-induced changes in primary cultures of mouse hepatocytes. *J. Gen. Virol.* **66**, 2171.

Leese, A. S. (1909). Two diseases of young camels. *J. Trop. Vet. Sci.* **4**, 1.

Levaditi, C., and Sanchis-Bayani, V. (1927). Infection spontanée du lapin pour le virus du vaccin jennerien. *C. R. Seances Soc. Biol. Ses Fil.* **97**, 371.

Levaditi, C., Harvier, P., and Nicolau, S. (1922). Étude expérimentale de l'encéphalite dite "léthargique." *Ann. Inst. Pasteur, Paris* **36**, 139.

Levaditi, C., Lépine, P., and Schoen, R. (1931). Virulence et neurotropisme du virus vaccinal. *C. R. Seances Soc. Biol. Ses Fil.* **107**, 802.

Levaditi, C., Fasquelle, R., Beguignon, R., and Reinié, L. (1938). Influence des sélecteurs sur le potential encéphalitogène du vaccin jennerian. *C. R. Hebd. Seances Acad. Sci., Ser. D* **207**, 688.

Lewis, J. A. (1986). Structure and expression of the Chinese hamster thymidine kinase gene. *Mol. Cell. Biol.* **6**, 1998.

Lin, P.-F., Lieberman, H. B., Yeh, D.-B., Tian, X., Zhao, S.-Y., and Ruddle, F. H. (1985). Molecular cloning and structural analysis of murine thymidine kinase genomic and cDNA sequences. *Mol. Cell. Biol.* **5**, 3149.

Lillie, R. D. (1930). Smallpox and vaccinia. The pathologic histology. *Arch. Pathol.* **10**, 241.

Loeffler, F., and Frosch, P. (1898). Berichte der Kommission zur Erforschung der Maul- und Klauenseuche bei dem Institut für Infektionskrankheiten in Berlin. *Zentralbl. Bakteriol., Parasitenkd., Infecktionskr. Hyg., Abt.1: Orig.* **23**, 371.

Loh, P. C., and Riggs, J. L. (1961). Demonstration of the sequential development of vaccinial antigens and virus in infected cells: Observations with cytochemical and differential fluorescent procedures. *J. Exp. Med.* **114**, 149.

Lourie, B., Bingham, P. G., Evans, H. H., Foster, S. O., Nakano, J. H., and Herrmann, K. L. (1972). Human infection with monkeypox virus: Laboratory investigation of six cases in West Africa. *Bull. W.H.O.* **46**, 633.

Lourie, B., Nakano, J. H., Kemp, G. E., and Setzer, H. W. (1975). Isolation of poxvirus from an African rodent. *J. Infect. Dis.* **132**, 677.

Low, R. B. (1918). "The Incidence of Small-pox Throughout the World in Recent Years," Reports to the Local Government Board on Public Health and Medical Subjects, NS: No. 117. H. M. Stationery Office, London.

Lum, G. S., Soriano, F., Trejos, A., and Llerena, J. (1967). Vaccinia epidemic and epizootic in El Salvador. *Am. J. Trop. Med. Hyg.* **16**, 332.

Lwoff, A., Anderson, T. F., and Jacob, F. (1959). Remarques sur les caractéristiques de la particule virale infectieuse. *Ann. Inst. Pasteur, Paris* **97**, 281.

McAuslan, B. R. (1963a). Control of induced thymidine kinase activity in the poxvirus infected cell. *Virology* **20**, 162.

McAuslan, B. R. (1963b). The induction and repression of thymidine kinase in the poxvirus-infected HeLa cell. *Viology* **21**, 383.

MacCallum, F. O., and McDonald, J. R. (1957). Survival of variola virus in raw cotton. *Bull. W.H.O.* **16**, 247.

MacCallum, F. O., Scott, T. F. M., Dalldorf, G., and Gifford, R. (1957). "Pseudo-lymphocytic choriomeningitis": A correction. *Br. J. Exp. Pathol.* **38**, 120.

MacCallum, W. G., and Moody, L. M. (1921). Alastrim in Jamaica. *Am. J. Hyg.* **1**, 388.

MacCallum, W. G., and Oppenheimer, E. H. (1922). Differential centrifugalization. A method for the study of filterable viruses, as applied to vaccinia. *J. Am. Med. Assoc.* **78**, 410.

McCarron, R. J., Cabrera, C. V., Esteban, M., McAllister, W. T., and Holowczak, J. A. (1978). Structure of vaccinia DNA: Analysis of the viral genome by restriction endonucleases. *Virology* **86**, 88.

McCarthy, K., and Downie, A. W. (1948). An investigation of immunological relationships between the viruses of variola, vaccinia, cowpox and ectromelia by neutralization tests on the chorioallantois of chick embryos. *Br. J. Exp. Pathol.* **29**, 501.

McCarthy, K., Downie, A. W., and Armitage, P. (1958a). The antibody response in man following infection with viruses of the pox group. I. An evaluation of the pock counting method for measuring neutralizing antibody. *J. Hyg.* **56**, 84.

McCarthy, K., Downie, A. W., and Bradley, W. H. (1958b). The antibody response in man following infection with viruses of the pox group. II. Antibody response following vaccination. *J. Hyg.* **56**, 466.

McClain, M. E. (1965). The host range and plaque morphology of rabbitpox virus (RPu^+) and its u mutants on chick fibroblast, PK-2a and L929 cells. *Aust. J. Exp. Biol. Med. Sci.* **43**, 31.

McConnell, S. J., Herman, Y. F., Mattson, D. E., and Erickson, L. (1962). Monkey pox disease in irradiated cynomolgus monkeys. *Nature (London)* **195**, 1128.

McConnell, S. J., Herman, Y. F., Mattson, D. E., Huxsoll, L., Lang, C. M., and Yager, R. H. (1964). Protection of rhesus monkeys against monkeypox by vaccinia virus immunization. *Am. J. Vet. Res.* **25**, 192.

McConnell, S. J., Hickman, R. L., Wooding, W. J., Jr., and Huxsoll, D. L. (1968). Monkeypox: Experimental infection in chimpanzee (*Pan satyrus*) and immunization with vaccinia virus. *Am. J. Vet. Res.* **29**, 1675.

McCrae, M. W., and Szilagyi, J. R. (1975). Preparation and characteristics of a subviral particle of vaccinia virus containing the DNA-dependent RNA polymerase activity. *Virology* **68**, 234.

McFadden, G., and Dales, S. (1979). Biogenesis of poxviruses: Mirror-image deletions in vaccinia virus DNA. *Cell (Cambridge, Mass.)* **18**, 101.

McGaughey, C. A., and Whitehead, R. (1933). Outbreaks of infectious ectromelia in laboratory and wild mice. *J. Pathol. Bacteriol.* **37**, 253.

Mack, T. M. (1972). Smallpox in Europe, 1950-1971. *J. Infect. Dis.* **125**, 161.

Mack, T. M., and Noble, J., Jr. (1970). Natural transmission of smallpox from man to performing monkeys. An ecological curiosity. *Lancet* **1**, 752.

Mackett, M. (1981). Restriction endonuclease analysis of orthopoxvirus DNA. Ph.D. Thesis, University of London.

Mackett, M., and Archard, L.C. (1979). Conservation and variation in orthopoxvirus genome structure. *J. Gen. Virol.* **45**, 683.

Mackett, M., and Arrand, J. R. (1985). Recombinant vaccinia virus induces neutralizing antibodies in rabbits against Epstein-Barr virus membrane antigen gp340. *EMBO J.* **4**, 3229.

Mackett, M., and Smith, G. L. (1986). Vaccinia virus expression vectors. *J. Gen. Virol.* **67**, 2067.

Mackett, M., Smith G. L., and Moss, B. (1982). Vaccinia virus: A selectable eukaryotic cloning and expression vector. *Proc. Natl. Acad. Sci. U.S.A.* **79**, 7415.

Mackett, M., Smith, G. L., and Moss, B. (1984). General method for the production and selection of infectious vaccinia virus recombinants expressing foreign genes. *J. Virol.* **49**, 857.

Mackett, M., Yilma, T., Rose, J., and Moss, B. (1985a). Vaccinia virus recombinants: Expression of VSV genes and protective immunization of mice and cattle. *Science* **227**, 433.

Mackett, M., Smith, G. L., and Moss, B. (1985b). The construction and characterization of vaccinia virus recombinants expressing foreign genes. *In* "DNA Cloning: A Practical Approach" (D. M. Glover, ed.), p. 191. I. R. L. Press, Oxford.

Mackie, T. J., and van Rooyen, C. E. (1937). John Brown Buist, MD (Edin) BSc (Edin) FRCP Ed, FRS Ed, 1846–1915; acknowledgement of his early contribution to the bacteriology of variola and vaccinia. *Edinburgh Med. J.* **44**, 72.

McNeill, T. A. (1968). The neutralization of pox viruses. II. Relations between vaccinia, rabbitpox, cowpox and ectromelia. *J. Hyg.* **66**, 549.

Madeley, C. R. (1968). The immunogenicity of heat-inactivated vaccinia virus in rabbits. *J. Hyg.* **66**, 89.

Mahnel, H. (1974). Labordifferenzierung der Orthopockenviren. *Zentralbl. Veterinaermed., Reihe B* **21**, 242.

Mahnel, H. (1983). Attenuierung von Mäusepocken (Ektromelie)-Virus. *Zentralbl. Veterinaermed., Reihe B* **30**, 701.

Mahnel, H. (1985a). Schutzimpfung gegen Mäusepocken. *Tieraerztl. Prax.* **13**, 403.

Mahnel, H. (1985b). Trinkwasserimpfung gegen mäusepocken (Ektromelie). *Zentralbl. Veterinaermed., Reihe B* **32**, 479.

Mahr, A., and Roberts, B. (1984). Arrangement of late mRNAs transcribed from a 7.1 kilobase *Eco*RI vaccinia virus DNA fragment. *J. Virol.* **49**, 510.

Maiboroda, A. D. (1982). Experimental infection of Norwegian rats (*Rattus norvegicus*) with ratpox virus. *Acta Virol. (Engl. Ed.)* **26**, 288.

Maiboroda, A. D., Lobanova, Z. I., and Dushkin, V. (1980). Pathogenicity of ratpox virus for laboratory mice. *Z. Versuchtierkd.* **22**, 25.

Maitland, H. B., and Maitland, M. C. (1928). Cultivation of vaccinia virus without tissue culture. *Lancet* **2**, 596.

Maltseva, N. N., and Marennikova, S. S. (1976). A method for serological differentiation of closely related poxviruses. *Acta Virol. (Engl. Ed.)* **20**, 250.

Maltseva, N. N., Akatova-Shelukhina, E. M., Yumasheva, M. A., and Marennikova, S. S. (1966). The aetiology of epizootics of certain smallpox-like infections in cattle and methods of differentiating cowpox, vaccinia and swine pox viruses. *J. Hyg., Epidemiol., Microbiol., Immunol.* **10**, 202.

Maltseva, N. N., Marennikova, S. S., Nakano, J., Matsevich, G. R., Khabakhpashaeva, N. A., Shelukhina, E. M., Arita, I., Gromyko, A. I., and Stepanova, L. G. (1984). [Data of the serological survey of the population of the Republic of Congo for the presence of antibodies to orthopoxviruses: Communication II. Specific identification of antibodies by means of solid-phase ELISA.] *J. Microbiol. Epidemiol. Immunol.* 64. (in Russian.)

Manning, P. J., and Fisk, C. S. (1981). Clinical, pathologic and serologic features of an epizootic of mousepox in Minnesota. *Lab. Anim. Sci.* **31**, 574.

Marchal, J. (1930). Infectious ectromelia. A hitherto undescribed virus disease of mice. *J. Pathol. Bacteriol.* **33**, 713.

Marennikova, S. S. (1979). Field and experimental studies of poxvirus infections in rodents. *Bull. W.H.O.* **57**, 461.

Marennikova, S. S., and Kaptsova, T. I. (1965). Age dependence of susceptibility of white mice to variola virus. *Acta Virol. (Engl. Ed.)* **9**, 230.

Marennikova, S. S., and Macevic, G. R. (1975). Experimental study of the role of inactivated vaccine in two-step vaccination against smallpox. *Bull. W.H.O.* **52**, 51.

Marennikova, S. S., and Shelukhina, E. M. (1976a). White rats as a source of pox infection in carnivora of the family *Felidae*. *Acta Virol. (Engl. Ed.)* **20**, 442.

Marennikova, S. S., and Shelukhina, E. M. (1976b). Susceptibility of some rodent species to monkeypox virus, and course of the infection. *Bull. W.H.O.* **53**, 13.

Marennikova, S. S., and Shelukhina, E. M. (1978). Whitepox virus isolated from hamsters inoculated with monkeypox virus. *Nature (London)* **276**, 291.

Marennikova, S. S., Gurvich, E. B., and Yumasheva, M. A. (1964). Laboratory diagnosis of smallpox and similar virus diseases by means of tissue culture methods. II. Differentiation of smallpox virus from varicella, vaccinia, cowpox and herpes viruses. *Acta Virol. (Engl. Ed.)* **8**, 135.

Marennikova, S. S., Gurvich, E. B., and Shelukhina, E. M. (1971). Comparison of the properties of five pox virus strains isolated from monkeys. *Arch. Gesamte Virusforsch.* **33**, 201.

Marennikova, S. S., Shelukhina, E. M., Maltseva, N. N., Cimiskjan, K. L., and Macevic, G. R. (1972a). Isolation and properties of the causal agent of a new variola-like disease (monkeypox) in man. *Bull. W.H.O.* **46**, 599.

Marennikova, S. S., Shelukhina, E. M., Maltseva, N. N., and Ladnyj, I. D. (1972b). Poxviruses isolated from clinically ill and asymptomatically infected monkeys and a chimpanzee. *Bull. W.H.O.* **46**, 613.

Marennikova, S. A., Shenkman, L. S., Shelukhina, E. M., and Maltseva, N. N. (1974). Isolation of camel pox virus and investigation of its properties. *Acta Virol. (Engl. Ed.)* **18**, 423.

Marennikova S. S., Shelukhina, E. M., Shenkman, L. S., Maltseva, N. N., and Matsevich, G. R. (1975a). [The results of a survey of wild monkeys for the presence of smallpox antibodies and pox-group viruses.] *Vopr. Virusol.* **20**(3), 321 (in Russian).

Marennikova, S. S., Maltseva, N. N., Korneeva, V. I., and Garanina, V. M. (1975b). Pox infection in carnivora of the family *Felidae*. *Acta Virol. (Engl. Ed.)* **19**, 260.

Marennikova, S. S., Shelukhina, E. M., and Shenkman, L. S. (1976a). "White-wild" (variola-like) poxvirus strain from rodents in Equatorial Africa. *Acta Virol. (Engl. Ed.)* **20**, 80.

Marennikova, S. S., Shelukhina, E. M., Matsevich, G. R., and Habahpasheva, N. A. (1976b). An antigenically atypical strain of variola virus. *J. Gen. Virol.* **33**, 513.

Marennikova, S. S., Maltseva, N. N., Korneeva, V. I., and Garanina, N. M. (1977). Pox outbreak among carnivora (Felidae) and Edentata. *J. Infect. Dis.* **135**, 358.

Marennikova, S. S., Shelukhina, E. M., and Finina, V. A. (1978a). Pox infection in white rats. *Lab. Anim.* **12**, 33.

Marennikova, S. S., Ladnyi, I. D., Ogorodnikova, Z. I., Shelukhina, E. M., and Maltseva, N. N. (1978b). Identification and study of a poxvirus isolated from wild rodents in Turkmenia. *Arch. Virol.* **56**, 7.

Marennikova, S. S., Shelukhina, E. M., Maltseva, N. N., and Matsevich, G. R. (1979). Monkeypox virus as a source of whitepox viruses. *Intervirology* **11**, 333.

Marennikova, S. S., Maltseva, N. N., and Habahpaseva, N. A. (1981). ELISA—a simple test for detecting and differentiating antibodies to closely related orthopoxviruses. *Bull. W.H.O.* **59**, 365.

Marennikova, S. S., Shelukhina, E. M., Maltseva, N. N., Efremova, E. V., Matsevich, G. R., Nikulina, V. G., Khabakhpasheva, N. A., Stepanova, L. G., Arita, I., and Gromyko, A. I. (1984). [Data obtained in the serological survey of the population of the Republic of Congo for the presence of antibodies to orthopoxviruses. Communication I. Comparative evaluation of different methods of investigation and total results.] *J. Microbiol. Epidem. Immunol.* **(3)**, 95. (in Russian).

Marquardt, J., Holm, S. E., and Lycke, E. (1965). Immunoprecipitating factors of vaccinia virus. *Virology* **27**, 170.

Marsden, J. P. (1936). "A Critical Review of the Clinical Features of 13,686 Cases of Smallpox (Variola Minor)," London County Council, Rep. No. 3209. London. (Reprinted as: Variola minor. A personal analysis of 13,686 cases. *Bull. Hyg.* **23**, 735, 1948)

Marshall, I. D. (1959). The influence of ambient temperature on the course of myxomatosis in rabbits. *J. Hyg.* **57**, 484.

Martin, S. A., and Moss, B. (1975). Modification of RNA by mRNA guanylyltransferase and mRNA (guanine-7-)methyltransferase from vaccinia virions. *J. Biol. Chem.* **250**, 9330.

Martin, S. A., and Moss, B. (1976). mRNA guanylyltransferase and mRNA (guanine-7-)methyltransferase from vaccinia virions. Donor and acceptor substrate activities. *J. Biol. Chem.* **251**, 7313.

Martin, S. A., Paoletti, E., and Moss, B. (1975). Purification of mRNA guanylyltransferase and mRNA (guanine-7-)methyltransferase from vaccinia virus. *J. Biol. Chem.* **250**, 9322.

Mathew, G. (1976). Comparative studies on the propagation of pox viruses in the chick embryos with special reference to buffalopox virus. *Kerala J. Vet. Sci.* **7**, 48.

Matsumoto, S. (1958). Electron microscope studies of ectromelia virus multiplication. *Annu. Rep. Inst. virus Res., Kyoto Univ.* **1**, 151.

Matthews, R. E. F. (1979). Classification and nomenclature of viruses. Third report of the International Committee on Taxonomy of Viruses. *Intervirology* **12**, 132.

Matthews, R. E. F. (1982). Classification and nomenclature of viruses. Fourth report of the International Committee on Taxonomy of Viruses. *Intervirology* **17**, 4.

Matthews, R. E. F., ed. (1983). "A Critical Appraisal of Viral Taxonomy." CRC Press, Boca Raton, Florida.

Mayr, A. (1966). Eine einfache und schnelle Methode zur Differenzierung zwischen

Vaccine - (poxvirus officinale) und Kuhpockenvirus (poxvirus bovis). *Zentralbl. Bakteriol., Parasitenkd., Infektionskr. Hyg., Abt. 1:Orig.* **199**, 144.

Mayr, A., and Herrlich, A. (1960). Zuchtung des variolavirus in der infantilen Maus. *Arch. Gesamte Virusforsch.* **10**, 226.

Mayr, A., Stickl, H., Müller, H. K., Danner, K., and Singer, H. (1978). The smallpox vaccination strain MVA: Marker, genetic structure, experience gained with the parenteral vaccination and behavior in organisms with a debilitated defence mechanism. *Zentralbl. Bakteriol. Parasitenkd. Infektionskr. Hyg., Abt. 1:Orig., Reihe B* **167**, 375.

Medzon, E. L., and Bauer, H. (1970). Structural features of vaccinia virus revealed by negative staining, sectioning and freeze-etching. *Virology* **40**, 860.

Melnick, J. E., and Gaylord, W. H., Jr. (1953). Problems with spontaneous ectromelia (mouse pox) in a virus laboratory. *Proc. Soc. Exp. Biol. Med.* **83**, 315.

Menna, A., Wittek, R., Bachmann, P. A., Mayr, A., and Wyler, R. (1979). Physical characterization of a stomatitis papulosa virus genome: A cleavage map for the restriction endonucleases *Hind*III and *Eco*RI. *Arch. Virol.* **59**, 145.

Merchlinsky, M., and Moss, B. (1986). Resolution of linear minichromosomes with hairpin ends for circular plasmids containing vaccinia virus concatemer junctions. *Cell (Cambridge, Mass.)* **45**, 879.

Michelson, H. E., and Ikeda, K. (1927). Microscopic changes in variola. *Arch. Dermatol. Syph.* **15**, 138.

Mika, L. A., and Pirsch, J. B. (1960). Differentiation of variola from other members of the poxvirus group by the plaque technique. *J. Bacteriol.* **81**, 861.

Milhaud, C., Klein, M., and Virat, J. (1969). Analyse d'un cas de variole du singe (monkey-pox) chez le chimpanzee (*Pan troglodytes*). *Exp. Anim.* **2**, 121.

Millard, C. K. (1914). "The Vaccination Question in the Light of Modern Experience." Lewis, London.

Mills, T., and Pratt, B. C. (1980). Differentiation of ectromelia virus haemagglutinin from haemaggutinins of other poxviruses. *Arch. Virol.* **63**, 153.

Mims, C. A. (1959). The response of mice to large intravenous injections of ectromelia virus. II. The growth of virus in the liver. *Br. J. Exp. Pathol.* **40**, 543.

Mims, C. A. (1960). Intracerebral injections and the growth of viruses in the mouse brain. *Br. J. Exp. Pathol.* **41**, 52.

Mims, C. A. (1964). Aspects of the pathogenesis of virus diseases. *Bacteriol. Rev.* **28**, 30.

Mims, C. A. (1966). The pathogenesis of rashes in virus diseases. *Bacteriol. Rev.* **30**, 739.

Mims, C. A. (1968). The response of mice to the intravenous injection of cowpox virus. *Br. J. Exp. Pathol.* **49**, 24.

Mims, C. A., and White, D. O. (1984). "Viral Pathogenesis and Immunology." Blackwell, Oxford.

Minnigan, H., and Moyer, R. W. (1985). Intracellular location of rabbit poxvirus nucleic acid within infected cells as determined by *in situ* hybridization. *J. Virol.* **55**, 634.

Mirchamsy, H., and Ahourai, P. (1971). Comparative adaptation of some poxviruses in two cell systems. *Arch. Inst. Razi* **23**, 93.

Mooser, H. (1943). Über die Mischinfection der weissen Maus mit einem Stamm Klassischen Fleckfiebers und dem Virus der infectiösen Ektromelie. *Schweiz. Z. Pathol. Bakteriol.* **6**, 463.

Morgan, C. (1976a). The insertion of DNA into vaccinia virus. *Science* **193**, 591.

Morgan, C. (1976b). Vaccinia virus reexamined: Development and release. *Virology* **73**, 43.

Morgan, C., Ellison, S. A., Rose, H. M., and Moore, D. H. (1954). Structure and development of viruses observed in the electron microscope. II. Vaccinia and fowlpox viruses. *J. Exp. Med.* **100**, 301.

Morita, M., Suzuki, K., Yasuda, A., Kojima, A., Sugimoto, M., Kobayashi, H., Watanabe, K., Kajiyama, K., and Hashizume, S. (1987). Recombinant vaccinia virus LC16mo or LC16m8 that expresses hepatitis B surface antigen while preserving the attenuation of the parental virus strain. *Vaccine* **5**, 65.

Morrison, D. K., and Moyer, R. W. (1986). Detection of a subunit of cellular polII within highly purified preparations of RNA polymerase isolated from rabbit poxvirus virions. *Cell (Cambridge, Mass.)* **44**, 587.

Morrison, D. K., Carter, J. K., and Moyer, R. W. (1985). Isolation and characterization of monoclonal antibodies directed against two subunits of rabbit poxvirus-associated DNA-directed RNA polymerase. *J. Virol.* **55**, 670.

Morse, H. C., III (1981). The laboratory mouse—a historical perspective. *In* "The Mouse in Biomedical Research" (H. L. Foster, J. D. Small, and J. G. Fox, eds.), Vol. 1, p. 1. Academic Press, New York.

Moss, B. (1974). Reproduction of poxviruses. *Compr. Virol.* **3**, 405.

Moss, B. (1985). Replication of poxviruses. *In* "Virology" (B. N. Fields *et al.*, eds.), p. 685. Raven Press, New York.

Moss, B., and Rosenblum, E. M. (1973). Protein cleavage and poxvirus morphogenesis: Tryptic peptide analysis of core precursors accumulated by blocking assembly with rifampicin. *J. Mol. Biol.* **19**, 65.

Moss, B., and Rosenblum, E. N. (1974). Vaccinia virus polyriboadenylate polymerase: Covalent linkage of the product with polyribonucleotide and polydeoxynucleotide primers. *J. Virol.* **14**, 86.

Moss, B., and Salzman, N. P. (1968). Sequential protein synthesis following vaccinia virus infection. *J. Virol.* **2**, 1016.

Moss, B., Rosenblum, E. N., and Gershowitz, A. (1975). Characterization of a polyriboadenylate polymerase from vaccinia virions. *J. Biol. Chem.* **250**, 4722.

Moss, B., Winters, E., and Cooper, N. (1981a). Instability and reiteration of DNA sequences within the vaccinia virus genome. *Proc. Natl. Acad. Sci. U.S.A.* **78**, 1611.

Moss, B., Winters, E., and Cooper, J. A. (1981b). Deletion of a 9,000 base pair segment of the vaccinia virus genome that encodes non-essential polypeptides. *J. Virol.* **40**, 387.

Moyer, R. W. (1987). The role of the host cell nucleus in vaccinia virus morphogenesis. *Virus Res.* **8**, 173.

Moyer, R. W., and Graves, R. L. (1981). The mechanism of cytoplasmic orthopoxvirus DNA replication. *Cell (Cambridge, Mass.)* **27**, 391.

Moyer, R. W., and Rothe, C. T. (1980). The white pock mutants of rabbit poxvirus. I. Spontaneous host range mutants contain deletions. *Virology* **102**, 199.

Moyer, R. W., Graves, R. L., and Rothe, C. T. (1980). The white pock (*u*) mutants of rabbit poxvirus. III. Terminal DNA sequence duplication and transposition in rabbit poxvirus. *Cell (Cambridge, Mass)* **22**, 545.

Müller, H. K., Wittek, R., Schaffner, W., Schümperli, D., Menna, A., and Wyler, R. (1978). Comparison of five poxvirus genomes by analysis with restriction endonucleases *Hind*III, *Bam*I and *Eco*RI. *J. Gen. Virol.* **38**, 135.

Munyon, W., Paoletti, E., and Grace, J. T., Jr. (1967). RNA polymerase activity in purified infectious vaccinia virus. *Proc. Natl. Acad. Sci. U.S.A.* **58**, 2280.

Munz, E., Reimann, M., Hoffman, R., and Göbel, E. (1974). Perorale Immunisierungversuche mit Vaccinia-Virus gegen Mäusepocken. *Zentralbl. Bakteriol., Parasitenkd., Infectionskr. Hyg., Abt. 1:Orig., Reihe B* **159**, 10.

Munz, E., Reimann, M., and Zschekel, W. D. (1976). Die Aerosol-Impfung gegen Mäusepocken (Infektiose Ektromelie) in Vergleich zu anderer Vakzinationsverfahren unter Verwerdung von Vaccinia-und Mausepockenvirushaltigen Impfstoffen. *Zentralbl. Veterinaermed., Reihe B* **23**, 431.

Murphy, F. A. (1985). Consideration of safety, efficacy, and potential application of vaccinia vectored vaccines for immune prophylaxis against animal diseases. *In* "Vaccinia Viruses as Vectors for Vaccine Antigens" (G. V. Quinnan, Jr., ed.), p. 237. Am. Elsevier, New York.

Murphy, W. J., Watkins, K. P., and Agabian, N. (1986). Identification of a novel Y branch structure as an intermediate in trypanosomes mRNA processing: Evidence for *trans* splicing. *Cell (Cambridge, Mass.)* **47**, 517.

Mutombo, M. W., Arita, I., and Jezek, Z. (1983). Human monkeypox transmitted by a chimpanzee in a tropical rain-forest area of Zaire. *Lancet* **1**, 735.

Nagington, J., and Horne, R. W. (1962). Morphological studies of orf and vaccinia viruses. *Virology* **16**, 248.

Nagler, F. P. O. (1942). Application of Hirst's phenomenon to the titration of vaccinia-immune serum. *Med. J. Aust.* **1**, 281.

Nagler, F. P. O. (1944). Red cell agglutination by vaccinia virus. Application to a comparative study of vaccination with egg vaccine and standard calf lymph. *Aust. J. Exp. Biol. Med. Sci.* **22**, 29.

Nakano, J. H. (1973). Evaluation of virological laboratory methods for smallpox diagnosis. *Bull. W.H.O.* **48**, 529.

Nakano, J. H. (1978). Comparative diagnosis of poxvirus diseases. *In* "Comparative Diagnosis of Viral Diseases: Human and Related Viruses" (E. Kurstak and C. Kurstak, eds.), Vol. 1, Part A, p. 287. Academic Press, New York.

Nakano, J. H. (1979). Poxviruses. *In* "Diagnostic Procedures for Viral, Rickettsial and Chlamydial Infections" (E. H. Lennette and N. J. Schmidt, eds.), 5th ed., p. 257. Am. Public Health Assoc., New York.

Nakano, J. H. (1985). Human poxvirus diseases. *In* "Laboratory Diagnosis of Viral Infections" (E. H. Lennette, ed.), p. 401. Dekker, New York.

Nakano, J. H. (1986). Poxviruses. *In* "Clinical Virology Manual" (S. Spector and G. J. Lancz, eds.), p. 417. Am. Elsevier, New York.

Needham, J. (1980). "China and the Origins of Immunology," Occas. Pap. Monogr. No. 41. Centre of Asian Studies, University of Hong Kong.

Needham, J., and Lu, G. -D. (1988). Smallpox in history. *In* "Science and Civilization in China," (J. Needham, ed.). Vol. 6, Part 4. Cambridge Univ. Press, London and New York (in press).

Negri, A. (1906). Ueber Filtration des Vaccinevirus. *Z. Hyg. Infektionskr.* **54**, 327.

Nelson, J. B. (1939). The behavior of pox viruses in respiratory tract. II. Response of mice to the nasal installation of variola virus. *J. Exp. Med.* **70**, 107.

Nelson, J. B. (1943). Stability of variola virus propagated in embryonated eggs. *J. Exp. Med.* **78**, 231.

Nevins, J. R., and Joklik, W. K. (1975). Poly(A) sequences of vaccinia virus messenger RNA: Nature, mode of addition and function during translation *in vitro* and *in vivo*. *Virology* **63**, 1.

Nevins, J. R., and Joklik, W. K. (1977). Isolation and properties of the vaccinia virus DNA-dependent RNA polymerase. *J. Biol. Chem.* **252**, 6930.

Newton, S., Mackett, M., Francis, M. J., and Brown, F. (1986). Expression of a foot and mouth disease immunogenic site in vaccinia virus. *In* "Vaccines 86" (F. Brown, R. M. Chanock, and R. A. Lerner, eds.). p. 303. Cold Spring Harbor Lab., Cold Spring Harbor, New York.

Nicolau, S., and Kopciowska, L. (1929). Epizootie maligne spontanée apparue chez des lapins. *C. R. Seances Soc. Biol. Ses Fil.* **101**, 551.

Niles, E. G., Condit, R. C., Caro, P., Davidson, K., Matusick, L., and Seto, J. (1986)

Nucleotide sequence and genetic map of the 16kb vaccinia virus *Hin*dIII D fragment. *Virology* **153**, 96.

Nizamuddin, M., and Dumbell, K. R. (1961). A single laboratory test to distinguish the virus of smallpox from that of alastrim. *Lancet* **1**, 68.

Noble, J., Jr. (1970). A study of New and Old World monkeys to determine the likelihood of a simian reservoir of smallpox. *Bull. W.H.O.* **42**, 509.

Noble, J., Jr., and Rich, J. A. (1969). Transmission of smallpox by contact and by aerosol routes in *Macaca irus*. *Bull. W.H.O.* **40**, 279.

Noguchi, H. (1915). Pure cultivation *in vivo* of vaccine virus free from bacteria. *J. Exp. Med.* **21**, 539.

North, E. A., Broben, J. A., and Mengoni, A. H. (1944). The use of the chorioallantois of the developing chick embryo in the diagnosis of smallpox. *Med. J. Aust.* **1**, 437.

Noyes, W. F. (1962). Further studies on the structure of vaccinia virus. *Virology* **18**, 511.

Obijeski, J. F., Palmer, E. C., Gafford, L. G., and Randall, C. C. (1973). Polyacrylamide gel electrophoresis of fowlpox and vaccinia virus proteins. *Virology* **51**, 512.

Oda, K., and Joklik, W. K. (1967). Hybridization and sedimentation studies on "early" and "late" vaccinia messenger RNA. *J. Mol. Biol.* **27**, 395.

Olitsky, P. K., and Long, P. H. (1929). Relation of vaccinial immunity to the persistence of the virus in rabbits. *J. Exp. Med* **50**, 263.

Olmsted, R. A., Elango, N., Prince, G. A., Murphy, B. R., Johnson, P. R., Moss, B., Chanock, R. M., and Collins, P. L. (1986). Expression of F glycoprotein of respiratory syncytial virus by a recombinant vaccinia virus: Comparisons of the individual contributions of the F and G glycoproteins to host immunity. *Proc. Natl. Acad. Sci. U.S.A.* **83**, 7462.

O'Neill, H. C., and Blanden, R. V. (1983). Mechanisms determining innate resistance to ectromelia in C57BL mice. *Infect. Immun.* **41**, 1391.

O'Neill, H. C., and Brenan, M. (1987). A role for cytotoxic T cells in resistance to ectromelia infection in mice. *J. Gen. Virol.* **68**, 2669.

O'Neill, H. C., Blanden, R. V., and O'Neill, T. J. (1983). H-2-linked control of resistance to ectromelia virus infection in B10 congenic mice. *Immunogenetics* **18**, 255.

Ono, K., and Kato, S. (1968). Lack of cell proliferation in the foci of variola virus-infected FL cells. *Biken J.* **11**, 333.

Osterhaus, A. D. M. E., Teppema, J. S., Wirahadiredja, R. M. S., and van Steenis, G. (1981). Mousepox in the Netherlands. *Lab. Anim. Sci.* **31**, 704.

Otten, L. (1927). Trockenlymphe. *Z. Hyg. Infektionskr.* **107**, 677.

Owen, D., Hill, A., and Argent, S. (1975). Reaction of mouse strains to skin test for ectromelia using an allied virus as inoculum. *Nature (London)* **254**, 598.

Packalén, T. (1945). Rickettsial agglutination and complement fixation studies in epidemic typhus fever. *Acta Pathol. Microbiol. Scand.* **22**, 573.

Packalén, T. (1947). An experimental study on a derivative of Laigret-Durand's strain of murine rickettsiae for the preparation of living typhus vaccine. II. Isolation of ectromelia virus from a vaccine strain. *Acta Pathol. Microbiol. Scand.* **24**, 375.

Padgett, B. L., and Tomkins, J. K. N. (1968). Conditional lethal mutants of rabbitpox virus. 3. Temperature-sensitive (*ts*) mutants; physiological properties, complementation and recombination. *Virology* **36**, 161.

Palmer, E. L., Ziegler, D. W., Kissling, R. E., Hutchinson, H. D., and Murphy, F. A. (1968). Poxvirus nature of the Motol virus. *J. Infect. Dis.* **118**, 500.

Panicali, D., and Paoletti, E. (1982). Construction of poxviruses as cloning vectors: Insertion of the thymidine kinase gene from herpes simplex into the DNA of infectious vaccinia virus. *Proc. Natl. Acad. Sci. U.S.A.* **79**, 4927.

Panicali, D., Davis, S. W., Mercer, S. R., and Paoletti, E. (1981). Two major DNA variants present in serially propagated stocks of the WR strain of vaccinia virus. *J. Virol.* **37,** 1000.

Panicali, D., Davis, S. W., Weinberg, R. L., and Paoletti, E. (1983). Construction of live vaccines by using genetically engineered poxviruses: Biological activity of recombinant vaccinia virus expressing influenza virus hemagglutinin. *Proc. Natl. Acad. Sci. U.S.A.* **80,** 5364.

Panicali, D., Grzelecki, A., and Huang, C. (1986). Vaccinia virus vectors utilizing the β-galactosidase assay for rapid selection of recombinant viruses and measurement of gene expression. *Gene* **47,** 193.

Paoletti, E. (1977a). *In vitro* synthesis of a high molecular weight virion-associated RNA by vaccinia. *J. Biol. Chem.* **252,** 866.

Paoletti, E. (1977b). High molecular weight virion-associated RNA of vaccinia. A possible precursor to 8–12 S mRNA. *J. Biol. Chem.* **252,** 872.

Paoletti, E., and Grady, L. J. (1977). Transcriptional complexity of vaccinia virus *in vivo* and *in vitro*. *J. Virol.* **23,** 608.

Paoletti, E., and Lipinskas, B. R. (1978a). Soluble endoribonuclease activity from vaccinia virus: Specific cleavage of virion-associated high molecular weight RNA. *J. Virol.* **26,** 822.

Paoletti, E., and Lipinskas, B. R. (1978b). The role of ATP in the biogenesis of vaccinia virus mRNA *in vitro*. *Virology* **87,** 317.

Paoletti, E., and Moss, B. (1974). Two nucleic acid-dependent nucleoside triphosphate phosphohydrolases from vaccinia virus. Nucleotide substrate and polynucleotide cofactor specificities. *J. Biol. Chem.* **249,** 3281.

Paoletti, E., Rosemond-Hornbeak, H., and Moss, B. (1974). Two nucleic acid-dependent nucleoside triphosphate phosphohydrolases from vaccinia virus. Purification and characterization. *J. Biol. Chem.* **249,** 3273.

Paoletti, E., Lipinskas, B. R., Samsanoff, C., Mercer, S., and Panicali, D. (1984). Construction of live vaccines using genetically engineered poxviruses: Biological activity of vaccinia virus recombinants expressing the hepatitis B surface antigen and the herpes simplex virus glycoprotein D. *Proc. Natl. Acad. Sci. U.S.A.* **81,** 193.

Parker, F., and Nye, R. N. (1925). Studies on filterable viruses. I. Cultivation of vaccine virus. *Am. J. Pathol.* **1,** 325.

Parker, R. F. (1938). Statistical studies of the nature of the infectious unit of vaccine virus. *J. Exp. Med.* **67,** 725.

Parker, R. F. (1939). The neutralization of vaccine virus by serum of vaccine-immune animals. *J. Immunol.* **36,** 147.

Parker, R. F., and Rivers, T. M. (1935). Immunological and chemical investigations of vaccine virus. I. Preparation of elementary bodies of vaccinia. *J. Exp. Med.* **62,** 65.

Parker, R. F., and Rivers, T. M. (1936). Immunological and chemical investigation of vaccine virus. IV. Statistical studies of elementary bodies in relation to infection and agglutination. *J. Exp. Med.* **64,** 439.

Parsons, B. L., and Pickup, D. J. (1987). Tandemly repeated sequences are present at the ends of the DNA of racoonpox virus. *Virology* **161,** 45.

Paschen, E. (1906). Was wissen wir über den Vakzineerreger? *Muench. Med. Wochenschr.* **53,** 2391.

Paschen, E. (1924). Die Pocken. *In* "Lehrbuch der Infektions-krankheiten" (Jochmann-Hegler, ed.), p. 843. Springer-Verlag, Berlin.

Paschen, E. (1936). Zuchtung der Ektromelievirus auf der Chorionallantois Membran von Huhnerembryonen. *Zentralbl. Bakeriol., Parasitenkd., Infektionskr. Hyg., Abt. 1:Orig.* **135,** 445.

Pasteur, L. (1881). Vaccination in relation to chicken-cholera and splenic fever. *Trans. Int. Med. Congr. London* **1**, 85. [Reprinted in "Oeuvres de Pasteur" (P. Vallery-Radot, ed.), Vol. 6, p. 378. Masson, Paris, 1933.]

Patel, D. D., and Pickup, D. J. (1987). Messenger RNAs of a strongly expressed late gene of cowpox virus contains 5'-terminal poly (A) sequences. *EMBO J.* **6**, 3787.

Patel, D. D., Pickup, D. J., and Joklik, W. K. (1986). Isolation of cowpox virus A-type inclusions and characterization of their major protein component. *Virology* **149**, 174.

Paul, G. (1915). Zur differentialdiagnose der Variola und der Varicellen. Die Erscheinungen an der variolarten Hornhaut des Kaninchens und ihre fruhzeitige Erkennung. *Zentralbl. Bakteriol., Parasitenkd., Infektionskr. Hyg., Abt. 1:Orig.* **75**, 518.

Paul, G. (1916). Objektiv Sichering der Varioladiagnose durch den Tierversuch. *Wien. Med. Wochenschr.* **29**, 862.

Payne, L. G. (1978). Polypeptide composition of extracellular enveloped vaccinia virus. *J. Virol.* **27**, 28.

Payne, L. G. (1979). Identification of the vaccinia hemagglutinin polypeptide from a cell system yielding large amounts of extracellular enveloped virus. *J. Virol.* **31**, 147.

Payne, L. G. (1980). Significance of extracellular enveloped virus in the *in vitro* and *in vivo* dissemination of vaccinia. *J. Gen. Virol.* **50**, 89.

Payne, L. G. (1986). The existence of an envelope on extracellular cowpox virus and its antigenic relationship to the vaccinia envelope. *Arch. Virol.* **90**, 125.

Payne, L. G., and Kristensson, K. (1979). Mechanism of vaccinia virus release and its specific inhibition by N_1-isonicotinoyl-N_2-3-methyl-4-chlorobenzoylhydrazine. *J. Virol.* **32**, 614.

Payne. L. G., and Kristensson, K. (1985). Extracellular release of enveloped vaccinia virus from mouse nasal epithelial cells *in vivo. J. Gen. Virol.* **66**, 643.

Payne, L. G., and Norrby, E. (1976). Presence of haemagglutinin in the envelope of extracellular vaccinia virus particles. *J. Gen. Virol.* **32**, 63.

Payne, L. D., and Norrby, E. (1978). Adsorption and penetration of enveloped and naked vaccinia virus particles. *J. Virol.* **27**, 19.

Pedley, C. B., and Cooper, R. J. (1987). The assay, purification and properties of vaccinia virus-induced uncoating protein. *J. Gen. Virol.* **68**, 1021.

Pedrali-Noy, G., and Weissbach, A. (1977). Evidence of a repetitive sequence in vaccinia virus DNA. *J. Virol.* **24**, 406.

Pelham, H. R. B. (1977). Use of coupled transcription and translation to study mRNA production by vaccinia cores. *Nature (London)* **269**, 532.

Pelham, H. R. B., Sykes, J. M. M., and Hunt, T. (1978). Characteristics of a coupled cell-free transcription and translation system directed by vaccinia cores. *Eur. J. Biochem.* **82**, 199.

Pennington, T. H. (1973). Vaccinia virus morphogenesis: A comparison of virus-induced antigens and polypeptides. *J. Gen. Virol.* **19**, 65

Pennington, T. H. (1974). Vaccinia virus polypeptide synthesis. Sequential appearance and stability of pre- and post-replicative polypeptides. *J. Gen. Virol.* **25**, 433.

Pennington, T. H., and Follett, E. A. C. (1974). Vaccinia virus replication in enucleate BSC-1 cells: Particle production and synthesis of viral DNA proteins. *J. Virol.* **13**, 488.

Perkus, M. E., Piccini, A., Lipinskas, B. R., and Paoletti, E. (1985). Recombinant vaccinia virus: Immunization against multiple pathogens. *Science* **229**, 981.

Peters, D. (1956). Morphology of resting vaccinia virus. *Nature (London)* **178**, 1453.

Peters, D., and Müller, G. (1963). The fine structure of the DNA-containing core of vaccinia virus. *Virology* **21**, 266.

Peters, J. C. (1966), A monkeypox-enzooty in the "Blijdorp" Zoo. *Tijdschr. Diergeneeskd.* **91**, 387.

Pfau, C. J., and McCrea, J. F. (1963). Studies on the deoxyribonucleic acid of vaccinia virus. III. Characterization of DNA isolated by different methods and its relation to virus structure. *Virology* **21**, 425.

Pickup, D. J., Bastia, D., Stone, H. O., and Joklik, W. K. (1982). Sequence of terminal regions of cowpox virus DNA: Arrangement of repeated and unique sequence elements. *Proc. Natl. Acad. Sci. U.S.A.* **79**, 7112.

Pickup, D. J., Ink, B. S., Parsons, B. L., Hu, W., and Joklik, W. K. (1984). Spontaneous deletions and duplications of sequences in the genome of cowpox virus. *Proc. Natl. Acad. Sci. U.S.A.* **81**, 6817.

Pickup, D. J., Ink, B. S., Hu, W., Ray, C. A., and Joklik, W. K. (1986). Hemorrhage in lesions caused by cowpox virus is induced by a viral protein that is related to plasma protein inhibitors of serine proteases. *Proc. Natl. Acad. Sci. U.S.A.* **83**, 7698.

Pilaski, J., Schaller, K., Matern, B., Kloppel, G., and Mayer, H. (1982). Pockenerkrankungen bei Elefanten und Nashornen. *Verh. Int. Symp. Erkrank. Zootiere, Veszprem, 24th, 1982,* p. 257.

Pincus, W. B., and Flick, J. A. (1963). The role of hypersensitivity in the pathogenesis of vaccinia virus infection in humans. *J. Pediatr.* **62**, 57.

Pirsch, J. B., and Purlson, E. H. (1962). A tissue culture assay for variola virus based upon the enumeration of hyperplastic foci. *J. Immunol.* **89**, 632.

Pirsch, J. B., Mika, L. A., and Purlson, E. H. (1963). Growth characteristics of variola virus in tissue culture. *J. Infect. Dis.* **113**, 170.

Plucienniczak, A., Schroeder, E., Zettlmeissl, G., and Streek, R. E. (1985). Nucleotide sequence of a cluster of early and late genes in a conserved segment of the vaccinia virus genome. *Nucleic Acids Res.* **13**, 985.

Pogo, B. G. T. (1977). Elimination of naturally occurring crosslinks in vaccinia virus DNA after viral penetration into cells. *Proc. Natl. Acad. Sci. U.S.A.* **74**, 1739.

Pogo, B. G. T. (1980). Changes in parental vaccinia virus DNA after viral pentration into cells. *Virology* **101**, 520.

Pogo, B. G. T., and Dales, S. (1969). Two deoxyribonuclease activities within purified vaccinia virus. *Proc. Natl. Acad. Sci. U. S. A.* **63**, 820.

Pogo, B. G. T., and O'Shea, M. T. (1977). Further characterization of deoxyribonucleases from vaccinia virus. *Virology* **77**, 55.

Pogo, B. G. T., and O'Shea, M. (1978). The mode of replication of vaccinia virus DNA. *Virology* **84**, 1.

Pogo, B. G. T., O'Shea, M., and Freimuth, P. (1981). Initiation and termination of vaccinia virus DNA replication. *Virology* **108**, 241.

Polisky, B., and Kates, J. (1972). Vaccinia virus intracellular DNA–protein complex. Biochemical characteristics of associated protein. *Virology* **49**, 168.

Porterfield, J. S., and Allison, A. C. (1960). Studies with poxviruses by an improved plaque technique. *Virology* **10**, 233.

Prier, J. E., Sauer, R. M., Malsberger, R. G., and Sillaman, J. M. (1960). Studies on a pox disease on monkeys. II. Isolation of the etiologic agent. *Am. J. Vet. Res.* **21**, 381.

Prowazek, S. (1905). Untersuchungen über die Vaccine. *Arb. Reichgesundh.* **22**, 535.

Puckett, C., and Moss, B. (1983). Selective transcription of vaccinia virus genes in template-dependent soluble extracts of infected cells. *Cell (Cambridge, Mass.)* **35**, 441.

Quinnan, G. V., Jr., ed. (1985). "Vaccinia Viruses as Vectors for Vaccine Antigens." Am. Elsevier, New York.

Ramyar, H., and Hessami, M. (1972). Isolation, cultivation and characterization of camelpox virus. *Zentralbl. Veterinaermed., Reihe B* **19**, 182.

Rao, A. R. (1972). "Smallpox." Kothari Book Depot, Bombay.

Rao, A. R., Savithri Sukuman, M., Paramasiram, T. V., Parasuraman, A. R., Kamalakshi, S., and Shantka, M. (1968). Experimental variola in monkeys. I. Studies on the disease-enhancing property of cortisone in smallpox. A preliminary report. *Indian J. Med. Res.* **56**, 1855.

Razzell, P. (1977a). "Edward Jenner's Cowpox Vaccine: The History of a Medical Myth." Caliban Books, Firle.

Razzell, P. (1977b). "The Conquest of Smallpox." Caliban Books, Firle.

Regnery, D. C. (1987). Isolation and partial characterization of an orthopoxvirus from a California vole *(Microtus californicus.). Arch. Virol.* **94,** 159.

Reisner, A. H. (1985). Similarity between the vaccinia virus 19 K early protein and epidermal growth factor. *Nature (London)* **313**, 801.

Report (1973). Procurement Specification (Contract Clause). IX. Defined Laboratory Rodents and Rabbits. Institute of Laboratory Animal Resources, Subcommittee on Procurement Standards for Defined Laboratory Rodents and Rabbits.

Report (1974). "Report of the Committee of Inquiry into the Smallpox Outbreak in London in March and April 1973." H. M. Stationery Office,London.

Report (1980). "Report of the Investigation into the Cause of the 1978 Birmingham Smallpox Occurrence." H. M. Stationery Office, London.

Rice, C. M., Franke, C. A., Strauss, J. H., and Hruby, D. E. (1985). Expression of Sindbis virus structural proteins via recombinant vaccinia virus: Synthesis, processing and incorporation into mature Sindbis virions. *J. Virol.* **56**, 227.

Richman, D. D., Cleveland, P. H., Oxman, M. N., and Johnson, K. M. (1982). The binding of staphyloccocal A protein by the sera of different animal species. *J. Immunol.* **128,** 2300.

Ricketts, T. F. (1908). "The Diagnosis of Smallpox." Cassell, London.

Rigby, P. W. J. (1983). Cloning vectors derived from animal viruses. *J. Gen. Virol.* **64**, 255.

Rivers, T. M. (1931). Cultivation of vaccine virus for Jennerian prophylaxis in man. *J. Exp. Med.* **54**, 453.

River, T. M., and Ward, S. M. (1935). Jennerian prophylaxis by means of intradermal injections of culture vaccine virus. *J. Exp. Med.* **62**, 549.

Rivers, T. M., Ward, S. M., and Baird, R. D. (1939). Amount and duration of immunity induced by intradermal inoculation of cultured vaccine virus. *J. Exp. Med.* **69**, 857.

Roberts, J. A. (1962a). Histopathogenesis of mousepox. I. Respiratory infection. *Br. J. Exp. Pathol.* **43**, 451.

Roberts, J. A. (1962b). Histopathogenesis of mousepox. II. Cutaneous infection. *Br. J. Exp. Pathol.* **43**, 462.

Roberts, J. A. (1964). Enhancement of the virulence of attenuated ectromelia virus in mice maintained in a cold environment. *Aust. J. Exp. Biol. Med. Sci.* **42**, 657.

Roberts, J. A. (1986). Susceptibility of Australian feral mice to ectromelia virus. *Aust. Wildl. Res.* **13**, 14.

Rohrmann, G., and Moss, B. (1985). Transcription of vaccinia virus early genes by template-dependent extract of purified virions. *J. Virol.* **56**, 349.

Rohrmann, G., Yuen, L., and Moss, B. (1986). Transcription of vaccinia virus early genes by enzymes isolated from vaccinia virions terminates downstream of a reulatory sequence. *Cell (Cambridge, Mass.)* **46**, 1029.

Rondle, C. J. M., and Dumbell, K. R. (1962). Antigens of cowpox virus. *J. Hyg.* **60**, 41.

Rondle, C. J. M., and Dumbell, K. R. (1982). A poxvirus antigen associated with pathogenicity for rabbits. *J. Hyg.* **89**, 383.

Rosahn, P. D., and Hu, C. K. (1935). Rabbitpox. Report of an epidemic. *J. Exp. Med.* **62**, 331.

Rosel, J., and Moss, B. (1985). Transcriptional and translational mapping and nucleotide sequence analysis of a vaccinia virus gene encoding the precursor of the major core polypeptide 4b. *J. Virol.* **56**, 830.

Rosel, J., Earl, P. L., Weir, J. P., and Moss, B. (1986). Conserved TAAATG sequence at the transcriptional and translational initiation sites of vaccinia virus late genes deduced by structural and functional analysis of the *Hind*III H genome fragment. *J. Virol.* **60**, 436.

Rosemond-Hornbeak, H., and Moss, B. (1974). Single-stranded deoxyribonucleic acid-specific nuclease from vaccinia virus. Endonucleolytic and exonucleolytic activities. *J. Biol. Chem.* **249**, 3292.

Rosemond-Hornbeak, H., Paoletti, E., and Moss, B. (1974). Single-stranded deoxyribonucleic acid specific nuclease from vaccinia virus. Purification and characterization. *J. Biol. Chem.* **249**, 3287.

Roslyakov, A. A. (1972). [Comparison of the ultrastructure of camelpox virus, the virus of a pox-like disease of camels ("ausdik") and contagious ecthyma virus.] *Vopr. Virusol.* **17**, 26 (in Russian).

Rouhandeh, H., Engler, R., Fouad, M. T. A., and Sells, L. D. (1967). Properties of monkeypox virus. *Arch. Gesamte Virusforsch.* **20**, 363.

Ruska, H., and Kausche, G. A. (1943). Über Form, Grossenverteilung und Struktur einiger Virus-Elementarkorper. *Zentralbl. Bakteriol., Parasitenkd., Infektionskr. Hyg., Abt. 1: Orig.* **150**, 311.

Sadykov, R. G. (1970). [Cultivation of camelpox virus in chick embryos.] *Virusng Bolezni Skh. Zhivotnykh* Part I, p. 55 (in Russian).

Salaman, M. H., and Tomlinson, A. J. H. (1957). Vaccination against mousepox: With a note on the suitability of vaccinated mice for work on chemical induction of tumours. *J. Pathol. Bacteriol.* **74**, 17.

Salzman, N. P., Shatkin, A. J., Sebring, E. D., and Munyon, W. (1962). On the replication of vaccinia virus. *Cold Spring Harbor Symp. Quant. Biol.* **27**, 237.

Sam, C. K., and Dumbell, K. R. (1981). Expression of poxvirus DNA in coinfected cells and marker rescue of thermosensitive mutants by subgenomic fragments of DNA. *Ann. Virol., Institut Pasteur, Paris* **132E**, 135.

Sambrook, J. F., Padgett, B. L., and Tomkins, J. K. N. (1966). Conditional lethal mutants of rabbitpox virus. I. Isolation of host cell-dependent and temperature-dependent mutants. *Virology* **28**, 592.

Sarkar, J. K., and Mitra, A. C. (1967). Virulence of variola viruses isolated from smallpox cases of varying severity. *Indian J. Med. Res.* **55**, 13.

Sarkar, J. K., and Mitra, A. C. (1968). A search for the causes of severity in smallpox. *J. Indian Med. Assoc.* **51**, 272.

Sarkar, J. K., Neogy, K. N., and Lahini, D. C. (1959). Behaviour of variola and vaccinia viruses in mice. *J. Indian Med. Assoc.* **32**, 279.

Sarov, I., and Joklik, W. K. (1972). Studies on the nature and location of the capsid polypeptides of vaccinia virions. *Virology* **50**, 579.

Sauer, R. M., Prier, J. E., Buchanan, R. S., Creamer, A. A., and Fegley, H. C. (1960). Studies on a pox disease of monkeys. I. Pathology. *Am. J. Vet. Res.* **21**, 377.

Schell, K. (1960a). Studies on the innate resistance of mice to infection with mousepox. I. Resistance and antibody production. *Aust. J. Exp. Biol. Med. Sci.* **38**, 271.

Schell, K. (1960b). Studies on the innate resistance of mice to infection with mousepox. II. Route of inoculation and resistance; and some observations on the inheritance of resistance. *Aust. J. Exp. Biol. Med. Sci.* **38**, 289.

Schell, K. (1964). On the isolation of ectromelia virus from the brains of mice from a "normal" mouse colony. *Lab. Anim. Care* **14**, 506.

Schlesinger, M. (1936). The Feulgen reaction of the bacteriophage substance. *Nature (London)* **138**, 508.

Schmidt, M. (1870). In "Die Krankheiten der Affen. Zoologische Klinik," (M. Schmidt, ed.), Vol. 1, Part 1, p. 97. Hirschwald, Berlin.

Schoen, R. (1938). Recherches sur une maladie spontanée mortelle des souris blanche. Transmission expérimentalle. *C. R. Seances Soc. Biol. Ses Fil. (Paris)* **128**, 695.

Schonbauer, M., Schonbauer-Langle, A., and Kolbl, S. (1982). Pockeninfektion bei einer Hauskatze. *Zentralbl. Veterinaermed., Reihe B* **29**, 434.

Schwer, B., Visca, P., Vos, J. C., and Stunnenberg, H. G. (1987). Discontinuous transcription or RNA processing of vaccinia virus late messengers results in a 5' poly(A) leader. *Cell* **50**, 163.

Scrimshaw, N. S., Taylor, C. E., and Gordon, J. E. (1968). Interactions of nutrition and infection. *W.H.O. Monogr. Ser.* **57**.

Seelemann, K. (1960). Zerebrale Komplikationen nach Pockenschutzimpfung mit besonderer Berucksichtigung der Alterdisposition in Hamburg 1939 bis 1958. *Dtsch. Med. Wochenschr.* **85**, 1081.

Sehgal, C. L., Ray, S. N., Ghosh, T. K., and Arora, R. R. (1977). An investigation of an outbreak of buffalopox in animals and human beings in the Dhulia district, Maharashtra. 1. Laboratory studies. *J. Commun. Dis.* **9**, 49.

Seymour-Price, M., Chachia, C., and Fendall, N. R. E. (1960). Smallpox in Kenya. *East Afr. Med. J.* **37**, 670.

Shaffer, R., and Traktman, P. (1987). Vaccinia virus encapsidates a novel topoisomerase with the properties of a eucaryotic type I enzyme. *J. Biol. Chem* **262**, 9309.

Sharma, G. K. (1934). An interesting outbreak of variola vaccinia in Lahore. *Misc. Bull. Imp. Counc. Agric. Res. India, Sel. Artic.* **8**, 1.

Sharp, D. G. (1963). Total multiplicity in the animal virus—cell reaction and its determination by electron microscopy. *Ergeb. Mikrobiol., Immunitaetsforsch. Exp. Ther.* **36**, 214.

Sharp, J. C. M., and Fletcher, W. B. (1973). Experience of antivaccinia immunoglubulin in the United Kingdom. *Lancet* **1**, 656.

Shedlovsky, T., and Smadel. J. E. (1942). The LS-antigen of vaccinia. II. Isolation of a single substance combining both L- and S-activity. *J. Exp. Med.* **75**, 165.

Shelukhina, E. M., Maltseva, N. N., Shenkman, L. S., and Marennikova, S. S. (1975). Properties of two isolates (MK-7-73 and MK-10-73) from wild monkeys. *Br. Vet. J.* **131**, 746.

Shelukhina, E. M., Shenkman, L. S., Rozina, E. E., and Marennikova, S. S. (1979). [Modes of maintenance of certain orthopoxviruses in nature.] *Vopr. Virusol.* **4**, 368 (in Russian).

Shida, H. (1986). Nucleotide sequence of the vaccinia virus hemagglutinin gene. *Virology* **150**, 451.

Shida, H., and Dales, S. (1981). Biogenesis of vaccinia: Carbohydrates of the hemagglutinin molecule. *Virology* **111**, 56.

Shida, H., and Dales, S. (1982). Biogenesis of vaccinia: Molecular basis for the hemagglutinin-negative phenotype of the IHD-W strain. *Virology* **117**, 219.

Shida, H., Tanabe, K., and Matsumoto, S. (1977). Mechanism of virus occlusion into A-type inclusion during poxvirus infection. *Virology* **76**, 217.

Shope, R. E. (1954). Report of a committee on infectious ectromelia of mice (mousepox). *J. Natl. Cancer Inst. (U.S.)* **15**, 405.

Shuman, S., and Hurwitz, J. (1981). Mechanism of mRNA capping by vaccinia virus guanylyltransferase: Characterization of an enzyme-guanylate intermediate. *Proc. Natl. Acad. Sci. U.S.A.* **78**, 187.

Shuman, S., Surks, M., Furneaux, H., and Hurwitz, J. (1980).Purification and characterization of a GTP-pyrophosphate exchange activity from vaccinia virions. Association of the GTP-pyrophosphate exchange activity with vaccinia mRNA guanylyltransferase·RNA (guanine-7-)methyltransferase complex capping enzyme. *J. Biol. Chem.* **255,** 11588.

Shuman, S., Broyles, S. S., and Moss, B. (1987). Purification and characterization of a transcription termination factor from vaccinia virus. *J. Biol. Chem.* **262,** 12372.

Silver, M., and Dales, S. (1982). Biogenesis of poxviruses: Interrelationships between post-translational cleavage, virus assembly and maturation. *Virology* **117,** 341.

Silver, M., McFadden, G., Wilton, S., and Dales, S. (1979). Biogenesis of poxviruses: Role for the DNA dependent RNA polymerase II of the host cell during expression of late functions. *Proc. Natl. Acad. Sci. U.S.A.* **76,** 4122.

Singh, I. P., and Singh, S. B. (1967). Isolation and characterization of the aetiologic agent of buffalo-pox. *J. Res. Ludhiana* **4,** 440.

Slabaugh, M. B., and Mathews, C. K. (1986). Hydroxyurea-resistant vaccinia virus: Overproduction of ribonucleotide reductase. *J. Virol.* **60,** 506.

Slabaugh, M. B., Johnson, T. L., and Mathews, C. K. (1984). Vaccinia virus induces ribonucleotide reductase in primate cells. *J. Virol.* **52,** 507.

Smadel, J. E., and Hoagland, C. L. (1942). Elementary bodies of vaccinia. *Bacteriol. Rev.* **6,** 79.

Smadel, J. E., Rivers, T. M., and Pickels, E. G. (1939). Estimation of the purity of preparations of elementary bodies of vaccinia. *J. Exp. Med.* **70,** 379.

Smadel, J. E., Lavin, G. I., and Dubos, R. J. (1940). Some constituents of elementary bodies of vaccinia virus. *J. Exp. Med.* **71,** 373.

Smadel, J. E., Rivers, T. M., and Hoagland, C. L. (1942). Nucleoprotein antigen of vaccine virus. I. A new antigen from elementary bodies of vaccinia. *Arch. Pathol.* **34,** 275.

Smith, G. L., and Moss, B. (1983). Infectious poxvirus vectors have capacity for at least 25,000 base pairs of foreign DNA. *Gene* **25,** 21.

Smith, G. L., Mackett, M., and Moss, B. (1983). Infectious vaccinia virus recombinants that express hepatitis B virus surface antigen. *Nature (London)* **302,** 490.

Smith, G. L., Godson, G. N., Nussenzweig, V., Nussenzweig, R. S., Barnwell, J., and Moss, B. (1984). *Plasmodium knowlesi* sporozoite antigen: Expression by infectious vaccinia virus. *Science* **224,** 397.

Smith, G. L., Cheng, K.-C., and Moss, B. (1986). Vaccinia virus: An expression vector for genes from parasites. *Parasitology* **92S,** 109.

Smith, H. O. (1979). Nucleotide sequence specificity of restriction endonucleases. *Science* **205,** 455.

Smith, H. O., and Wilcox, K. W. (1970). A restriction enzyme from *Haemophilus influenzae* I. Purification and general properties. *J. Mol. Biol.* **51,** 379.

Smith, M. M. (1974). The "Real Expedición Marítima de la Vacuna" in New Spain and Guatemala. *Trans. Am. Philos. Soc.* [N. S.] **64,** Part 1.

Society of General Microbiology (1953). "The Nature of Virus Multiplication." Cambridge Univ. Press, London and New York.

Society of General Microbiology (1959). "Virus Growth and Variation." Cambridge Univ. Press, London and New York.

Soekawa, M., Matsumoto, K., Izawa, H., Iwabuchi, H., Takahashi, K., and Saito, Y. (1964a). Studies on the cowpox-like disease outbreak in Hokkaido, Japan. I. Isolation of the causative agent. *Jpn. J. Vet. Sci.* **26,** 25.

Soekawa, M., Izawa, H., Iwabuchi, H., Matsumoto, K., Takahashi, K., and Saito, Y. (1964b). Studies on the cowpox-like disease outbreak in Hokkaido, Japan. II. Properties and identification of the isolates. *Jpn. J. Vet. Sci.* **26,** 295.

Soloski, M. J., and Holowczak, J. A. (1981). Characterization of supercoiled nucleoprotein complexes released from detergent-treated vaccinia virions. *J. Virol.* **37**, 770.

Spencer, E., Loring, D., Hurwitz, J., and Monroy, G. (1978). Enzymatic conversion of 5'-phosphate terminated RNA to 5'-di and triphosphate-terminated RNA. *Proc. Natl. Acad. Sci. U.S.A.* **75**, 4793.

Spencer, E., Shuman, S., and Hurwitz, J. (1980). Purification and properties of vaccinia virus DNA-dependent RNA polymerase. *J. Biol. Chem.* **255**, 5388.

Stagles, M. J., Watson, A. A., Boyd, J. F. More, I. A. R., and McSeveney, D. (1985). The histopathology and electron microscopy of a human monkeypox lesion. *Trans R. Soc. Trop. Med. Hyg.* **79**, 192.

Stanley, W. M. (1935). Isolation of a crystalline protein possessing the properties of tobacco mosaic virus. *Nature (London)* **81**, 644.

Stearn, E. W., and Stearn, A. E. (1945). "The Effect of Smallpox on the Destiny of the Amerindian." Humphries, Boston, Massachusetts.

Steinhardt, E., Israeli, C., and Lambert, R. A. (1913). Studies on the cultivation of the virus of vaccinia. *J. Infect. Dis.* **13**, 294.

Stent, G. S. (1963). "Molecular Biology of Bacterial Viruses." Freeman, San Francisco, California.

Stern, W., and Dales, D. (1974). Biogenesis of vaccinia: Concerning the origin of the envelope phospholipids. *Virology* **62**, 293.

Stern, W., and Dales, S. (1976a). Biogenesis of vaccinia: Isolation and characterization of a surface component that elicits antibody suppressing infectivity and cell–cell fusion. *Virology* **75**, 232.

Stern, W., and Dales, S. (1976b). Biogenesis of vaccinia: Relationship of the envelope to virus assembly. *Virology* **75**, 242.

Stern, W., Pogo, B. G. T., and Dales, S. (1977). Biogenesis of poxviruses: Analysis of the morphogenetic sequence using a conditional lethal mutant defective in envelope self-assembly. *Proc. Natl. Acad. Sci. U.S.A.* **74**, 2162.

Stickl, H., Hochstein-Mintzel, V., Mayr, A., Huber, H. C., Schafer, H., and Holzner, A. (1974). MVA-Stufenimpfung gegen Pocken. *Dtsch. Med. Wochenschr.* **99**, 2386.

Stoeckenius, W., and Peters, D. (1955). Untersuchungen am Virus der Variolavaccine. IV Mitt. Die Morphologie des Innenkorpers. *Naturforsch., B:Anorg, Chem., Org. Chem., Biochem., Biophys., Biol.* **10B**, 77.

Stone, J. D. (1946). Inactivation of vaccinia and ectromelia haemagglutinins by lecithinase. *Aust J. Exp. Biol. Med. Sci.* **24**, 191.

Stroobant, P., Rice, A. P., Gullick, W. J., Cheng, D. J., Kerr, I. M., and Waterfield, M. D. (1985). Purification and characterization of vaccinia virus growth factor. *Cell (Cambridge, Mass.)* **42**, 383.

Stuart, G. (1947). Memorandum on post-vaccinal encephalitis. *Bull. W.H.O.* **1**, 36.

Subrahmanyan, T. P. (1968). A study of the possible basis of age-dependent resistance of mice to poxvirus diseases. *Aust. J. Exp. Biol. Med. Sci.* **46**, 251.

Symmers, W. St. C. (1978). The lymphoreticular system. *In* "Systemic Pathology" (W. St. C. Symmers, ed.), 2nd ed., Vol. 2, p. 644. Churchill-Livingstone, Edinburgh.

Syzbalski, W., Erikson, R. L., Gentry, G. A., Gafford, L. G., and Randall, C. C. (1963). Unusual properties of fowlpox virus DNA. *Virology* **19**, 586.

Takahashi-Nishimaki, F., Suzuki, K., Morita, M., Maruyama, T., Miki, K., Miki, S., Hashizume, S., and Sugimoto, M. (1987). Genetic analysis of vaccinia Lister strain and its attenuated variant LC16m8: Production of intermediate variants by homologous recombination. *J. Gen. Virol.* **68**, 2705.

Tantawi, H. H. (1974). Comparative studies on camel pox, sheep pox and vaccinia viruses. *Acta Virol. (Engl. Ed.)* **18**, 347.

Tantawi, H. H., Sabban, M. S., Reda, I. M., and El Dahaby, H. (1974). Camelpox virus in Egypt. I. Isolation and characterization. *Bull. Epizoot. Dis. Afr.* **22**, 315.

Tantawi, H. H., Fayed, A. A., Shalaby, M. A., and Skalinsky, E. I. (1977). Isolation, cultivation and characterization of poxvirus from Egyptian water buffaloes. *J. Egypt. Vet. Med. Assoc.* **37**, 15.

Tattersall, P., and Ward, D. C. (1976). Rolling hairpin model for replication of parvovirus and linear chromosomal DNA. *Nature (London)* **263**, 106.

Taylor, J. L., and Rouhandeh, H. (1977). Yaba virus-specific DNA in the host cell nucleus. *Biochim. Biophys. Acta* **478**, 59.

Theiler, M., and Smith, H. H. (1937). Use of yellow fever virus modified by *in vitro* cultivation for human immunization. *J. Exp. Med.* **65**, 787.

Thomas, E. K., Palmer, E. L., Obijeski, J. F., and Nakano, J. H. (1975). Further characterization of raccoonpox virus. *Arch. Virol.* **49**, 217.

Thomas, G. (1970). Sampling rabbit pox aerosols of natural origin. *J. Hyg.* **68**, 511.

Thomsett, L. R., Baxby, D., and Denham, E. M. H. (1978). Cowpox in the domestic cat. *Vet. Rec.* **108**, 567.

Tint, H. (1973). The rationale for elective pre-vaccination with attenuated vaccinia (CV1-78) in preventing some vaccination complications. *Symp. Ser. Immunobiol. Stand.* **19**, 281.

Topciu, V., Rosiu, N., Csaky, N., and Argan, P. (1972). Isolements accidentales de virus ectromelique latent chez des souris blanches de laboratoire, l'occasion d'études sur les arbovirus. *Arch. Roum. Pathol. Exp. Microbiol.* **31**, 75.

Topciu, V., Luca, I., Modovan, E., Stoianovici, V., Plavosin, L., Milin, D., and Welter, E. (1976). Transmission of vaccinia virus from vaccinated milkers to cattle. *Rev. Roum. Med. Virol.* **27**, 279.

Topley, W. W. C. (1923). The spread of bacterial infection: Some general considerations. *J. Hyg.* **21**, 226.

Torres, C. M., and Teixeira, J. de C. (1935). Culture du virus de l'alastrim sur les membranes de l'embryon de poulet. *C. R. Seances Soc. Biol. Ses Fil.* **118**, 1023.

Traktman, P., Sridhar, R., Condit, R. C., and Roberts, B. (1984). Transcriptional mapping of the DNA polymerase gene of vaccinia virus. *J. Virol.* **49**, 125.

Trentin, J. J. (1953). An outbreak of mouse-pox (infectious ectromelia) in the United States. I. Presumptive diagnosis. *Science* **117**, 226.

Trentin, J. J., and Ferrigno, M. A. (1957). Control of mouse pox (infectious ectromelia) by immunization with vaccinia virus. *J. Natl. Cancer Inst. (U. S.)* **18**, 757.

Tsao, H., Ren, G. F., and Chu, C. M. (1986). Gene coding for the late 11,000-dalton polypeptide of the Tian Tan strain of vaccinia virus and its 5'-flanking region: Nucleotide sequence. *J. Virol.* **57**, 693.

Tsuchiya, Y., and Tagaya, I. (1972). Isolation of a multinucleated giant cell-forming and haemagglutinin-negative variant of variola virus. *Arch. Gesamte Virusforsch.* **39**, 292.

Turner, A., and Baxby, D. (1979). Structural polypeptides of *Orthopoxvirus*; their distribution in various members and within the virion. *J. Gen. Virol.* **45**, 537.

Turner, G. S., and Squires, E. J. (1971). Inactivated smallpox vaccine: Immunogenicity of inactivated intracellular and extracellular virus. *J. Gen. Virol.* **13**, 19.

Turner, G. S., Squires, E. J., and Murray, H. G. S. (1970). Inactivated smallpox vaccine. A comparison of inactivation methods. *J. Hyg.* **68**, 197.

Tutas, D. J., and Paoletti, E. (1977). Purification and characterization of core-associated polynucleotide 5'-triphosphatase from vaccinia virus. *J. Biol. Chem.* **252**, 3092.

Twardzik, D. R., Brown, J. P., Ranchalis, J. E., Todaro, G. J., and Moss, B. (1985). Vaccinia virus-infected cells release a novel polypeptide functionally related to transforming and epidermal growth factors. *Proc. Natl. Acad. Sci. U.S.A.* **82**, 5300.

Ueda, Y., Ito, M., and Tagaya, I. (1969). A specific surface antigen induced by poxvirus. *Virology* **38**, 180.

Ueda, Y., Tagaya, I., Amano, H., and Ito, M. (1972). Studies on the early antigens induced by vaccinia virus. *Virology* **49**, 794.

Upton, C., and McFadden, G. (1986). Identification and molecular sequence of the thymidine kinase gene of Shope fibroma virus *J. Virol.* **60**, 920.

van den Berg, C. A. (1946). L'encéphalite post-vaccinale aux Pays-Bas. *Bull. Off. Int. Hyg. Publ.* **38**, 847.

van Rooyen, C. E., and Rhodes, A. J. (1948). "Virus Diseases of Man." Nelson, New York.

van Tongeren, H. A. E. (1952). Spontaneous mutation of cowpox-virus by means of eggpassage. *Arch. Gesamte Virusforsch.* **5**, 35.

Varich, N. L., Sychova, I. V., Kaverin, N. V., Autonova, T. P., and Chernos, V. I. (1979). Transcription of both DNA strands of vaccinia virus genome *in vivo. Virology* **96**, 412.

Vassef, A. (1987). Conserved sequences near the early transcription start of vaccinia virus. *Nucleic Acids Res.* **15**, 1427.

Vassef, A., Mars, M., Dru, A., Plusienniczak A., Streeck, R. E., and Beaud, G. (1985). Isolation of *cis*-acting vaccinia virus DNA fragments promoting the expression of herpes simplex virus thymidine kinase by recombinant viruses. *J. Virol.* **55**, 163.

Venkatesan, S., and Moss, B. (1981). *In vitro* transcription of the inverted terminal repetition of the vaccinia virus genome: Correspondence of initiation and cap sites. *J. Virol.* **37**, 738.

Venkatesan, S., Gershowitz, A., and Moss, B. (1980). Modification of the 5'-end of mRBA: Association of RNA triphosphatase with the RNA guanylyltransferase RNA·(guanine-7-)methyltransferase complex from vaccinia virus. *J. Biol. Chem.* **255**, 903.

Venkatesan, S., Baroudy, B. M., and Moss, B. (1981). Distinctive nucleotide sequences adjacent to multiple initiation and termination sites of an early vaccinia virus gene. *Cell (Cambridge, Mass.)* **125**, 805.

Venkatesan, S., Gershowitz, A., and Moss, B. (1982). Complete nucleotide sequences of two adjacent early vaccinia virus genes located within the inverted terminal repetition. *J. Virol.* **44**, 637.

Verlinde, J. D. (1951). Koepokken bij de mens. *Tidjschr. Diergeneeskd.* **76**, 334.

Verlinde, J. D., and Wensinck, F. (1951). Manifestations of a laboratory epizootic of rabbit pox by non-specific stimuli. *Antonie van Leeuwenhoek* **17**, 232.

Vichniakov, V. E. (1968). A study of immunity to smallpox in persons who have experienced a previous attack. *Bull. W.H.O.* **39**, 433.

Vieuchange, J., Brion, G., and Gruest, J. (1958). Virus variolique en culture de cellules. *Ann. Inst. Pasteur, Paris* **95**, 681.

Villarreal, E. C., and Hruby, D. E. (1986). Mapping the genomic location of the gene encoding αamantin resistance in vaccinia virus mutants. *J. Virol.* **57**, 65.

Villarreal, E. C., Roseman, N. A., and Hruby, D. E. (1984). Isolation of vaccinia virus mutants capable of replicating independently of the host cell nucleus. *J. Virol.* **51**, 359.

von Magnus, P., Andersen, E. K., Petersen, B. K., and Birch-Andersen, A. (1959). A pox-like disease in cynomolgus monkeys. *Acta Pathol. Microbiol. Scand.* **46**, 156.

Wagner, H., and Rollinghoff, M. (1978). T-T-cell interactions during *in vitro* cytotoxic allograft responses. I. Soluble products from activated Ly1⁺ T cells trigger autonomously antigen-primed Ly23⁺ T cells to proliferation and cytolytic activity. *J. Exp. Med.* **148**, 1523.

Wallace, G. D. (1981). Mouse pox threat. *Science* **211**, 438.

Wallace, G. D., and Buller, R. M. L. (1985). Kinetics of ectromelia virus (mousepox) transmission and clinical response in C57BL/6J, BALB/cByJ and AKR/J inbred mice. *Lab. Anim. Sci.* **35**, 41.

Wallace, G. D., and Buller, R. M. L. (1986). Ectromelia virus (mousepox): Biology, epizootiology, prevention and control. In "Viral and Mycoplasmal Infections of Laboratory Rodents: Effects on Biomedical Research" (P. N Bhatt, R. O. Jacoby, H. C. Morse, III, and A. E. New, eds.), p. 539. Academic Press, Orlando, Florida.

Wallace, G. D., Werner, R. M., Golway, P. L., Hernandez, D. M., Alling, D. W., and George, D. A. (1981). Epizootiology of an outbreak of mousepox at the National Institutes of Health. *Lab. Anim. Sci.* **31**, 609.

Wallace, G. D., Buller, R. M. L., and Morse, H. M., III (1985). Genetic determinants of resistance to ectromelia (mousepox) virus-induced mortality. *J. Virol.* **55**, 890.

Wallnerova, Z., and Mims, C. A. (1970). Thoracic duct cannulation and hemal node formation in mice infected with cowpox virus. *Br. J. Exp. Pathol.* **51**, 118.

Wang, T. L. (1947). Recherches sur le virus de la peste des lapins. *Ann. Inst. Pasteur, Paris* **73**, 1207.

Watson, J. D. (1972). Origin of concatemeric T7 DNA. *Nature (London)* New Biol. **239**, 197.

Weber, G., and Lange, J. (1961). Zur Variationsbreite der "Inkubationszeiten" postvakzinaler zerebraler Erkrankungen. *Dtsch. Med. Wochenschr.* **86**, 1461.

Wehrle, P. F., Posch, J., Richter, K. H., and Henderson, D. A. (1970). An airborne outbreak of smallpox in a German hospital and its significance with respect to other recent outbreaks in Europe. *Bull. W.H.O.* **43**, 669.

Wei, C. M., and Moss, B. (1975). Methylated nucleotides block 5'-terminus of vaccinia virus mRNA. *Proc. Natl. Acad. Sci. U.S.A.* **72**, 318.

Weinrich, S. L., and Hruby, D. E. (1986). A tandemly-orientated late gene cluster within the vaccinia virus genome. *Nucleic Acids Res.* **14**, 3003.

Weintraub, S., and Dales, S. (1974). Biogenesis of poxviruses: Genetically controlled modifications of structural and funtional components of plasma membrane. *Virology* **60**, 96.

Weir, J. P., and Moss, B. (1983). Nucleotide sequence of the vaccinia virus thymidine kinase gene and the nature of spontaneous frameshift mutations. *J. Virol.* **46**, 530.

Weir, J. P., and Moss, B. (1984). Regulation of expression and nucleotide sequence of a late vaccinia virus gene. *J. Virol.* **51**, 662.

Weir, J. P., Bajszar, G., and Moss, B. (1982). Mapping of the vaccinia virus thymidine kinase gene by marker rescue and by cell-free translation of selected mRNA. *Proc. Natl. Acad. Sci. U.S.A.* **79**, 1210.

Wells, D. G. T. (1967). Studies on variola virus. Ph.D. Thesis, London University.

Wenner, H. A., Bolano, C. R., Cho, C. T., and Kamitsuka, S. (1969). Studies on the pathogenesis of monkeypox. III. Histopathological lesions and sites of immunofluorescence. *Arch. Gesamte Virusforsch.* **27**, 179.

Werner, G. T. (1982). Transmission of mouse-pox in colonies of mice. *Zentralbl. Veterinaermed., Reihe B* **29**, 401.

Werner, G. T., Jentzsch, V., Metzger, E., and Simon, J. (1980). Studies on poxvirus infection in irradiated animals. *Arch. Virol.* **64**, 247.

Westwood, J. C. N. (1963). Virus pathogenicity In "Mechanisms of Virus Infection" (W. Smith, ed.), p. 255. Academic Press, London.

Westwood, J. C. N., Harris, W. J., Zwartouw, H. T., Titmuss, D. H. J., and Applyard, G. (1964). Studies on the structure of vaccinia virus. *J. Gen. Microbiol.* **34**, 67.

Westwood, J. C. N., Zwartouw, H. T., Applyard, G., and Titmus, D. H. J. (1965). Comparison of soluble antigens and virus particle antigens of vaccinia virus. *J. Gen. Microbiol.* **38**, 47.

Westwood, J. C. N., Boulter, E. A., Bowen, E. T. A., and Maber, H. B. (1966).

Experimental respiratory infection with pox viruses. 1. Clinical, virological and epidemiological studies.*Br. J. Exp. Pathol.* **47**, 453.

Whellock, E. F. (1964). Interferon in dermal crusts of human vaccinia virus vaccinations. Possible explanation of relative benignity of variolation smallpox. *Proc. Soc. Exp. Biol. Med.* **117**, 650.

Whitney, R. A. (1974). Ectromelia in mouse colonies. *Science* **184**, 609.

Whitney, R. A., Jr., Small, J. D., and New, A. (1981). Mousepox—National Institutes of Health experiences. *Lab. Anim. Sci.* **31**, 570.

Wiktor, T. J., Macfarlan, R. I., Reagan, K. J., Dietzschold, B., Curtis, P. J., Wunner, W. H., Kieny, M.-P., Lathe, R., Lecocq, J.-P., Mackett, M., Moss, B., and Koprowski, H. (1984). Protection from rabies by a vaccinia recombinant containing the rabies virus glycoprotein gene. *Proc. Natl. Acad. Sci. U.S.A.* **81**, 7194.

Wildy, P. (1971). Classification and nomenclature of viruses. First report of the International Committee on Nomenclature of Viruses. *Monogr. Virol.* **5**, 1.

Wilkinson, L. (1979). The development of the virus concept as reflected in the corpora of studies on individual pathogens. 5. Smallpox and the evolution of ideas on acute (viral) infections. *Med. Hist.* **23**, 1.

Williamson, J. D., and Mackett, M. (1982). Arginine deprivation and the generation of white variants in cowpox virus-infected cell cultures. *J. Hyg.* **89**, 373.

Wilson, G. S. (1967). "The Hazards of Immunization." Athlone, London.

Wittek, R., and Moss, B. (1980). Tandem repeats within the inverted terminal repetition of vaccinia virus. *Cell (Cambridge, Mass.)* **21**, 277–284.

Wittek, R., Menna, A., Schümperli, D., Stoffel, S., Müller, H. K., and Wyler, R. (1977). *Hind*III and *Sst*1 restriction sites mapped on rabbit poxvirus and vaccinia virus DNA. *J. Virol.* **23**, 669.

Wittek, R., Müller, H. K., Menna, A., and Wyler, R. (1978a). Length heterogeneity in the DNA of vaccinia virus is eliminated on cloning the virus. *FEBS Lett.* **90**, 41.

Wittek, R., Menna, A., Müller, H. K., Schümperli, D., Boseley, P. G., and Wyler, R. (1978b). inverted terminal repeats in rabbitpox virus and vaccinia virus DNA. *J. Virol.* **28**, 171.

Wittek, R., Barbosa, E., Cooper, J. A., Garon, C. F., Chan H., and Moss, B. (1980a). Inverted terminal repetition in vaccinia virus DNA encodes early mRNAs. *Nature (London)* **285**, 21.

Wittek, R., Cooper, J., Barbosa, E., and Moss, B. (1980b). Expression of the vaccinia virus genome—analysis and mapping of mRNAs encoded with the inverted terminal repetition. *Cell (Cambridge, Mass.)* **21**, 487.

Wittek, R., Cooper, J., and Moss, B. (1981). Transcriptional and translational mapping of a 6.6 kilobase-pair DNA fragment containing the junction of the terminal repetition and unique sequence at the left end of the vaccinia virus genome. *J. Virol.* **39**, 722.

Wittek, R., Hänggi, M., and Hiller, G. (1984a). Mapping of a gene coding for a major late structural polypeptide on the vaccinia virus genome. *J. Virol.* **49**, 371.

Wittek, R., Richner, B., and Hiller, G. (1984b). Mapping of the genes coding for the two major vaccinia virus core polypeptides. *Nucleic Acids Res.* **12**, 4835.

Wokatsch, R. (1972). Vaccinia virus. *In* "Strains of Human Viruses" (M. Majer and S. A. Plotkin, eds.), p. 241. Karger, Basel.

Wolff, H. L., and Croon, J. J. A. B. (1968). The survival of smallpox virus (variola minor) in natural circumstances. *Bull W.H.O.* **38**, 492.

Woodroofe, G. M., and Fenner, F. (1960). Genetic studies with mammalian poxviruses IV. Hybridization between several different poxviruses. *Virology* **12**, 272.

Woodroofe, G. M., and Fenner, F. (1962). Serological relationships within the poxvirus group; An antigen common to all members of the group. *Virology* **16**, 334.

Woodruff, A. M., and Goodpasture, E. W. (1931). The susceptibility of the chorioallantoic membrane of chick embryos to infection with the fowlpox virus. *Am. J. Pathol.* **7**, 209.

World Health Organization (1965). Requirements for Biological Substances. Report of WHO Expert Group. *W.H.O. Tech. Rep. Ser.* **323.**

World Health Organization (1969). "Guide to the Laboratory Diagnosis of Smallpox for Smallpox Eradication Programmes." W.H.O., Geneva.

World Health Organization (1980a). "The Global Eradication of Smallpox. Final Report of the Global Commission for the Certification of Smallpox Eradication," Hist. Int. Public Health, No. 4. W.H.O., Geneva.

World Health Organization (1980b). "Management of Reserve Stocks of Vaccine in the Post-smallpox Eradication Era," WHO/SE/80.158 Rev. 1. Archives of Smallpox Eradication Unit, W.H.O., Geneva.

World Health Organization (1985). "Guidelines for the Development of Recombinant Vaccinia Viruses for Use as Vaccines." Report of a WHO Informal Meeting, Geneva, September 6 and 7 1985. W.H.O., Geneva.

Wulff, H., Chin, T. D. Y., and Wenner, H. A. (1969). Serologic responses of children after primary vaccination and revaccination against smallpox. *Am. J. Epidemiol.* **90**, 312.

Yewdell, J. W., Bennink, J. R., Smith, G. L., and Moss, B. (1985). Influenza A virus nucleoprotein is a major target antigen for cross-reactive anti-influenza A cytotoxic T lymphocytes. *Proc. Natl. Acad. Sci. U.S.A.* **82**, 1785.

Yuen, L, and Moss, B. (1986). Multiple 3' ends of mRNA encoding vaccinia virus growth factor occur within a series of repeated sequences downstream of T clusters. *J. Virol.* **60**, 320.

Zaslavsky, V., and Yakobson, E. (1975). Control of thymidine kinase synthesis in IHD vaccinia virus-infected thymidine kinase-deficient LM cells. *J. Virol.* **16**, 210.

Zeller, H., and Reckzeh, G. (1965a). Zur Immunisierung der weissen Maus gegen Ektromelie mit aktiven Vaccine Virus. I. Mitteilung: Methoden and Grundlagen. *Zentralbl. Bakteriol., Parasitenkd., Infektionsk Hyg., Abt. I: Orig., Reihe B* **195**, 282.

Zeller, H., and Reckzeh, G. (1965b). Zur Immunisierung der weissen Maus gegen Ektromelie mit aktiven Vaccine Virus. II. Mitteilung: Praktische Erfahrungen. *Zentralbl. Bakteriol., Parasitenkd., Infektionsk Hyg., Abt. 1: Orig., Reihe B* **197**, 34.

Zinkernagel, R. M., and Althage, A. (1977). Anti-viral protection by virus-immune cytotoxic T cells: Infected target cells are lysed before infectious virus progeny are assembled. *J. Exp. Med.* **145**, 644.

Zuelzer, W. (1874). Zur aetiologie der variola. *Zentralbl. Med. Wiss.* **12**, 82.

Zwart, P., Gispen, R., and Peters, J. C. (1971). Cowpox in *Okapi johnstoni* at Rotterdam Zoo. *Br. Vet. J.* **127**, 20.

Zwartouw, H. T. (1964). The chemical composition of vaccinia virus. *J. Gen. Microbiol.* **34**, 115.

Index

423

host range, 9, 310
meningoencephalomyelitis in mice,
    113–114, 310
Radioimmunoassay, 121
    adsorption, 121, 239–240, 260–261, 262–264
Rash,
    monkeypox,
        human, 243–244
        simian, 95, 231, 232
    mousepox, 95, 279–282
    smallpox, 95, 97, 318–320
        histopathology, 106–110
        inclusion bodies, 107, 108
Rat, infection with,
    cowpox virus, 115, 116, 174, 191, 193–194
    ectromelia virus, 273–274
Resistance to,
    mousepox,
        genetic, 285–288
        immunological, 125–126, 300–302
    smallpox,
        immunological, 133, 134–136
        non-specific, 140, 141
Rodents, infection with,
    cowpox virus, 10, 11, 115, 116, 117
    monkeypox virus, 11, 116, 117, 233–234
    see also Gerbil; Mouse; Mousepox; Rat

S

Smallpox, 317–252
    case-fatality rate, 318, 321, 422
    clinical features, 317–323
    deaths from, 326, 328–329, 336
    elimination, 1900–1958, 333–340
        1958–1966, 341
        1967–1977, 341–347
    epidemiology, 197, 323–324
    Eradication Unit of WHO, 230, 244, 261,
        341, 342, 343
    global eradication, 26–27, 333–352
        certification, 348–349, 350
        factors involved, 349, 351–352
        strategies, 343–346
    history, 1, 2, 324–330
    laboratory diagnosis, 344–345
    mass vaccination, 341, 343, 344
    laboratory-associated outbreaks, 220–221
    pathology,
        internal organs, 111–112

mucosal lesions, 109–110
    skin lesions, 106–109
pockmarks, 318, 324
portals of entry of virus, 86–91
rash, 95, 318–320
recognition card, 319
recurrent infectivity, 117, 118
sequelae, 323
shedding of variola virus, 318, 323–324
spread,
    airborne, 324
    colonists, 327–330
    contact, 324
    fomites, 324
    variola minor, 332–333
spread of infection in body, 91–95
subclinical infection, 323
surveillance and containment, 343–344
toxemia, 97–98
Squirrels, monkeypox virus, 263–266

T

Tatera poxvirus, 12, 312–313
    biological characteristics, 7, 312–313
    discovery, 27, 312
    DNA restriction map, 8, 313
    differentiation from variola virus, 210,
        312–313
    geographic range, 9, 312
    host range, 9, 312
Thymidine kinase of poxviruses, 56–57
    gene,
        effect on virulence, 372
        sequences, 56–57, 211
        use for recombinant vaccines, 355, 357
    sensitivity to thymidine triphosphate, 7,
        183, 184, 206–207
Tissue culture,
    cellular changes after orthopoxvirus
        infection, 98–99
    growth of orthopoxviruses, 20–21
    hemadsorption, 99
    hyperplastic foci, 99
    see also under names of individual
        viruses
Toxemia,
    mousepox, 98
    rabbitpox, 97–98
    smallpox, 95, 97–98, 323